Springer Series in
SOLID-STATE SCIENCES 173

Springer Series in
SOLID-STATE SCIENCES

Series Editors:
M. Cardona P. Fulde K. von Klitzing R. Merlin H.-J. Queisser H. Störmer

The Springer Series in Solid-State Sciences consists of fundamental scientific books prepared by leading researchers in the field. They strive to communicate, in a systematic and comprehensive way, the basic principles as well as new developments in theoretical and experimental solid-state physics.

Please view available titles in *Springer Series in Solid-State Sciences* on series homepage http://www.springer.com/series/682

Pierre A. Deymier

Editor

Acoustic Metamaterials and Phononic Crystals

 Springer

Editor
Pierre A. Deymier
Department of Materials Science and Engineering
University of Arizona
Tucson, Arizona
USA

Series Editors:
Professor Dr., Dres. h. c. Manuel Cardona
Professor Dr., Dres. h. c. Peter Fulde*
Professor Dr., Dres. h. c. Klaus von Klitzing
Professor Dr., Dres. h. c. Hans-Joachim Queisser
Max-Planck-Institut für Festkörperforschung, Heisenbergstrasse 1, 70569 Stuttgart, Germany
* Max-Planck-Institut für Physik komplexer Systeme, Nöthnitzer Strasse 38
  01187 Dresden, Germany

Professor Dr. Roberto Merlin
Department of Physics, University of Michigan
450 Church Street, Ann Arbor, MI 48109-1040, USA

Professor Dr. Horst Störmer
Dept. Phys. and Dept. Appl. Physics, Columbia University, New York, NY 10027 and
Bell Labs., Lucent Technologies, Murray Hill, NJ 07974, USA

ISSN 0171-1873
ISBN 978-3-642-31231-1   ISBN 978-3-642-31232-8 (eBook)
DOI 10.1007/978-3-642-31232-8
Springer Heidelberg New York Dordrecht London

Library of Congress Control Number: 2012953273

Printed on acid-free paper

Springer is part of Springer Science+Business Media (www.springer.com)

Dedication
To Christine, Alix, and Martin

Preface

Phononic crystals and acoustic metamaterials have generated rising scientific interests for very diverse technological applications ranging from sound abatement to ultrasonic imaging to telecommunications to thermal management and thermo-electricity. Phononic crystals and acoustic metamaterials are artificially structured composite materials that enable manipulation of the dispersive properties of vibrational waves. Phononic crystals are made of periodic distributions of inclusions (scatterers) embedded in a matrix. Phononic crystals are designed to control the dispersion of waves through Bragg scattering, the scattering of waves by a periodic arrangement of scatterers with dimensions and periods comparable to the wavelength. Acoustic metamaterials have the added feature of local resonance, and although often designed as periodic structures, their properties do not rely on periodicity. The structural features of acoustic metamaterials can be significantly smaller than the wavelength of the waves they are affecting. Local resonance may lead to negative effective dynamic mass density and bulk modulus and therefore to their unusual dispersion characteristics. Whether these materials impact wave dispersion (i.e., band structure) through Bragg's scattering or local resonances, they can achieve a wide range of unusual spectral (ω-space), wave vector (k-space), and phase (φ-space) properties. For instance, under certain conditions, absolute acoustic band gaps can form. These are spectral bands where propagation of waves is forbidden independently of the direction of propagation. Mode localization in phononic crystals or acoustic metamaterials containing defects (e.g., cavities, linear defects, stubs, etc.) can produce a hierarchy of spectral features inside the band gap that can lead to a wide range of functionalities such as frequency filtering, wave guiding, wavelength multiplexing, and demultiplexing. The wave vector properties result from passing bands with unique refractive characteristics, such as negative refraction, when the wave group velocity (i.e., the direction of propagation of the energy) is antiparallel to the wave vector. Negative refraction can be exploited to achieve wave focusing with flat lenses. Under specific conditions involving amplification of evanescent waves, super-resolution imaging can also be obtained, that is, forming images that beat the Rayleigh limit of resolution. Phononic crystals and acoustic metamaterials with anisotropic band structures may exhibit zero-angle refraction and can lead to wave guiding/collimation without the need for linear defects. The dominant mechanisms behind the control of phase of propagating acoustic waves at some specific frequency is associated with the noncollinearity of the wave vector and the group velocity leading to phase shift. More recent developments have considered phononic crystals and acoustic metamaterials composed of materials that go beyond the regime of linear continuum elasticity theory. These include strongly nonlinear phononic structures such as granular media, the effect of damping and viscoelasticity on band structure, phononic structures composed of at least one active medium, and phononic crystals made of discrete anharmonic lattices. Phononic structures composed of strongly nonlinear media can show phenomena with no linear analogue and can exhibit unique behaviors associated with solitary waves, bifurcation, and tunability. Tunability of the band structure can also be achieved with constitutive media with mixed properties such as acousto-optic or acousto-magnetic properties. Dissipation, often seen as having a negative effect on wave propagation, can be turned into a mean of controlling band structure.

Finally, the study of phononic crystals and acoustic metamaterials has also extensively relied on a combination of experiments and theory that have shown extraordinary complementarity.

In light of the strong interest in phononic crystals and acoustic metamaterials, we are trying in this book to respond to the need for a pedagogical treatment of the fundamental concepts necessary to understand the properties of these artificial materials. For this, we use simple models to ease the reader into understanding the fundamental concepts underlying the behavior of these materials. We also expose the reader to the current state of knowledge through results from established and cutting-edge research. We also present recent progresses in our understanding of these materials. The chapters in this book are written by some of the pioneers in the field as well as emerging young talents who are redirecting that field. These chapters try to strike a balance, when possible, between theory and experiments. We have made a coordinated effort to harmonize some of the contents of the chapters and we have tried to follow a common thread based on variations on a simple model, namely the one-dimensional (1-D) chain of spring and masses. In Chap. 1, we present a non-exhaustive state of the field with some attention paid to its chronological development. Chapter 2 serves as a pedagogical introduction to many of the fundamental concepts and tools that are needed to understand the properties of phononic crystals and acoustic metamaterials. Particular attention is focused on the contrast between scattering by periodic structures and local resonances. In that chapter, we use the 1-D harmonic chain as a simple metaphor for wave propagation in more complex structures. This simple model will recur in many of the other chapters of this book. Logically, Chap. 3 treats the vibrational properties of 1-D phononic crystals (superlattices) of both discrete and continuous media. A comparison of the theoretical results with experimental data available in the literature is also presented. Chapter 4 then considers two-dimensional (2-D) and three-dimensional (3-D) phononic crystals. A combination of experimental and theoretical methods are presented and used to shed light not only on the spectral properties of phononic crystals but also importantly on refractive properties. Particular attention is paid to the phenomenon of negative refraction. Chapter 5 considers acoustic metamaterials whose properties are determined by local resonators. These properties are related to the unusual behavior of the dynamic mass density and bulk modulus in materials composed of locally resonant structures. Chapters 6–9 introduce new directions for the field of phononic crystal and acoustic metamaterials. The more recent topics of phononic structures composed of dissipative media (Chap. 6), of strongly nonlinear media (Chap. 7), and media enabling tunability of the band structure (Chap. 8) are presented. Chapter 9 illustrates the richness of behavior of phononic structures that may be encountered at the nanoscale when accounting for the anharmonicity of interatomic forces. Finally, Chap. 10 serves again a pedagogical purpose and is a compilation of the different theoretical and computational methods that are used to study phononic crystals and acoustic metamaterials. It is intended to support the other chapters in providing additional details on the theoretical and numerical methods commonly employed in the field.

We hope that this book will stimulate future interest in the field of phononic crystals and acoustic metamaterials and will initiate new developments in their study and design.

Tucson, AZ   Pierre A. Deymier

Contents

List of Contributors

N. Boechler Engineering and Applied Science, California Institute of Technology, Pasadena, CA, USA, boechler@caltech.edu

O. Bou Matar International Associated Laboratory LEMAC, Institut d'Electronique, Microelectronique et Nanotechnologie (IEMN), UMR CNRS 8520, PRES Lille Nord de France, Ecole Centrale de Lille, Villeneuve d'Ascq, France, olivier.boumatar@iemn.univ-lille1.fr

C. Daraio Engineering and Applied Science, California Institute of Technology, Pasadena, CA, USA, daraio@caltech.edu

Pierre A. Deymier Department of Materials Science and Engineering, University of Arizona, Tucson, AZ, USA, deymier@email.arizona.edu

B. Djafari-Rouhani Institut d'Electronique, de Microélectronique et de Nanotechnologie (IEMN), UFR de Physique, Université de Lille 1, Villeneuve d'Ascq Cédex, France, bahram.djafari-rouhani@univ-lille1.fr

L. Dobrzynski Equipe de Physique, des Ondes, des Nanostructures et des Interfaces, Groupe de Physique, Institut d'Electronique, de Microélectronique et de Nanotechnologie (IEMN), Université de Lille 1, Villeneuve d'Ascq Cédex, France, Leonard.Dobrzynski@Univ-Lille1.fr

El Houssaine El Boudouti LDOM, Departement de Physique, Faculté des Sciences, Université Mohamed I, Oujda, Morocco, elboudouti@yahoo.fr

R. Erdmann Department of Materials Science and Engineering, University of Arizona, Tucson, AZ, USA, fluid.thought@gmail.com

Michael J. Frazier Department of Aerospace Engineering Sciences, University of Colorado Boulder, Boulder, CO, USA, michael.frazier@colorado.edu

A.-C. Hladky-Hennion Institut d'Electronique, de Micro-électronique et de Nanotechnologie, UMR CNRS 8520, Dept. ISEN, Lille Cedex, France, anne-christine.hladky@isen.fr

Mahmoud I. Hussein Department of Aerospace Engineering Sciences, University of Colorado Boulder, Boulder, CO, USA, Mih@Colorado.EDU

Guancong Ma Department of Physics, The Hong Kong University of Science and Technology, Kowloon, Hong Kong, China

Jun Mei Department of Physics, The Hong Kong University of Science and Technology, Kowloon, Hong Kong, China

B. Merheb Department of Materials Science and Engineering, University of Arizona, Tucson, AZ, USA, bassam@merheb.net

K. Muralidharan Department of Materials Science and Engineering, University of Arizona, Tucson, AZ, USA, Krishna@email.arizona.edu

J. H. Page Department of Physics and Astronomy, University of Manitoba, Winnipeg, MB, Canada, jhpage@cc.umanitoba.ca

J. F. Robillard Institut d'Electronique, de Micro-électronique et de Nano-technologie (IEMN), UMR CNRS 8520, Cité Scientifique, Villeneuve d'Ascq Cedex, France, jean-francois.robillard@isen.fr

Ping Sheng Department of Physics, The Hong Kong University of Science and Technology, Kowloon, Hong Kong, China, sheng@ust.hk

A. Sukhovich Laboratoire Domaines Océaniques, UMR CNRS 6538, UFR Sciences et Techniques, Université de Bretagne Occidentale, Brest, France, alexei_suhov@yahoo.co.uk

N. Swinteck Department of Materials Science and Engineering, University of Arizona, Tucson, AZ, USA, swinteck@email.arizona.edu

G. Theocharis Engineering and Applied Science, California Institute of Technology, Pasadena, CA, USA, georgiostheocharis@gmail.com

J. O. Vasseur Institut d'Electronique, de Micro-électronique et de Nanotechnologie (IEMN), UMR CNRS 8520, Cité Scientifique, Villeneuve d'Ascq Cedex, France, Jerome.Vasseur@univ-lille1.fr

Jason Yang Department of Physics, The Hong Kong University of Science and Technology, Kowloon, Hong Kong, China

Min Yang Department of Physics, The Hong Kong University of Science and Technology, Kowloon, Hong Kong, China

Chapter 1
Introduction to Phononic Crystals and Acoustic Metamaterials

Pierre A. Deymier

Abstract The objective of this chapter is to introduce the broad subject of phononic crystals and acoustic metamaterials. From a historical point of view, we have tried to refer to some of the seminal contributions that have made the field. This introduction is not an exhaustive review of the literature. However, we are painting in broad strokes a picture that reflects the biased perception of this field by the authors and coauthors of the various chapters of this book.

1.1 Properties of Phononic Crystals and Acoustic Metamaterials

The field of phononic crystals (PCs) and acoustic metamaterials emerged over the past two decades. These materials are composite structures designed to tailor elastic wave dispersion (i.e., band structure) through Bragg's scattering or local resonances to achieve a range of spectral (ω-space), wave vector (k-space), and phase (ϕ-space) properties.

1.1.1 Spectral Properties

The development of phononic crystals for the control of vibrational waves followed by a few years the analogous concept of photonic crystals (1987) for electromagnetic waves [1]. Both concepts are based on the idea that a structure composed of a periodic arrangement of scatterers can affect quite strongly the propagation of classical waves

P.A. Deymier (✉)
Department of Materials Science and Engineering, University of Arizona, Tucson,
AZ 85721, USA
e-mail: deymier@email.arizona.edu

P.A. Deymier (ed.), *Acoustic Metamaterials and Phononic Crystals*,
Springer Series in Solid-State Sciences 173, DOI 10.1007/978-3-642-31232-8_1,
© Springer-Verlag Berlin Heidelberg 2013

such as acoustic/elastic or electromagnetic waves. The names photonic and phononic crystals are based on the elementary excitations associated with the particle description of vibrational waves (phonon) and electromagnetic waves (photon). The first observation of a periodic structure, a GaAs/AlGaAs superlattice, used to control the propagation of high-frequency phonons was reported by Narayanamurti et al. in 1979 [2]. Although not called a phononic crystal then, a superlattice is nowadays considered to be a one-dimensional phononic crystal. The actual birth of two-dimensional and three-dimensional phononic crystals can be traced back to the early 1990s. Sigalas and Economou demonstrated the existence of band gaps in the phonon density of state and band structure of acoustic and elastic waves in three-dimensional structures composed of identical spheres arranged periodically within a host medium [3] and in two-dimensional fluid and solid systems constituted of periodic arrays of cylindrical inclusions in a matrix [4]. The first full band structure calculation for transverse polarization of vibration in a two-dimensional periodic elastic composite was subsequently reported [5]. In 1995, Francisco Meseguer and colleagues determined experimentally the aural filtering properties of a perfectly real but fortuitous phononic crystal, a minimalist sculpture by Eusebio Sempere standing in a park in Madrid, Spain [6] (Fig. 1.1). This sculpture is a two-dimensional periodical square arrangement of steel tubes in air. They showed that attenuation of acoustic waves occurs at certain frequencies due not to absorption since steel is a very stiff material but due to multiple interferences of sound waves as the steel tubes behave as very efficient scatterers for sound waves. The periodic arrangement of the tubes leads to constructive or destructive interferences depending on the frequency of the waves. The destructive interferences attenuate the amplitude of transmitted waves, and the phononic structure is said to exhibit forbidden bands or band gaps at these frequencies. The properties of phononic crystals result from the scattering of acoustic or elastic waves (i.e., band folding effects) in a fashion analogous to Bragg scattering of X-rays by periodic crystals. The mechanism for the formation of band gaps in phononic crystals is a Bragg-like scattering of acoustic waves with wavelength comparable to the dimension of the period of the crystal i.e., the crystal lattice constant. The first experimentally observed ultrasonic full band gap for longitudinal waves was reported for an aluminum alloy plate with a square array of cylindrical holes filled with mercury [7]. The first experimental and theoretical demonstration of an absolute band gap in a two-dimensional solid-solid phononic crystal (triangular array of steel rods in an epoxy matrix) was demonstrated 3 years later [8]. The absolute band gap spanned the entire Brillouin zone of the crystal and was not limited to a specific type of vibrational polarization (i.e., longitudinal or transverse).

In 2000, Liu et al. [9] presented a class of sonic crystals that exhibited spectral gaps with lattice constants two orders of magnitude smaller than the relevant sonic wavelength. The formation of band gaps in these acoustic metamaterials is based on the idea of locally resonant structures. Because the wavelength of sonic waves is orders of magnitude larger than the lattice constant of the structure, periodicity is not necessary for the formation of a gap. Disordered composites made from such localized resonant structures behave as a material with effective negative elastic

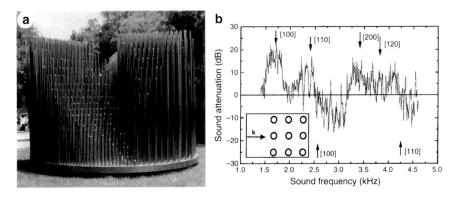

Fig. 1.1 (a) Eusebio Sempere's sculpture in Madrid, Spain, (b) Measured sound attenuation as a function of frequency. The *inset* illustrates the direction of propagation of sound waves. The *brackets* [hkl] represent, in the vocabulary of X-ray diffraction, crystallographic planes for which Bragg interferences will occur (after [6])

constants and a total wave reflector within certain tunable sonic frequency ranges. This idea was implemented with a simple cubic crystal consisting of a heavy solid core material (lead) coated with elastically soft material (silicone elastomer) embedded in a hard matrix material (epoxy). Centimeter size structures produced narrow transmission gaps at low frequencies corresponding to that of the resonances of the lead/elastomer resonator (Fig. 1.2).

While early phononic crystals and acoustic metamaterials research on spectral properties focused on frequencies in the sonic (10^2–10^3 Hz) and ultrasonic (10^4–10^6 Hz) range, phononic crystals with hypersonic (GHz) properties have been fabricated by lithographic techniques and analyzed using Brillouin Light Scattering [10]. It has also been shown theoretically and experimentally that phononic crystals may be used to reduce thermal conductivity by impacting the propagation of thermal phonons (THz) [11, 12].

Wave localization phenomena in defected phononic crystals containing linear and point defects have been also considered [13]. Kafesaki et al. [14] calculated the transmission of elastic waves through a straight waveguide created in a two-dimensional phononic crystal by removing a row of cylinders. The guidance of the waves is due to the existence of extended linear defect modes falling in the band gap of the phononic crystal. The propagation of acoustic waves through a linear waveguide, created inside a two-dimensional phononic crystal, along which a stub resonantor (point defect) was attached to its side has also been studied [15]. The primary effect of the resonator is to induce zeros of transmission in the transmission spectrum of the perfect waveguide. The transmittivity exhibits very narrow dips whose frequencies depend upon the width and the length of the stub. When a gap exists in the transmittivity of the perfect waveguide, the stub may also permit selective frequency transmission in this gap.

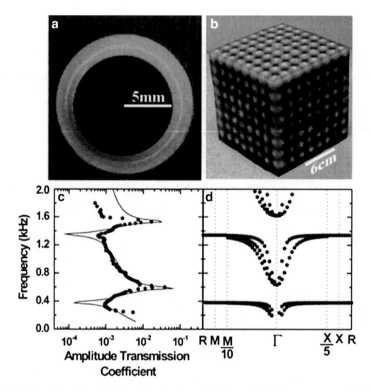

Fig. 1.2 (a) Cross section of a coated lead sphere that forms the basic structure unit (b) for an 8 × 8 × 8 sonic crystal. (c) Calculated (*solid line*) and measured (*circles*) amplitude transmission coefficient along the [100] direction as a function of frequency, (d) calculated band structure of a simple cubic structure of coated spheres in very good agreement with measurements (the directions to the left and the right of the Γ point are the [110] and [100] directions of the Brillouin zone, respectively (after [9])

In addition to bulk elastic waves, various authors have studied theoretically the existence of surface acoustic waves (SAW) localized at the free surface of a semi-infinite two-dimensional phononic crystal [16–19]. For this geometry, the parallel inclusions are of cylindrical shape and the surface considered is perpendicular to their axis. Various arrays of inclusions [16, 17], crystallographic symmetries of the component materials [9], and also the piezoelectricity of one of the constituent [19] were considered. The band structures of 2D phononic crystal plates with two free surfaces [20, 21] were also calculated. This includes the symmetric Lamb mode band structure of 2D phononic crystal plates composed of triangular arrays of W cylinders in a Si background. Charles et al. [21] reported on the band structure of a slab made of a square array of iron cylinders embedded in a copper matrix. Hsu and Wu [22] determined the lower dispersion curves in the band structure of 2D gold-epoxy phononic crystal plates. Moreover, Manzanares-Martinez and Ramos-Mendieta have also considered the propagation of acoustic waves along a surface parallel to the cylinders in a 2D phononic crystal [23]. Sainidou and Stefanou investigated the guided elastic waves in a glass plate coated on one side with a periodic monolayer of polymer spheres immersed in water [24]. On the experimental point of view, Wu et al. [25] observed high-frequency SAW with a pair of interdigital transducers placed on both sides of a very thick silicon plate in which a square array of holes was drilled. Similar experiments were conducted by Benchabane et al. on a 2D square lattice piezoelectric phononic crystal etched in lithium niobate [26]. Zhang et al. [27] have shown the existence of gaps for acoustic waves propagating at the surface of an air-aluminum 2D phononic crystal plate through laser ultrasonic measurements.

1.1.2   Wave Vector Properties

The wave vector (k-space) properties of phononic crystals and acoustic metamaterials result from passing bands with unique refractive characteristics, such as negative refraction or zero-angle refraction. Negative refraction of acoustic waves is analogous to negative refraction of electromagnetic waves also observed in electromagnetic and optical metamaterials [28]. Negative refraction is achieved when the wave group velocity (i.e., the direction of propagation of the energy) is antiparallel to the wave vector. In electromagnetic metamaterials, the unusual refraction is associated with materials that possess negative values of the permittivity and permeability , so-called double negative materials [29]. Negative refraction of acoustic waves may be achieved with double negative acoustic metamaterials in which both the effective mass density and bulk modulus are negative [30]. The double negativity of the effective dynamical mass and bulk modulus results from the coexistence in some specific range of frequency of monopolar and dipolar resonances [31]. The monopolar resonance may be due to a breathing mode of inclusions resonating out of phase with an incident acoustic wave leading to an effective negative bulk modulus. The dipolar resonance of heavy inclusions coated with a soft material embedded in a stiff matrix can result in a displacement of the center of mass of the metamaterial that is out of phase with the acoustic wave, leading to an effective negative dynamical mass density. Negative refraction may also be achieved through band-folding effect due to Bragg's scattering using phononic crystals. Band folding can produce bands with negative slopes (i.e., negative group velocity and positive phase velocity), a prerequisite for negative refraction. A combined theoretical and experimental study of a three-dimensional phononic crystal composed of tungsten carbide beads in water has shown the existence of a strongly anisotropic band with negative refraction [32]. A slab of this crystal was used to make a flat lens [33] to focus a diverging sound beam without curved interfaces typically employed in conventional lenses. A two-dimensional phononic crystal constituted of a triangular lattice of steel rods immersed in a liquid exhibited negative refraction and was used to focus ultrasound [34, 35]. High-fidelity imaging is obtained when all-angle negative refraction conditions are satisfied, that is, the equifrequency contour of the phononic crystal is circular and matches that of the medium in which it is embedded. A flat lens of this latter crystal achieved focusing and subwavelength imaging of acoustic waves [36]. This lens beat the diffraction limit of conventional lenses by transmitting the evanescent components of a sound point source via the excitation of a vibrational mode bound to the phononic crystal slab. In contrast, a conventional lens transmits only the propagating component of the source. Negative refraction of surface acoustic waves [37] and Lamb waves [38] has also been reported.

A broader range of unusual refractive properties was also reported in a study of a phononic crystal consisting of a square array of cylindrical polyvinylchloride (PVC) inclusions in air [39]. This crystal exhibits positive, negative, or zero refraction depending on the angle of the incident sound beam. Zero angle refraction can lead to wave guiding/localization without defects. The refraction in this crystal is highly anisotropic due to the nearly square shape of the fourth vibrational band. For all three cases of refraction, the transmitted beam undergoes splitting upon exiting the crystal because the equifrequency contour on the incident medium (air) in which a slab of the phononic crystal is immersed is larger than the Brillouin zone of the crystal. In this case Block modes in the extended Brillouin zone are excited inside the crystal and produce multiple beams upon exit.

1.1.3 Phase properties

Only recently has progress been made in the extension of properties of phononic crystals beyond ω-k space and into the space of acoustic wave phase (ϕ-space). The concept of phase control between propagating waves in a phononic crystal can be realized through analysis of its band structure and equi-frequency contours [40]. The dominant mechanism behind the control of phase between propagating acoustic waves in two-dimensional phononic crystals arises from the non-colinearity of the wave vector and the group velocity.

1.2 Beyond Macroscopic, Linear Elastic, Passive Structures and Media

Until recently, phononic crystals and acoustic metamaterials have been constituted of passive media satisfying continuum linear elasticity. A richer set of properties is emerging by utilizing dissipative media or media obeying nonlinear elasticity. Lossy media can be used to modify the dispersive properties of phononic crystals. Acoustic structures composed of nonlinear media can support nondispersive waves. Composite structures constituted of active media, media responding to internal or external stimuli, enable the tunability of their band properties.

1.2.1 Dissipative Media

Psarobas studied the behavior of a composite structure composed of close packed viscoelastic rubber spheres in air [41]. He reported the existence of an appreciable omnidirectional gap in the transmission spectrum in spite of the losses. The existence of band gaps in phononic crystals constituted of viscoelastic silicone rubber and air was also reported [42]. It was also shown that viscoelasticity did not only attenuate acoustic waves traversing a rubber-based phononic crystal but also modified the frequency of passing bands in the transmission spectrum [43]. A theory of damped Bloch waves [44] was employed to show that damping alters the shape of dispersion curves and reduces the size of band gaps as well as opens wave vector gaps via branch cutoff [45]. Loss has an effect on the complete complex band structure of phononic systems including the group velocity [46].

1.2.2 Nonlinear Media

In this subsection, we introduce only the nonlinear behavior of granular-type acoustic structures. The nonlinearity of vibrational waves in materials at the atomic scales due to the anharmonic nature of interatomic forces will be addressed in Sect. 1.2.4. The nonlinearity of contact forces between grains in granular materials has inspired the design of strongly nonlinear phononic structures. Daraio has demonstrated that a one-dimensional phononic crystal assembled as a chain of polytetrafluoroethylene (PTE-Teflon) spheres supports strongly nonlinear solitary waves with very low speed [47]. Using a similar system composed of a chain of stainless-steel spheres, Daraio has also shown the tunability of wave propagation properties [48]. Precompression of the chain of spheres lead to a significant increase in solitary wave speed. The study of noncohesive granular phononic crystals lead to the prediction of translational modes but also, due to the rotational degrees of freedom, of rotational modes and coupled rotational and translational modes [49]. The dispersion laws of these modes may also be tuned by an external loading on the granular structure.

1.2.3 Tunable Structures

To date the applications of phononic crystals and acoustic metamaterials have been limited because their constitutive materials exhibit essentially passive responses. The ability to control and tune the phononic/acoustic properties of these materials may overcome these limitations. Tunability may be achieved by changing the geometry of the inclusions [50] or by varying the elastic characteristics of the constitutive materials through application of contact and noncontact external stimuli [51]. For instance, some authors have proposed the use of electro-rheological materials in conjunction with application of an external electric field [52]. Some authors have considered the effect of temperature on the elastic moduli [53, 54]. Other authors [55] have controlled the band structure of a phononic crystal by applying an external stress that alters the crystal's structure. Tunability can also be achieved by using active constitutive materials. Following this approach, some authors [56, 57] have studied how the piezoelectric effect can influence the elastic properties of a PC and subsequently change its dispersion curves and gaps. Several studies have also reported noticeable changes in the band structures of magneto-electro-elastic phononic crystals when the coupling between magnetic, electric, and elastic phenomena are taken into account [58, 59] or when external magnetic fields are applied [60].

1.2.4   Scalability

The downscaling of phononic structures to nanometric dimensions requires an atomic treatment of the constitutive materials. At the nanoscale, the propagation of phonons may not be completely ballistic (wave-like) and nonlinear phenomena such as phonon–phonon scattering (Normal and Umklapp processes) occur. These nonlinear phenomena are at the core of the finiteness of the thermal conductivity of materials. Gillet et al. investigated the thermal-insulating behavior of atomic-scale three-dimensional nanoscale phononic crystals [11]. The phononic crystal consists of a matrix of diamond-cubic Silicon with a periodic array of nano-particles of Germanium (obtained by substitution of Si atoms by Ge atoms inside the phonoic crystal unit cell). These authors calculated the band structure of the nanoscale phononic crystal with classical lattice dynamics. They showed a flat-tening of the dispersion curves leading to a significant decrease in the phonon group velocities. This decrease leads to a reduction in thermal conductivity. In addition to these linear effects associated with Bragg scattering of the phonons by the periodic array of inclusions, another reduction in thermal conductivity is obtained from multiple inelastic scattering of the phonons using Boltzmann transport equation. The nanomaterial thermal conductivity can be reduced by several orders of magnitude compared with bulk Si. Atomistic computational methods such as molecular dynamics and the Green-Kubo method were employed to shed light on the transport behavior of thermal phonons in models of graphene-based nanophononic crystals comprising periodic arrays of holes [61]. The pho-non lifetime and thermal conductivity as a function of the crystal filling fraction and temperature were calculated. These calculations suggested a competition between elastic Bragg's scattering and inelastic phonon–phonon scattering and an effect of elastic scattering via modification of the band structure on the phonon lifetime (i.e., inelastic scattering).

1.3 Phoxonic Structures

Recent effort has been aimed at designing periodic structures that can control simultaneously the propagation of phonons and photons. Such periodic materials possess band structure characteristics such as the simultaneous existence of photonic and phononic band gaps. For this reason, these materials are named "phoxonic" materials. Maldovan and Thomas have shown theoretically that simultaneous two-dimensional phononic and photonic band gaps exist for in-plane propagation in periodic structures composed of square and triangular arrays of cylindrical holes in silicon [62]. They have also shown localization of photonic and phononic waves in defected phoxonic structures. Simultaneous photonic and phononic band gaps have also been demonstrated computationally in two-dimensional phoxonic crystal structures constituted of arrays of air holes in lithium niobate [63]. Planar structures such as phoxonic crystal composed of arrays of void cylindrical holes in silicon slabs with a finite thickness have been shown to possess simultaneous photonic and phononic band gaps [64]. Other examples of phoxonic crystals include three-dimensional lattices of metallic nanospheres embedded into a dielectric matrix [65]. Phoxonic crystals with spectral gaps for both optical and acoustic waves are particularly suited for applications that involve acousto-optic interactions to control photons with phonons. The confinement of photons and phonons in a one-dimensional model of a phoxonic cavity incorporating nonlinear acousto-optic effects was shown to lead to enhanced modulation of light by acoustic waves through multiphonon exchange mechanisms [66].

References

1. E. Yablonovitch, Inhibited spontaneous emission in solid state physics and electronics. Phys. Rev. Lett. **58**, 2059 (1987)
2. V. Narayanamurti, H.L. Störmer, M.A. Chin, A.C. Gossard, W. Wiegmann, Selective transmission of high-frequency phonons by a superlattice: the "Dielectric" phonon filter. Phys. Rev. Lett. **2**, 2012 (1979)
3. M.M. Sigalas, E.N. Economou, Elastic and acoustic wave band structure. J. Sound Vib. **158**, 377 (1992)
4. M. Sigalas, E. Economou, Band structure of elastic waves in two-dimensional systems. Solid State Commun. **86**, 141 (1993)
5. M.S. Kushwaha, P. Halevi, L. Dobrzynski, B. Djafari-Rouhani, Acoustic band structure of periodic elastic composites. Phys. Rev. Lett. **71**, 2022 (1993)
6. R. Martinez-Salar, J. Sancho, J.V. Sanchez, V. Gomez, J. Llinares, F. Meseguer, Sound attenuation by sculpture. Nature **378**, 241 (1995)
7. F.R. Montero de Espinoza, E. Jimenez, M. Torres, Ultrasonic band gap in a periodic two-dimensional composite. Phys. Rev. Lett. **80**, 1208 (1998)
8. J.O. Vasseur, P. Deymier, B. Chenni, B. Djafari-Rouhani, L. Dobrzynski, D. Prevost, Experimental and theoretical evidence for the existence of absolute acoustic band gap in two-dimensional periodic composite media. Phys. Rev. Lett. **86**, 3012 (2001)

9. Z. Liu, X. Zhang, Y. Mao, Y.Y. Zhu, Z. Yang, C.T. Chan, P. Sheng, Locally resonant sonic crystal. Science **289**, 1734 (2000)
10. T. Gorishnyy, C.K. Ullal, M. Maldovan, G. Fytas, E.L. Thomas, Hypersonic phononic crystals. Phys. Rev. Lett. **94**, 115501 (2005)
11. J.N. Gillet, Y. Chalopin, S. Volz, Atomic-scale three-dimensional phononic crystals with a very low thermal conductivity to design crystalline thermoelectric devices. J. Heat Transfer **131**, 043206 (2009)
12. P.E. Hopkins, C.M. Reinke, M.F. Su, R.H. Olsson III, E.A. Shaner, Z.C. Leseman, J.R. Serrano, L.M. Phinney, I. El-Kady, Reduction in the thermal conductivity of single crystalline silicon by phononic crystal patterning. Nano Lett. **11**, 107 (2011)
13. M. Torres, F.R. Montero de Espinosa, D. Garcia-Pablos, N. Garcia, Sonic band gaps in finite elastic media: surface states and localization phenomena in linear and point defects. Phys. Rev. Lett. **82**, 3054 (1999)
14. M. Kafesaki, M.M. Sigalas, N. Garcia, Frequency modulation in the transmittivity of wave guides in elastic-wave band-gap materials. Phys. Rev. Lett. **85**, 4044 (2000)
15. A. Khelif, B. Djafari-Rouhani, J.O. Vasseur, P.A. Deymier, P. Lambin, L. Dobrzynski, Transmittivity through straight and stublike waveguides in a two-dimensional phononic crystal. Phys. Rev. B **65**, 174308 (2002)
16. Y. Tanaka, S.I. Tamura, Surface acoustic waves in two-dimensional periodic elastic structures. Phys. Rev. B **58**, 7958 (1998)
17. Y. Tanaka, S.I. Tamura, Acoustic stop bands of surface and bulk modes in two-dimensional phononic lattices consisting of aluminum and a polymer. Phys. Rev. B **60**, 13294 (1999)
18. T.T. Wu, Z.G. Huang, S. Lin, Surface and bulk acoustic waves in two-dimensional phononic crystal consisting of materials with general anisotropy. Phys. Rev. B **69**, 094301 (2004)
19. V. Laude, M. Wilm, S. Benchabane, A. Khelif, Full band gap for surface acoustic waves in a piezoelectric phononic crystal. Phys. Rev. E **71**, 036607 (2005)
20. J.J. Chen, B. Qin, J.C. Cheng, Complete band gaps for lamb waves in cubic thin plates with periodically placed inclusions. Chin. Phys. Lett. **22**, 1706 (2005)
21. C. Charles, B. Bonello, F. Ganot, Propagation of guided elastic waves in two-dimensional phononic crystals. Ultrasonics **44**, 1209(E) (2006)
22. J.C. Hsu, T.T. Wu, Efficient formulation for band-structure calculations of two-dimensional phononic-crystal plates. Phys. Rev. B **74**, 144303 (2006)
23. B. Manzanares-Martinez, F. Ramos-Mendieta, Surface elastic waves in solid composites of two-dimensional periodicity. Phys. Rev. B **68**, 134303 (2003)
24. R. Sainidou, N. Stefanou, Guided and quasiguided elastic waves in phononic crystal slabs. Phys. Rev. B **73**, 184301 (2006)
25. T.T. Wu, L.C. Wu, Z.G. Huang, Frequency band gap measurement of two-dimensional air/silicon phononic crystals using layered slanted finger interdigital transducers. J. Appl. Phys. **97**, 094916 (2005)
26. S. Benchabane, A. Khelif, J.-Y. Rauch, L. Robert, V. Laude, Evidence for complete surface wave band gap in a piezoelectric phononic crystal. Phys. Rev. E **73**, 065601(R) (2006)
27. X. Zhang, T. Jackson, E. Lafond, P. Deymier, J. Vasseur, Evidence of surface acoustic wave band gaps in the phononic crystals created on thin plates. Appl. Phys. Lett. **88**, 0419 (2006)
28. N. Engheta, R.W. Ziolkowski, *Metamaterials: Physics and Engineering Explorations* (Wiley, New York, 2006)
29. V.G. Veselago, The electrodynamics of substances with simultaneous negative values of ε and μ. Sov. Phys. Usp. **10**, 509 (1967)
30. J. Li, C.T. Chan, Double negative acoustic metamaterials. Phys. Rev. E **70**, 055602 (2004)
31. Y. Ding, Z. Liu, C. Qiu, J. Shi, Metamaterials with simultaneous negative bulk modulus and mass density. Phys. Rev. Lett. **99**, 093904 (2007)
32. S. Yang, J.H. Pahe, Z. Liu, M.L. Cowan, C.T. Chan, P. Sheng, Focusing of sound in a 3D phononic crystal. Phys. Rev. Lett. **93**, 024301 (2004)
33. J.B. Pendry, Negative refraction makes a perfect lens. Phys. Rev. Lett. **85**, 3966 (2000)

34. M. Ke, Z. Liu, C. Qiu, W. Wang, J. Shi, W. Wen, P. Sheng, Negative refraction imaging with two-dimensional phononic crystals. Phys. Rev. B **72**, 064306 (2005)
35. A. Sukhovich, L. Jing, J.H. Page, Negative refraction and focusing of ultrasound in two-dimensional phononic crystals. Phys. Rev. B **77**, 014301 (2008)
36. A. Sukhovich, B. Merheb, K. Muralidharan, J.O. Vasseur, Y. Pennec, P.A. Deymier, J.H. Pae, Experimental and theoretical evidence for subwavelength imaging in phononic crystals. Phys. Rev. Lett. **102**, 154301 (2009)
37. B. Bonello, L. Beillard, J. Pierre, J.O. Vasseur, B. Perrin, O. Boyko, Negative refraction of surface acoustic waves in the subgigahertz range. Phys. Rev. B **82**, 104108 (2010)
38. M.K. Lee, P.S. Ma, I.K. Lee, H.W. Kim, Y.Y. Kim, Negative refraction experiments with guided shear horizontal waves in thin phononic crystal plates. Appl. Phys. Lett. **98**, 011909 (2011)
39. J. Bucay, E. Roussel, J.O. Vasseur, P.A. Deymier, A.-C. Hladky-Hennion, Y. Penec, K. Muralidharan, B. Djafari-Rouhani, B. Dubus, Positive, negative, zero refraction and beam splitting in a solid/air phononic crystal: theoretical and experimental study. Phys. Rev. B **79**, 214305 (2009)
40. N. Swinteck, J.-F. Robillard, S. Bringuier, J. Bucay, K. Muralidaran, J.O. Vasseur, K. Runge, P.A. Deymier, Phase controlling phononic crystal. Appl. Phys. Lett. **98**, 103508 (2011)
41. I.E. Psarobas, Viscoelastic response of sonic band-gap materials. Phys. Rev. B **64**, 012303 (2001)
42. B. Merheb, P.A. Deymier, M. Jain, M. Aloshyna-Lessuffleur, S. Mohanty, A. Berker, R.W. Greger, Elastic and viscoelastic effects in rubber/air acoustic band gap structures: a theoretical and experimental study. J. Appl. Phys. **104**, 064913 (2008)
43. B. Merheb, P.A. Deymier, K. Muralidharan, J. Bucay, M. Jain, M. Aloshyna-Lesuffleur, R.W. Greger, S. Moharty, A. Berker, Viscoelastic Effect on acoustic band gaps in polymer-fluid composites. Model. Simul. Mat. Sci. Eng. **17**, 075013 (2009)
44. M.I. Hussein, Theory of damped bloch waves in elastic media. Phys. Rev. B **80**, 212301 (2009)
45. M.I. Hussein, M.J. Frazier, Band structure of phononic crystals with general damping. J. Appl. Phys. **108**, 093506 (2010)
46. R.P. Moiseyenko, V. Laude, Material loss influence on the complex band structure and group velocity of phononic crystals. Phys. Rev. B **83**, 064301 (2011)
47. C. Daraio, V.F. Nesterenko, E.B. Herbold, S. Jin, Strongly non-linear waves in a chain of Teflon beads. Phys. Rev. E **72**, 016604 (2005)
48. C. Daraio, V. Nesterenko, E. Herbold, S. Jin, Tunability of solitary wave properties in one-dimensional strongly non-linear phononic crystals. Phys. Rev. E **73**, 26610 (2006)
49. A. Merkel, V. Tournat, V. Gusev, Dispersion of elastic waves in three-dimensional noncohesive granular phononic crystals: properties of rotational modes. Phys. Rev. E **82**, 031305 (2010)
50. C. Goffaux, J.P. Vigneron, Theoretical study of a tunable phononic band gap system. Phys. Rev. B **64**, 075118 (2001)
51. J. Baumgartl, M. Zvyagolskaya, C. Bechinger, Tailoring of phononic band structure in colloidal crystals. Phys. Rev. Lett. **99**, 205503 (2007)
52. J.-Y. Yeh, Control analysis of the tunable phononic crystal with electrorheological material. Physica B **400**, 137 (2007)
53. Z.-G. Huang, T.-T. Wu, Temperature effect on the band gaps of surface and bulk acoustic waves in two-dimensional phononic crystals. IEEE Trans. Ultrason. Ferroelectr. Freq. Control **52**, 365 (2005)
54. K.L. Jim, C.W. Leung, S.T. Lau, S.H. Choy, H.L.W. Chan, Thermal tuning of phononic bandstructure in ferroelectric ceramic/epoxy phononic crystal. Appl. Phys. Lett. **94**, 193501 (2009)
55. K. Bertoldi, M.C. Boyce, Mechanically triggered transformations of phononic band gaps in periodic elastomeric structures. Phys. Rev. B **77**, 052105 (2008)

56. Z. Hou, F. Wu, Y. Liu, Phononic crystals containing piezoelectric material. Solid State Commun. **130**, 745 (2004)
57. Y. Wang, F. Li, Y. Wang, K. Kishimoto, W. Huang, Tuning of band gaps for a two-dimensional piezoelectric phononic crystal with a rectangular lattice. Acta Mech. Sin. **25**, 65 (2008)
58. Y.-Z. Wang, F.-M. Li, W.-H. Huang, X. Jiang, Y.-S. Wang, K. Kishimoto, Wave band gaps in two-dimensional piezoelectric/piezomagnetic phononic crystals. Int. J. Solids Struct. **45**, 4203 (2008)
59. Y.-Z. Wang, F.-M. Li, K. Kishimoto, Y.-S. Wang, W.-H. Huang, Elastic wave band gaps in magnetoelectroelastic phononic crystals. Wave Motion **46**, 47 (2009)
60. J.-F. Robillard, O. Bou Matar, J.O. Vasseur, P.A. Deymier, M. Stippinger, A.-C. Hladky-Hennion, Y. Pennec, B. Djafari-Rouhani, Tunable magnetoelastic phononic crystals. Appl. Phys. Lett. **95**, 124104 (2009)
61. J.-F. Robillard, K. Muralidharan, J. Bucay, P.A. Deymier, W. Beck, D. Barker, Phononic metamaterials for thermal management: an atomistic computational study. Chin. J. Phys. **49**, 448 (2011)
62. M. Maldovan, E.L. Thomas, Simultaneous complete elastic and electromagnetic band gaps in periodic structures. Appl. Phys. B **83**(4), 595 (2006)
63. M. Maldovan, E.L. Thomas, Simultaneous localization of photons and phonons in two-dimensional periodic structures. Appl. Phys. Lett. **88**(25), 251 (2006)
64. S. Sadat-Saleh, S. Benchabane, F. Issam Baida, M.-P. Bernal, V. Laude, Tailoring simultaneous photonic and phononic band gaps. J. Appl. Phys. **106**, 074912 (2009)
65. N. Papanikolaou, I.E. Psarobas, N. Stefanou, Absolute spectral gaps for infrared ligth and hypersound in three-dimensional metallodielectric phoxonic crystals. Appl. Phys. Lett, **96**, 231917 (2010)
66. I.E. Psarobas, N. Papanikolaou, N. Stefanou, B. Djafari-Rouhani, B. Bonello, V. Laude, Enhanced acousto-optic interactions in a one-dimensional phoxonic cavity. Phys. Rev. B **82**, 174303 (2010)

Chapter 2
Discrete One-Dimensional Phononic and Resonant Crystals

Pierre A. Deymier and L. Dobrzynski

Abstract The objective of this chapter is to introduce the broad range of concepts necessary to appreciate and understand the various aspects and properties of phononic crystals and acoustic metamaterials described in subsequent chapters. These concepts range from the most elementary concepts of vibrational waves, propagating waves, and evanescent waves, wave vector, phase and group velocity, Bloch waves, Brillouin zone, band structure and band gaps, and bands with negative group velocities in periodic or locally resonant structures. Simple models based on the one-dimensional harmonic crystal serve as vehicles for illustrating these concepts. We also illustrate the application of some of the tools used to study and analyze these simple models. These analytical tools include eigenvalue problems ($\omega(k)$ or $k(\omega)$) and Green's function methods. The purpose of this chapter is primarily pedagogical. However, the simple models discussed herein will also serve as common threads in each of the other chapters of this book.

2.1 One-Dimensional Monoatomic Harmonic Crystal

The one-dimensional (1-D) monoatomic harmonic crystal consists of an infinite chain of masses, m, with nearest neighbor interaction modeled by harmonic springs with spring constant, β. The separation distance between the masses at rest is defined as a. This model system is illustrated in Fig. 2.1.

P.A. Deymier (✉)
Department of Materials Science and Engineering, University of Arizona, Tucson, AZ 85721, USA
e-mail: deymier@email.arizona.edu

L. Dobrzynski
Equipe de Physique, des Ondes, des Nanostructures et des Interfaces, Groupe de Physique, Institut d'Electronique, de Microélectronique et de Nanotechnologie, Université des Sciences et Technologies de Lille, Centre National de la Recherche Scientifique (UMR 8520), 59655 Villeneuve d'Ascq Cédex, France
e-mail: Leonard.Dobrzynski@Univ-Lille1.fr

P.A. Deymier (ed.), *Acoustic Metamaterials and Phononic Crystals*,
Springer Series in Solid-State Sciences 173, DOI 10.1007/978-3-642-31232-8_2,
© Springer-Verlag Berlin Heidelberg 2013

n-1   **n**   **n+1**

Fig. 2.1 Schematic illustration of one 1-D mono-atomic harmonic crystal

In the absence of external forces, the equation describing the motion of atom "n" is given by

$$m\frac{d^2u_n}{dt^2} = \beta(u_{n+1} - u_n) - \beta(u_n - u_{n-1}).$$ (2.1)

In this equation, u_n represents the displacement of the mass "n" with respect to its position at rest. The first term on the right-hand side of the equal sign is the harmonic force on mass "n" resulting from the spring on its right. The second term is the force due to the spring on the left of "n." The dynamics of the 1-D monoatomic harmonic crystal can, therefore, be studied by solving (2.2):

$$m\frac{d^2u_n}{dt^2} = \beta(u_{n+1} - 2u_n + u_{n-1}).$$ (2.2)

The next subsections aim at seeking solutions of (2.2).

2.1.1 Propagating Waves

We seek solutions to (2.2) in the form of propagating waves:

$$u_n = Ae^{ikna}e^{i\omega t},$$ (2.3)

where k is a wave number and ω is an angular frequency. Inserting solutions of the form given by (2.3) into (2.2) and simplifying by $Ae^{ikna}e^{i\omega t}$, one obtains the relation between angular frequency and wave number:

$$\omega^2 = -\frac{\beta}{m}\left(e^{\frac{ika}{2}} - e^{-\frac{ika}{2}}\right)^2.$$ (2.4)

We use the relation $2i\sin\theta = e^{i\theta} - e^{-i\theta}$ and the fact that ω is a positive quantity to obtain the so-called dispersion relation for propagating waves in the 1-D harmonic crystal:

$$\omega(k) = \omega_0\left|\sin k\frac{a}{2}\right|,$$ (2.5)

with $\omega_0 = 2\sqrt{\frac{\beta}{m}}$ representing the upper limit for angular frequency. Since the monoatomic crystal is discrete and waves with wave-length $\lambda = \frac{2\pi}{k}$ larger than $2a$

Fig. 2.2 Illustration of the
dispersion relation for
propagating waves in 1-D
mono-atomic harmonic
crystal

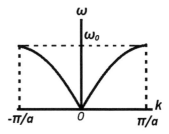

are physically equivalent to those with wave-length smaller than $2a$, the dispersion relation of (2.5) needs only be represented in the symmetrical interval $k \in \left[-\frac{\pi}{a}, \frac{\pi}{a}\right]$ (see Fig. 2.2). This interval is the first Brillouin zone of the 1-D monoatomic periodic crystal.

2.1.2   Phase and Group Velocity

The velocity at which the phase of the wave with wave vector, k, and angular frequency, ω, propagates is defined as

$$v_\varphi = \frac{\omega}{k}. \tag{2.6}$$

The group velocity is defined as the velocity at which a wave packet (a superposition of propagating waves with different values of wave number ranging over some interval) propagates. It is easier to understand this concept by considering the superposition of only two waves with angular velocities, ω_1 and ω_2, and wave vectors, k_1 and k_2. Choosing, $\omega_1 = \omega - \frac{\Delta\omega}{2}$ and $\omega_2 = \omega + \frac{\Delta\omega}{2}$, and, $k_1 = k - \frac{\Delta k}{2}$ and $k_2 = k + \frac{\Delta k}{2}$. The superposition of the two waves, assuming that they have the same amplitude, A, leads to the displacement field at mass "n":

$$u_n^s = 2A e^{ikna} e^{i\omega t} \cos\left(\frac{\Delta k}{2} na + \frac{\Delta\omega}{2} t\right). \tag{2.7}$$

The first part of the right-hand side of (2.7) is a traveling wave that is modulated by the cosine term. This later term represents a beat pulse. The velocity at which this modulation travels is the group velocity and is given by

$$v_g = \frac{\Delta\omega}{\Delta k}. \tag{2.8}$$

In the limit of infinitesimally small differences in wave number and frequency, the group velocity is expressed as a derivative of the dispersion relation:

$$v_g = \frac{d\omega(k)}{dk}.$$

(2.9)

In the case of the 1-D harmonic crystal, the group velocity is given by $v_g = \omega_0 \frac{a}{2} \cos k \frac{a}{2}$.

We now open a parenthesis concerning the group velocity and show that it is also equal to the velocity of the energy transported by a propagating wave. To that effect, we calculate the average energy density as the sum of the potential energy and the kinetic energy averaged over one cycle of time. The average energy is given by

$$\langle E \rangle = \frac{1}{2}\beta(u_n - u_{n-1})(u_n - u_{n-1})^* + \frac{1}{2}m\dot{u}_n\dot{u}_n^*.$$

(2.10)

In (2.10), the * denotes the complex conjugate and $\dot{u}$ the time derivative of the displacement (i.e., the velocity of the mass "n"). Inserting into (2.10) the displacements given by (2.3) and the dispersion relation given by (2.5) yields the average energy density:

$$\langle e \rangle = \frac{\langle E \rangle}{a} = 4A^2 \frac{\beta}{a} \sin^2 k \frac{a}{2}.$$

(2.11)

We now calculate the energy flow through one unit cell of the 1-D crystal in the form of the real part of the power, Φ, defined as the product of the force on mass "n" due to one spring and the velocity of the mass:

$$\Phi = \mathrm{Re}\{\beta(u_{n+1} - u_n)\dot{u}_n^*\} = \beta A^2 \omega_0 \left| \sin k \frac{a}{2} \right| \sin ka.$$

(2.12)

The velocity of the energy, v_e, is therefore the ratio of the energy flow to the average energy density, which after using trigonometric relations yields: $v_e = \omega_0 \frac{a}{2} \cos k \frac{a}{2}$. This expression is the same as that of the group velocity. In summary, the group velocity represents also the velocity of the energy transported by the propagating waves in the crystal.

2.1.3 Evanescent Waves

In Sect. 2.1.1, we sought solutions to the equation of motion (2.2) in the form of propagating waves (Eq. (2.3)). We may also seek solutions in the form of nonpropagating waves with exponentially decaying amplitude:

$$u_n = A e^{-k''na} e^{ik'na} e^{i\omega t}.$$

(2.13)

Equation (2.13) can be obtained by inserting a complex wave number $k = k' + ik''$ into (2.3). Combining solutions of the form given by (2.13) and the equation of motion (2.2), one gets

$$- m\omega^2 = \beta \left(e^{ik'\frac{a}{2}} e^{-k''\frac{a}{2}} - e^{-ik'\frac{a}{2}} e^{k''\frac{a}{2}} \right)^2. \tag{2.14}$$

Since the mass and the angular frequency are positive numbers, (2.14) possesses solutions only when the difference inside the parenthesis is imaginary. This condition is met at the edge of the Brillouin zone, when, $k' = \frac{\pi}{a}$. In this case, (2.14) yields the dispersion relation:

$$\omega = \omega_0 \cosh k'' \frac{a}{2}. \tag{2.15}$$

This condition is only met for angular frequencies greater than ω_0, that is, for frequencies above the dispersion curves of propagating waves illustrated in Fig. 2.2.

The solutions of (2.2) in the form of propagating and evanescent waves did not need to be postulated as was done above and in Sect. 2.1.1. We illustrate below a different path to solving (2.2). Instead of solving for the frequency as a function of wave number, this approach solves for the wave number as a function of frequency. This approach is particularly interesting as it will enable us to determine iso-frequency maps in wave vector space when dealing with 2-D or 3-D phononic structures.

We start with (2.4) and rewrite it in the form

$$- m\omega^2 = \beta(e^{ika} - 2 + e^{-ika}). \tag{2.16}$$

We now define the new variable: $X = e^{ika}$. Consequently, equation (2.16) becomes a quadratic equation in terms of X:

$$X^2 + \left(\frac{m}{\beta} \omega^2 - 2 \right) X + 1 = 0. \tag{2.17}$$

This equation has two solutions, which in terms of ω_0 are

$$X = \frac{1}{\omega_0^2} \left(\omega_0^2 - 2\omega^2 \right) \pm \frac{2}{\omega_0^2} \sqrt{\omega^2 \left(\omega^2 - \omega_0^2 \right)}. \tag{2.18a}$$

The solutions given by (2.18a) are real or complex depending on the value of the angular frequency. Let us consider first the case, $\omega \leq \omega_0$, for which

$$X = \frac{1}{\omega_0^2} \left(\omega_0^2 - 2\omega^2 \right) \pm \frac{2i}{\omega_0^2} \sqrt{\omega^2 \left(\omega_0^2 - \omega^2 \right)}. \tag{2.18b}$$

We now generalize the problem to complex wave numbers $k = k' + ik''$. In this case, X should take the form

$$X = e^{-k''a}\cos k'a + ie^{-k''a}\sin k'a. \qquad (2.19)$$

We identify the real and imaginary parts of equations (2.18b) and (2.19) and solve for k'' and k'. We find using standard trigonometric relations that $k'' = 0$ and $\sin^2 k' \frac{a}{2} = \frac{\omega^2}{\omega_0^2}$. This solution corresponds to propagating waves with a dispersion relation equivalent to that previously found in Sect. 2.1.1 (Eq. (2.5)).

In contrast, when we consider $\omega > \omega_0$, (2.18a) remains purely real. The real part of (2.19) should then be equal to the right-hand side term of (2.18a). We will denote this term $h^{\pm}(\omega)$. The imaginary part of (2.19) is zero. A trivial solution exists for $k' = 0$. However, in this case, the function $h^{\pm}(\omega)$ is always negative and one cannot find a corresponding value for k''. There exists another solution, namely, $k' = \frac{\pi}{a}$ (there is also a similar solution $k' = -\frac{\pi}{a}$), for which, we obtain

$$k''^{\pm}(\omega) = -\frac{1}{a}\ln(-h^{\pm}(\omega)). \qquad (2.20)$$

One of the solutions given by (2.20) is positive and the other negative. In the former case, the displacement is representative of an exponentially decaying evanescent wave. In the latter case, the displacement grows exponentially. This second solution is unphysical. This unphysical solution is a mathematical artifact of the approach used here as it leads to a quadratic equation in X (i.e., k) for a 1-D monoatomic crystal. Since this crystal has only one mass per unit cell "a" it should exhibit only one solution for $\omega(k)$ in the complex plane. We illustrate in Fig. 2.3 the dispersion relations for the propagating and evanescent waves in the complex plane $k = k' + ik''$.

2.1.4 Green's Function Approach

In anticipation of subsequent sections where Green's function approaches will be used to shed light on the vibrational behavior of more complex harmonic structures, we present here the Green's function formalism applied to the 1-D monoatomic crystal. Considering harmonic solution with angular frequency ω, the equation of motion (2.2) can be recast in the form

$$\frac{1}{m}[\beta u_{n+1} + (m\omega^2 - 2\beta)u_n + \beta u_{n-1}] = 0. \qquad (2.21)$$

Fig. 2.3 Dispersion curves
for the 1-D mono-atomic
harmonic crystal extended to
the wave-number complex
plane. The *black solid curves*
are for propagating waves,
and the *grey solid curve* is for
the evanescent waves

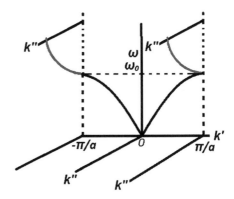

We now rewrite (2.21) in matrix form when applying it to all masses in the 1-D monoatomic crystal:

$$\overleftrightarrow{H_0}\,\vec{u} = \frac{1}{m}
\begin{bmatrix}
\ddots & \vdots & \vdots & \vdots & \vdots & \vdots & \vdots & \vdots & \vdots \\
\cdots & 0 & \beta & -\gamma & \beta & 0 & 0 & 0 & \cdots \\
\cdots & 0 & 0 & \beta & -\gamma & \beta & 0 & 0 & \cdots \\
\cdots & 0 & 0 & 0 & \beta & -\gamma & \beta & 0 & \cdots \\
& \vdots & \vdots & \vdots & \vdots & \vdots & \vdots & \vdots & \ddots
\end{bmatrix}
\begin{bmatrix}
\vdots \\ u_{n-1} \\ u_n \\ u_{n+1} \\ \vdots
\end{bmatrix}
=
\begin{bmatrix}
\vdots \\ 0 \\ 0 \\ 0 \\ \vdots
\end{bmatrix}, \quad (2.22)$$

where $\gamma = 2\beta - m\omega^2$. The operator, $\overleftrightarrow{H_0}$, is a more compact representation of the dynamic matrix in (2.22), and $\vec{u}$ is the vector whose components are the displacements of the masses in the crystal. With this notation, the Green's function, $\overleftrightarrow{G_0}$, associated with $\overleftrightarrow{H_0}$ is defined by the relation

$$\overleftrightarrow{H_0}\overleftrightarrow{G_0} = \overleftrightarrow{I}. \qquad (2.23)$$

In this equation, $\vec{I}$ is the identity matrix. Equation (2.23) is written in component form as

$$\sum_{n''} H_0(n, n'')G_0(n'', n') = \delta_{nn'}. \qquad (2.24)$$

Here, we have used the Kroenecker symbol δ'_{nn} to represent the components of the identity matrix, that is 1 when $n = n'$ and 0 when $n \neq n'$. Since $\overleftrightarrow{H_0}$ is tridiagonal (harmonic interactions are limited to first nearest neighbors), (2.23) becomes

$$\frac{1}{m}[\beta G_0(n+1, n') - \gamma G_0(n, n') + \beta G_0(n-1, n')] = \delta_{nn'}. \qquad (2.25)$$

From a physical point of view, the Green's function $G_0(n, n')$ is the displacement of mass "n" when a unit external force is applied at the site of mass "n'". The solution of (2.25) is known [1] and has the general form

$$G_0(n, n') = \frac{m}{\beta} \frac{t^{|n-n'|+1}}{t^2 - 1}.$$ (2.26)

The quantity, t, is determined by inserting this general solution into (2.25) and choosing $n = n'$. In this case, we obtain the simple quadratic equation:

$$t^2 - 2\xi t + 1 = 0,$$ (2.27)

with $\xi = \frac{\ddot{\gamma}}{2\beta} = 1 - \frac{m\omega^2}{2\beta} = 1 - \frac{2\omega^2}{\omega_0^2}$. The resolution of the quadratic equation yields

$$t = \begin{cases} \xi - (\xi^2 - 1)^{1/2} & \text{if} \quad \xi > 1 \\ \xi + (\xi^2 - 1)^{1/2} & \text{if} \quad \xi < -1 \\ \xi + i(1 - \xi^2)^{1/2} & \text{if} \quad -1 \leq \xi \leq 1 \end{cases}.$$ (2.28)

We note that for $\omega \epsilon [0, \omega_0]$ and $\xi \epsilon [-1, 1]$ t is a complex quantity. We introduce some wave number, k, and write this complex quantity, $t = e^{ika}$. We equate the real part and the imaginary part of this quantity with those of the third form of the solution in (2.28) and using standard trigonometric relations, we obtain the dispersion relation given by (2.5). We therefore recover the propagating wave solution in the crystal. For, $\omega > \omega_0$ and $\xi < -1$, $t \epsilon [-1, 0]$. Introducing a wave number, k, we can therefore rewrite $\xi = -\cosh ka$ and $t = -e^{-ka}$ represents an evanescent wave.

As a final note, we recast the operator, $\overleftrightarrow{H_0}$, as the difference, $\overleftrightarrow{H'}_0 - \omega^2 \overleftrightarrow{I}$, where the operator, H'_0, depends on the spring constant β only. Equation (2.23) then states that

$$\overleftrightarrow{G_0} = \overleftrightarrow{I} \left(\overleftrightarrow{H'}_0 - \omega^2 \right)^{-1},$$ (2.29)

meaning that the poles (zeros of the denominator) of the Green's function are the eigenvalues of the operator, $\overleftrightarrow{H'}_0$. According to (2.26), the poles of the Green's function of the 1-D monoatomic harmonic crystal are, therefore, given by the equation

$$t^2 - 1 = 0.$$ (2.30)

This condition is met when $t = e^{ika} = \cos ka + i \sin ka$. In the case, $\omega \epsilon [0, \omega_0]$, $t = \xi + i(1 - \xi^2)^{1/2}$ if $-1 \leq \xi \leq 1$. We can subsequently write $\cos ka = \xi = 1 - \frac{2\omega^2}{\omega_0^2}$, which, using trigonometric relation, reduces to the dispersion relation of propagating waves in the crystal (Eq. (2.5)).

2.2   Periodic One-Dimensional Harmonic Crystals

2.2.1   One-Dimensional Monoatomic Crystal and Super-Cell Approach

We consider again the 1-D monoatomic harmonic crystal but treat it as a periodic system with a super-period, $R = Na$ i.e., a super-cell representation of the crystal. This system is represented in Fig. 2.4.

We will solve the equation of motion of the mass, "l" in the first super-cell, that is, $l \in [0, N-1]$. Equation (2.21) applied to "l" is

$$- m\omega^2 u_l = \beta(u_{l+1} - 2u_l + u_{l-1}). \tag{2.31}$$

In contrast to Sect. 2.1, we now assume that the displacement obeys Block's theorem [2]. The solutions of (2.31) are the product of plane waves and a periodic function of the super-cell structure:

$$u_1(k) = e^{ikla}\tilde{u}_l(k). \tag{2.32}$$

The periodic function, $\tilde{u}_l(k)$, satisfies the condition: $\tilde{u}_l(k) = \tilde{u}_{l+N}(k)$. The wave number, k, is now limited to the interval: $\left[-\frac{\pi}{R}, \frac{\pi}{R}\right]$. The periodic function $\tilde{u}_l(k)$ is subsequently written in the form of a Fourier series:

$$\tilde{u}_l(k) = \sum_g u_g(k) e^{igla}, \tag{2.33}$$

where the reciprocal lattice vector of the periodic structure of super-cells $g = \frac{2\pi}{Na} m$ with m being an integer. Inserting (2.33) and (2.32) into (2.31) gives after some algebra

$$\sum_g u_g(k) e^{i(k+g)la} \left[-m\omega^2 - \beta\left(e^{i(k+g)a} - 2 + e^{-i(k+g)a}\right)\right] = 0. \tag{2.34}$$

In addition to the trivial solution, $u_g(k) = 0$, (2.34) admits nontrivial dispersion relations:

$$\omega(k) = \omega_0 \left| \sin(k+g)\frac{a}{2} \right|. \tag{2.35}$$

We illustrate this dispersion relation for a super-cell 2a long and containing two masses. For $N = 2$, the reciprocal space vectors, $g = \frac{\pi}{a}n$. Equation (2.35) becomes

$$\omega(k) = \omega_0 \left| \frac{\sin\left(k + \frac{\pi}{a}n\right)a}{2} \right|.$$

WOWWOWWOWWOWWOWWOWWOWWOWWOWWOWWOWWOWWOWWON
0   1   ···· | ···· R₁=Na ····   ···· R₂=2Na

Fig. 2.4 Schematic representation of the 1-D mono-atomic crystal as a periodic structure with super-period Na

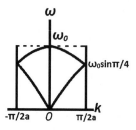

Fig. 2.5 Schematic illustration of the dispersion relation of the 1-D monoatomic harmonic crystal in the super-cell representation, $N = 2$

When $n = 0$, this dispersion relation is identical to that given by (2.5) that was illustrated in Fig. 2.2. The dispersion relation when $n = 1$ is equivalent to that of (2.5) translated along the wave number axis by $\frac{\pi}{a}$. For $n = 2a$ and other even values, one obtains again the same result than for $n = 0$. The case $n = 3$ and other odd values are identical to the case $n = 1$. There are therefore only two possible nonequivalent representations of the dispersion relation (2.35). These dispersion relations are only valid in the interval of wave number: $\left[-\frac{\pi}{2a}, \frac{\pi}{2a}\right]$. They are illustrated in Fig. 2.5.

In the super-cell representation, the dispersion relation consists of two branches that can be obtained graphically by folding the dispersion curve of Fig. 2.2 about two vertical lines at wave numbers $-\frac{\pi}{2a}$ and $\frac{\pi}{2a}$. The super-cell representation of the band structure of the monoatomic crystal is a purely mathematical representation. In general, one can construct the band structure of a super-cell with period $R = Na$ by folding the single dispersion curve of Fig. 2.2 N times inside a reduced Brillouin zone: $\left[0, \frac{\pi}{Na}\right]$. We will show in the next section that this representation may be useful in interpreting the band structure of the 1-D diatomic harmonic crystal.

2.2.2   One-Dimensional Diatomic Harmonic Crystal

The 1-D diatomic harmonic crystal is illustrated in Fig. 2.6.

The equations of motion of two adjacent odd and even atoms are

$$\begin{cases} m_1 \ddot{u}_{2n} = \beta(u_{2n+1} - u_{2n} + u_{2n-1}) \\ m_2 \ddot{u}_{2n+1} = \beta(u_{2n+2} - u_{2n+1} - u_{2n}) \end{cases}. \tag{2.36}$$

Fig. 2.6 Schematic illustration of the 1-D diatomic harmonic crystal. The atoms with an even label have a mass m_1, and the odd atoms have a mass m_2. The force constant of the springs is β. The periodicity of the crystal is $2a$

We seek solutions in the form of propagating waves with different amplitudes for odd or even atoms as their masses are different:

$$\begin{cases} u_{2n} = Ae^{i\omega t}e^{ik2na} \\ u_{2n+1} = Be^{i\omega t}e^{ik(2n+1)a} \end{cases}.$$ (2.37)

Inserting these solutions into (2.36) leads, after some algebraic manipulations and using the definition of the cosine in terms of complex exponentials, to the set of two linear equations in A and B:

$$\begin{cases} (2\beta - m_1\omega^2)A - 2\beta\cos ka\, B = 0 \\ -2\beta\cos ka\, A + (2\beta - m_1\omega^2)B = 0 \end{cases}.$$ (2.38)

This is an eigenvalue problem in ω^2. This set of equations admits nontrivial solutions (i.e., $A \neq 0, B \neq 0$) when the determinant of the matrix composed of the linear coefficients in equation (2.38) is equal to zero, that is,

$$\begin{vmatrix} 2\beta - m_1\omega^2 & -2\beta\cos ka \\ -2\beta\cos ka & 2\beta - m_2\omega^2 \end{vmatrix} = 0.$$ (2.39)

Setting $\alpha = \omega^2$, (2.39) takes the form of the quadratic equation:

$$\alpha^2 - 2\beta\left(\frac{1}{m_1} + \frac{1}{m_2}\right)\alpha + \frac{4\beta^2}{m_1 m_2}\sin^2 ka = 0,$$ (2.40)

which admits two solutions:

$$\omega^2 = \alpha = \beta\left(\frac{1}{m_1} + \frac{1}{m_2}\right) \pm \sqrt{\beta^2\left(\frac{1}{m_1} + \frac{1}{m_2}\right)^2 - \frac{4\beta^2}{m_1 m_2}\sin^2 ka}.$$ (2.41)

These two solutions are periodic in wave number, k, with a period of $\frac{\pi}{a}$. These solutions are represented graphically in the band structure of Fig. 2.7 over the interval, $k \in \left[0, \frac{\pi}{2a}\right]$. This interval is the smallest interval, the so-called irreducible Brillouin zone, for representing the band structure. The complete band structure is reconstructed by mirror symmetry with respect to a vertical line passing though the origin.

Fig. 2.7 Schematic
representation of the band
structure of the 1-D diatomic
harmonic crystal in the
irreducible Brillouin zone

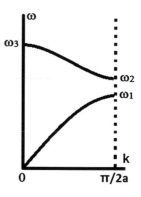

The frequencies ω_1, ω_2 and ω_3 are given by $\omega_1 = \sqrt{\frac{2\beta}{m_1}}$, $\omega_2 = \sqrt{\frac{2\beta}{m_2}}$, and $\omega_3 = \sqrt{2\beta\left(\frac{1}{m_1} + \frac{1}{m_2}\right)}$ if one chooses $m_1 > m_2$. The band structure of Fig. 2.7 exhibits two branches since the unit cell of the 1-D diatomic crystal contains two atoms. These branches are separated by a gap in the interval of frequency $[\omega_1, \omega_2]$. The low-frequency branch is called the acoustic branch. The high-frequency branch is called the optical branch. In the limit $m_1 = m_2 = m$, the diatomic crystal reduces to a monoatomic crystal. The band structure of Fig. 2.7 becomes that of the 1-D monoatomic harmonic crystal in the super-cell representation with $N = 2$ (see Fig. 2.5). The construction of the band structure of the diatomic crystal may then be understood conceptually by first considering the folded band structure of the monoatomic crystal with a super-period $R = 2a$. The waves with wave number $k = \frac{\pi}{2a}$ have a wavelength $\lambda = \frac{2\pi}{k} = 4a$. The wavelength is twice the period of the diatomic crystal. Then, we label alternating atoms with odd and even numbers in the monoatomic crystal. If at some instant an even atom undergoes a zero displacement, then the displacement of all other even atoms will also be zero. At the same time, all odd number atoms will be subjected to a maximum displacement. The even atom and odd atom sublattices support the $\lambda = 4a$ wave with the same frequency as long as their masses are the same. However, if now one perturbs the monoatomic crystal by making the mass of atoms on one sub-lattice different from the atoms on the other (leading to the formation of a diatomic crystal), the frequency of the $\lambda = 4a$ wave will be lower for the heavier atoms than for the lighter ones. Approaching the diatomic crystal by perturbing the masses of the monoatomic crystals separates the folded branches of the monoatomic crystal at $k = \frac{\pi}{2a}$ leading to the formation of a gap.

It is interesting to note that in contrast to the acoustic branch, the optical branch has a negative slope, i.e., a negative group velocity. The group velocity and energy velocity point in a direction opposite to the direction of the wave vector and of the phase velocity. This observation is particularly important when dealing with the concept of negative refraction. However, since the diatomic crystal is one dimensional, we cannot address the phenomenon of refraction yet. However, we rewrite the real part of the displacement of a superposition of waves given by (2.7) in the form

$$u_n^s = 2A \, \cos(k(na + v_\varphi t)) \cos\left(\frac{\Delta k}{2}(na + v_g t)\right). \tag{2.42}$$

Equation (2.42) shows that the envelope of the wave packet appears to propagate in the opposite direction of the superposition of waves when the phase velocity and the group velocity have opposite signs.

2.2.3 Evanescent Waves in the Diatomic Crystal

In this section, we use the method introduced in Sect. 2.1.3 to shed some light on the nature of waves with frequencies corresponding to the gap of Fig. 2.7. For this, we start with (2.38) and recast it in the form

$$\begin{cases} -m_1\omega^2 A = \beta B e^{ika} - 2\beta A + \beta B e^{-ika} \\ -m_2\omega^2 B = \beta A e^{ika} - 2\beta B + \beta A e^{-ika} \end{cases} \tag{2.43}$$

We set $X = e^{ika}$ and insert it into the equations of motion (2.43) to obtain after some algebraic manipulations the set of two quadratic equations:

$$\begin{cases} X^2 A = -A + X\left(2 - \dfrac{m_2\omega^2}{\beta}\right)B \\ X^2 B = -B + X\left(2 - \dfrac{m_1\omega^2}{\beta}\right)A \end{cases} \tag{2.44}$$

Equation (2.44) is recast further in the form of an eigenvalue problem taking the matrix form

$$X\begin{pmatrix} A \\ B \\ XA \\ XB \end{pmatrix} = \begin{pmatrix} 0 & 0 & 1 & 0 \\ 0 & 0 & 0 & 1 \\ -1 & 0 & 0 & 2 - \dfrac{m_2\omega^2}{\beta} \\ 0 & -1 & 2 - \dfrac{m_1\omega^2}{\beta} & 0 \end{pmatrix}\begin{pmatrix} A \\ B \\ XA \\ XB \end{pmatrix}. \tag{2.45}$$

There exists a nontrivial solution when

$$\begin{vmatrix} -\alpha & 0 & 1 & 0 \\ 0 & -\alpha & 0 & 1 \\ -1 & 0 & -\alpha & 2 - \dfrac{m_2\omega^2}{\beta} \\ 0 & -1 & 2 - \dfrac{m_1\omega^2}{\beta} & -\alpha \end{vmatrix} = 0 \tag{2.46}$$

In (2.46), the eigenvalues are $\alpha = e^{ika}$. This equation yields a fourth-order equation:

$$\alpha^4 + \alpha^2 \left[2 - \left(2 - \frac{m_2 \omega^2}{\beta} \right) \left(2 - \frac{m_1 \omega^2}{\beta} \right) \right] + 1 = 0. \tag{2.47}$$

By setting $\zeta = \alpha^2 = e^{i2ka}$, we transform (2.47) in a quadratic equation whose solutions are

$$\zeta = \frac{1}{\omega_1^2 \omega_2^2} \left(\omega_1^2 \omega_2^2 + 2\omega^4 - 2\omega^2 \omega_3^2 \right) \pm \frac{2}{\omega_1^2 \omega_2^2} \sqrt{\omega^2 \left(\omega^2 - \omega_1^2 \right) \left(\omega^2 - \omega_2^2 \right) \left(\omega^2 - \omega_3^2 \right)}. \tag{2.48}$$

To obtain (2.48), we have used the relations $\omega_1 = \sqrt{\frac{2\beta}{m_1}}$, $\omega_2 = \sqrt{\frac{2\beta}{m_2}}$ and $\omega_3 = \sqrt{2\beta \left(\frac{1}{m_1} + \frac{1}{m_2} \right)} = \sqrt{\omega_1^2 + \omega_2^2}$. If $0 < \omega < \omega_1$ or $\omega_2 < \omega < \omega_3$, then the argument of the square root in equation (2.48) is negative and ζ is a complex function of frequency corresponding to propagating waves (i.e., real wave number k). These cases represent the acoustic and optical branches of the band structure of the diatomic crystal. Inside the gap ($\omega_1 < \omega < \omega_2$), ζ is a real function. Introducing a complex wave number $k = k' + ik''$, we redefine ζ as the quantity:

$$\zeta = e^{-2k''a} \cos 2k'a + i e^{-2k''a} \sin 2k'a. \tag{2.49}$$

ζ is therefore real only when $\sin 2k'a = 0$, that is when $k' = \frac{\pi}{2a}$. Equating the real part of (2.49) to (2.48) leads to two solutions for k''. The positive solution is unphysical as it represents an exponentially increasing wave. Again, the emergence of this unphysical solution results from the fact that in the current eigenvalue problem we used a 4×4 matrix (Eq. (2.45)) that is two times larger than the actual 2×2 dynamical matrix of the diatomic harmonic crystal. The negative solution for k'' corresponds to an evanescent wave with exponentially decaying amplitude. Similarly, the vibrational modes for frequencies beyond ω_3 also correspond to evanescent waves. The complete band structure of the 1-D diatomic harmonic crystal is illustrated schematically in Fig. 2.8.

2.2.4 Monoatomic Crystal with a Mass Defect

To shed additional light on the origin of the band gap in the band structure of the diatomic harmonic crystal, we investigate the propagation of waves in a 1-D monoatomic harmonic crystal with a single mass defect. This is accomplished by substituting one atom with mass m by another atom with mass m'. The diatomic crystal may subsequently be created as a periodic substitution of atoms with different masses. We address the following question: does the gap originate from

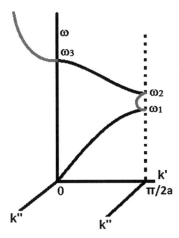

Fig. 2.8 Complete band structure of the 1-D diatomic harmonic crystal. The *black solid lines* correspond to propagating waves. The *gray lines* correspond to evanescent waves

Fig. 2.9 Schematic illustration of the 1-D mono-atomic harmonic crystal with a single mass defect at site 0. The springs are all identical with the same spring constant

the scattering of propagating waves by mass defects independently of their periodicity or does the gap originate from the periodic arrangement of the mass defects?

The defected monoatomic crystal is illustrated in Fig. 2.9.

The equations of motion of the atoms in the defected crystal are

$$\begin{cases} -m\omega^2 u_n = \beta(u_{n+1} - 2u_n + u_{n-1}) & \text{for} \quad n \neq 0 \\ -m'\omega^2 u_o = \beta(u_1 - 2u_0 + u_{-1}) \end{cases} \tag{2.50}$$

Let us consider an incident wave (i) propagating from the left of the crystal:

$$u_n^{(i)} = A_i e^{ikna} \quad \text{for} \quad n \leq -1. \tag{2.51}$$

Part of this wave will be reflected by the mass defect. Another part of the incident wave will be transmitted through the defect. We write the displacements associated with these reflected and transmitted waves in the form

$$u_n^{(r)} = A_r e^{-ikna} \quad \text{for} \quad n \leq -1$$
$$u_n^{(t)} = A_t e^{ikna} \quad \text{for} \quad n \geq 1. \tag{2.52}$$

In (2.52), the upper-scripts (r) and (t) stand for reflected and transmitted waves, respectively. The total displacement on the left of the defect is the sum of the incident and reflected displacement. The displacement on the right of the defect consists only of the transmitted wave. The total displacement is therefore given by

$$u_n = u_n^{(i)} + u_n^{(r)} \quad \text{for} \quad n \leq -1$$
$$u_n = u_n^{(t)} \quad \text{for} \quad n \geq 1. \tag{2.53}$$

The continuity of the displacement at the defected site "0" imposes the condition

$$u_0 = u_0^{(i)} + u_0^{(r)} = u_0^{(t)}. \tag{2.54}$$

Substituting (2.51) and (2.52) into the condition (2.54) yields a relation between the amplitudes of the incident, the reflected, and the transmitted waves:

$$A_i + A_r = A_t. \tag{2.55}$$

We now substitute equations (2.51), (2.52), and (2.54) into (2.50) for the motion of the mass m'. After some algebraic steps, this equation becomes

$$\left(-m'\omega^2 + 2\beta - \beta e^{ika}\right)A_T = A_i \beta e^{-ika} + A_R \beta e^{ika}. \tag{2.56}$$

Equations (2.55) and (2.56) constitute a set of linear equations in the amplitudes of the incident, reflected, and transmitted waves. We can express the amplitude of the reflected and transmitted waves in terms of the amplitude of the incident wave to define a transmission coefficient and a reflection coefficient:

$$T = \frac{A_t}{A_i} = \frac{\beta 2i \sin ka}{(m' - m)\omega^2 + \beta 2i \sin ka}$$
$$R = \frac{A_r}{A_i} = \frac{-(m' - m)\omega^2}{(m' - m)\omega^2 + \beta 2i \sin ka}. \tag{2.57}$$

To obtain (2.57), we have used the fact that for the 1-D monoatomic harmonic crystal, the dispersion relation of (2.5) can be recast in the form $m\omega^2 = 2\beta(1 - \cos ka)$. To analyze the behavior of the defected crystal further, we calculate the square of the modulus of the transmission coefficient:

$$T^2 = TT^* = \frac{4\beta^2 \sin^2 ka}{(m' - m)^2 \omega^4 + 4\beta^2 \sin^2 ka}. \tag{2.58}$$

We note that when $m' = m$, the incident wave propagates without reflection, i.e., the transmission coefficient ((2.58)) is equal to 1. We also note that for $k = \frac{\pi}{2a}$, i.e., the edge of the Brillouin zone for the diatomic crystal, the transmission coefficient simplifies to $T^2 = \frac{4\beta^2}{(m'-m)^2\omega^4 + 4\beta^2}$. The transmission coefficient decreases monotonically as a function of frequency showing no sign of resonance or any other localized vibration phenomenon. In the absence of such a resonant phenomenon, the band structure of the diatomic harmonic crystal can, therefore, be ascribed to the periodicity of the structure, only. The presence of an acoustic branch and of an optical branch separated by a gap results from scattering of waves by the periodic crystal, namely, Bragg's scattering.

2.2.5 Monoatomic Harmonic Crystal with a General Perturbation

The approach of Sect. 2.2.4 is generalized by introducing a frequency dependent perturbation, $V(\omega)$, of the 1-D monoatomic crystal at site 0. The equations of motion of the atoms in this defected crystal are

$$\begin{cases} -m\omega^2 u_n = \beta(u_{n+1} - 2u_n + u_{n+1}) & \text{for } n \neq 0 \\ -m\omega^2 u_0 = \beta(u_1 - 2u_0 + u_{-1}) + V(\omega)u_0 \end{cases}. \tag{2.59}$$

Following the derivation of the transmission and reflection coefficients in the previous section, we obtain

$$T = \frac{\beta 2i \sin ka}{V(\omega) + \beta 2i \sin ka}$$

$$R = \frac{V(\omega)}{V(\omega) + \beta 2i \sin ka}. \tag{2.60}$$

We note that if $V(\omega) = \infty$, then an incident wave is totally reflected. Such a condition may arise from a local resonance. This case is discussed in the next section.

2.2.6 Locally Resonant Structure

In this section, we are interested in the behavior of a monoatomic crystal with a structural perturbation taking the form of a side branch. The side branch is composed of L' atoms of mass m' interacting via harmonic springs with force constant β'. The side branch is attached to the monoatomic crystal at site "0" via a spring with stiffness β_l. We assume that the lattice parameter is the same in the

Fig. 2.10 Illustration of the
1-D mono-atomic crystal
perturbed by a side branch

side branch and the infinite monoatomic crystal. This structure is illustrated in
Fig. 2.10.

The derivation of an expression for the perturbation potential V begins with the
equations of motion of atoms in the side branch:

$$\begin{cases} -m'\omega^2 u_{n'} = \beta'(u_{n'+1} - 2u_{n'} + u_{n'-1}) & \text{for } n' \neq 1', L' \quad (a) \\ -m'\omega^2 u_{L'} = -\beta'(u_{L'} - u_{L'-1}) & (b). \\ -m'\omega^2 u_{1'} = -\beta_I(u_{1'} - u_0) + \beta'(u_{2'} - u_{1'}) & (c) \end{cases} \quad (2.61)$$

This set of equations is complemented by the equation of motion of site "0":

$$-m\omega^2 u_o = \beta(u_1 - 2u_0 + u_{-1}) + \beta_I(u_{1'} - u_0). \quad (2.62)$$

To find the perturbation potential, we are interested in coupling (2.61) and (2.62)
to obtain an effective equation taking the form of equation (2.69) for site "0."
Rewriting (2.62) as $\left(-m\omega^2 + \beta_I\left(1 - \frac{u_{1'}}{u_0}\right)\right)u_0 = \beta(u_1 - 2u_0 + u_{-1})$ yields

$$V = -\beta_I\left(1 - \frac{u_{1'}}{u_0}\right). \quad (2.63)$$

The ratio of displacements in equation (2.63) is found by considering the general
solution to (2.61)(a):

$$u_{n'} = A'e^{ik'n'a} + B'e^{-ik'n'a}. \quad (2.64)$$

Inserting this solution in (2.61)(a) gives

$$m'\omega^2 = 2\beta'(1 - \cos k'a). \quad (2.65)$$

Should site N' have been in an infinite monoatomic crystal, its equation of motion would have been

$$- m' \omega^2 u_{L'} = \beta'(u_{L'+1} - 2u_{L'} + u_{L'-1}).$$ (2.66)

Subtracting (2.66) and (2.61)(b) gives

$$\beta'(u_{L'+1} - u_{L'}) = 0.$$ (2.67)

This equation serves as a boundary condition on site N' in the side branch. We define the displacement $u_{L'+1}$ at a fictive site "$L' + 1$" as support for the boundary condition (2.67). Similarly subtracting the equation of motion (2.61)(c) and that of site "1'" if it were embedded in an infinite monoatomic crystal leads to the boundary condition

$$- \beta_I(u_{1'} - u_0) + \beta'(u_{1'} - u_{0'}) = 0.$$ (2.68)

Fictive site "0'" is only used to impose the boundary condition. The two boundary conditions at sites "1'" and "L'" form the set of equations:

$$\begin{cases} u_{L'} - u_{L'+1} = 0 \\ (\beta_I - \beta')u_{1'} + \beta'u_{0'} = 0. \end{cases}$$ (2.69)

We insert the general solution (2.64) into (2.69) and obtain the set of linear equations

$$\begin{cases} A'e^{ik'L'a}\left(1 - e^{ik'a}\right) + B'e^{-ik'L'a}\left(1 - e^{-ik'a}\right) = 0 \\ A'\left[(\beta_I - \beta')e^{ik'a} + \beta'\right] + B'\left[(\beta_I - \beta')e^{-ik'a} + \beta'\right] = \beta_I u_0 \end{cases}.$$ (2.70)

Solving (2.70) gives

$$A' = -\beta_I u_0 e^{-ik'L'a}\left(1 - e^{-ik'a}\right)/\Delta$$

$$B = \beta_I u_0 e^{ik'L'a}\left(1 - e^{ik'a}\right)/\Delta,$$ (2.71)

where

$$\Delta = -4i\sin\frac{k'a}{2}\left[(\beta_I - \beta')\cos k'\left(L' - \frac{1}{2}\right)a + \beta'\cos k'\left(L' + \frac{1}{2}\right)a\right].$$ (2.72)

To obtain (2.72), we have used a variety of trigonometric relations.

It is worth noting that in the limit of $\beta_I = 0$, the set of (2.70) can be used to find the displacement of an isolated finite segment of monoatomic crystal. The existence of nontrivial solutions for the amplitudes A' and B' is ensured by the condition $\Delta = 0$. This condition is rewritten as $\cos k' (L' - \frac{1}{2})a + \cos k' (L' + \frac{1}{2})a = -2\sin k' L' a \sin \frac{k'a}{2}$ $= 0$ or $\sin k' L' a = 0$. These solutions correspond to vibrational modes of the finite crystal of length L', i.e., standing waves with wave vectors: $k' = \frac{p\pi}{L'a}$, where p is an integer.

Finally, to find the perturbation V, we use (2.70) and (2.64) to obtain the displacement of atom "1'," which we subsequently insert into (2.63). After several algebraic and trigonometric manipulations, the perturbation becomes

$$V(\omega) = \frac{2\beta' \beta_I \sin \frac{k'a}{2} \sin L' k' a}{(\beta_I - \beta')\cos k' (L' - \frac{1}{2})a + \beta' \cos k' (L' + \frac{1}{2})a}. \qquad (2.73)$$

The effect of the side branch on the propagation of waves along the infinite crystal is most easily understood by considering the limiting case: $\beta = \beta_I = \beta'$ and $m = m'$ such that $k = k'$. In this case, the side branch is constituted of the same material as the infinite crystal and equation (2.73) becomes

$$V(\omega) = \frac{2\beta \sin \frac{ka}{2} \sin L' ka}{\cos k (L' + \frac{1}{2})a} \qquad (2.74)$$

with the dispersion relation $\omega(k) = \omega_0 |\sin \frac{a}{2}|$ (i.e., (2.5)). At the frequency (wave number) corresponding to the standing wave modes of the side branch, the perturbation $V = 0$. The transmission and reflection coefficients given by (2.60) are equal to 1 and 0, respectively. Zeros of transmission and complete reflection occur when $V = \infty$, that is, when $\cos k (L' + \frac{1}{2})a = 0$ or $k = (2p + 1) \frac{\pi}{(2L'+1)a}$. These conditions correspond to resonances with the side branch. For instance for a single atom side branch, i.e., $L' = 1$, there is one zero of transmission in the irreducible Brillouin zone of the monoatomic crystal at $k = \frac{\pi}{3a}$. For a two-atom side branch, $L' = 2$, there are two zeros of transmission in the irreducible Brillouin zone of the monoatomic crystal at $k = \frac{\pi}{5a}$ and $k = \frac{3\pi}{5a}$. The number of zeros of transmission scales with the number of atoms in the side branch. Therefore, in contrast to the result of Sect. 2.2.4 where the mass defect did not introduce any zeros of transmission, the side branch leads to perturbations of the band structure of the supporting infinite 1-D monoatomic crystal. These perturbations arise from resonances $(V = \infty)$ of the side branch. The alterations to the band structure of the monoatomic crystal due to the side branch may be visualized as infinitesimally narrow band gaps. The crystal with a single side branch is not periodic, and the perturbed band structure results only from local resonances. In the next sections, we develop the formalism necessary to shed light on the interplay between Bragg's scattering and local resonances on the band structure of a 1-D monoatomic crystal with periodic arrangements of side branches. This formalism is based on the Green's function approach called the Interface Response Theory.

2.3 Interface Response Theory

2.3.1 Fundamental Equations of the Interface response Theory

In this section, we review the fundamental equations of the Interface Response Theory (IRT) for discrete systems [3]. This formalism allows the calculation of the Green's function of a perturbed system in terms of Green's functions of unperturbed systems. We recall (2.23) and (2.22) defining the Green's function, $\overleftrightarrow{G_0}$, by

$$\overleftrightarrow{H_0}\overleftrightarrow{G_0} = I.$$

The operator H_0 is the infinite tridiagonal dynamic matrix:

$$\overleftrightarrow{H_0} = \frac{1}{m}\begin{bmatrix} \ddots & \vdots & \vdots & \vdots & \vdots & \vdots & \vdots & \vdots & \vdots \\ \cdots & 0 & \beta & -\gamma & \beta & 0 & 0 & \cdots & \cdots \\ \cdots & \cdots & 0 & \beta & -\gamma & \beta & 0 & \cdots & \cdots \\ \cdots & \cdots & \cdots & 0 & \beta & -\gamma & \beta & 0 & \cdots \\ \vdots & \vdots & \vdots & \vdots & \vdots & \vdots & \vdots & \vdots & \ddots \end{bmatrix}, \tag{2.75}$$

where $\gamma = 2\beta - m\omega^2$. We initially consider a type of perturbation that cleaves the 1-D monoatomic harmonic crystal by severing a bond between two neighboring atoms (Fig. 2.11).

The equations of motion of the atoms 0 and 1 are

$$\begin{cases} \frac{1}{m}(-\alpha u_0 + \beta u_{-1}) = 0 \\ \frac{1}{m}(-\alpha u_1 + \beta u_2) = 0 \end{cases}, \tag{2.76}$$

with $\alpha = m\omega^2 - \beta$.

The dynamical operator for the cleaved crystal is written as

$$\overleftrightarrow{h_0} = \begin{bmatrix} \overleftrightarrow{h_{S1}} & \overleftrightarrow{0} \\ \overleftrightarrow{0} & \overleftrightarrow{h_{S2}} \end{bmatrix}$$

$$= \frac{1}{m}\begin{bmatrix} \cdots & -3 & -2 & -1 & 0 & 1 & 2 & 3 & \cdots \\ \beta & -\gamma & \beta & 0 & 0 & 0 & 0 & 0 & 0 \\ 0 & \beta & -\gamma & \beta & 0 & 0 & 0 & 0 & 0 \\ 0 & 0 & \beta & -\gamma & \beta & 0 & 0 & 0 & 0 \\ 0 & 0 & 0 & \beta & -\alpha & 0 & 0 & 0 & 0 \\ 0 & 0 & 0 & 0 & 0 & -\alpha & \beta & 0 & 0 \\ 0 & 0 & 0 & 0 & 0 & \beta & -\gamma & \beta & 0 \\ 0 & 0 & 0 & 0 & 0 & 0 & \beta & -\gamma & \beta \end{bmatrix} \begin{matrix} \vdots \\ -3 \\ -2 \\ -1 \\ 0 \\ 1 \\ 2 \\ \vdots \end{matrix} . \tag{2.77}$$

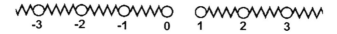

-3   -2   -1   0   1   2   3

Fig. 2.11 Schematic illustration of the 1-D mono-atomic harmonic crystal cleaved between atoms 0 and 1

In (2.77), $\overleftrightarrow{h_0}$ is a block matrix composed of two independent matrices $\overline{h}_{S1}$ and $\overline{h}_{S2}$, corresponding to the two semi-infinite crystals on the left and right of the cleaved bond, respectively. The Green's function of the perturbed system, $\overleftrightarrow{g}_0$, is therefore defined through the relation

$$\overleftrightarrow{h_0}\overleftrightarrow{g_0} = \overleftrightarrow{I}. \tag{2.78}$$

Since the dynamical matrix of the cleaved system is a block matrix, its associated Green's function is also a Block matrix:

$$\overleftrightarrow{g_0} = \begin{bmatrix} \overleftrightarrow{g}_{S1} & \overleftrightarrow{0} \\ \overleftrightarrow{0} & \overleftrightarrow{g}_{S2} \end{bmatrix}. \tag{2.79}$$

We define the perturbation operator or cleavage operator as the difference between the dynamical matrices of the cleaved and unperturbed crystals:

$$\overleftrightarrow{V_0} = \overleftrightarrow{h_0} - \overleftrightarrow{H_0}. \tag{2.80}$$

Using the matrix representation, the cleavage operator is a 2×2 matrix limited to the sites 0 and 1 of the crystal:

$$\overleftrightarrow{V_0} = \begin{pmatrix} V_0(0,0) & V_0(1,0) \\ V_0(0,1) & V_0(1,1) \end{pmatrix} = \frac{1}{m} \begin{pmatrix} \beta & -\beta \\ -\beta & \beta \end{pmatrix}. \tag{2.81}$$

We rewrite (2.78) in the form $\overleftrightarrow{g_0}\overleftrightarrow{h_0} = \overleftrightarrow{I}$ by using the commutative property of the product of a matrix with its inverse. Introduction (2.80) into this later relation, multiplying both sides of the equal sign by $\overleftrightarrow{G_0}$, applying the distributive property of the product of matrices, and finally using (2.23) yields

$$\overleftrightarrow{g_0}\left(\overleftrightarrow{I} + \overleftrightarrow{V_0}\overleftrightarrow{G_0}\right) = \overleftrightarrow{g_0}\left(\overleftrightarrow{I} + \overleftrightarrow{A_0}\right) = \overleftrightarrow{G_0}. \tag{2.82}$$

Equation (2.82) is called Dyson's equation. It enables the determination of the Green's function of a perturbed system in terms of the perturbation operator and the Green's function of the unperturbed system. In (2.82), we have defined the surface operator:

$$\overleftrightarrow{A_0} = \overleftrightarrow{V_0}\overleftrightarrow{G_0}. \tag{2.83}$$

The Green's function of the perturbed system is then given by

$$\ddot{g}_0 = \ddot{G}_0 \left(\ddot{I} + \ddot{A}_0 \right)^{-1}. \tag{2.84}$$

The poles of $\ddot{g}_0$ (i.e., the eigenvalues of the operator $\ddot{h}_0$) are the zeros of $\ddot{I} + \ddot{A}_0$.

2.3.2 Green's Function of the Cleaved 1-D Monoatomic Crystal

We apply (2.84) to the calculation of the Green's function of the semi-infinite crystal on the right of the cleaved bond in Fig. 2.11 (i.e., $n \geq 1$). The components of the surface operator defined by (2.83) are written as

$$A_{S2}(n, n') = \sum_{n''} V_0(n, n'')G_0(n'', n') \quad \text{with} \quad n, n' \geq 1. \tag{2.85}$$

The only nonzero components of the cleavage operator are for $n, n' \epsilon [0, 1]$, so (2.85) reduces to

$$A_{S2}(1, n') = V_0(1, 0)G_0(0, n') + V_0(1, 1)G_0(1, n'), \quad n' \geq 1. \tag{2.86}$$

Inserting the terms in (2.81) and (2.26) into (2.86) results in

$$A_{S2}(1, n') = \frac{t^{n'} - t^{n'+1}}{t^2 - 1}. \tag{2.87}$$

We now write equation (2.82) in component form:

$$g_{S2}(n, n') + g_{S2}(n, 1)A_{S2}(1, n') = G_0(n, n'), \quad n, n' \geq 1. \tag{2.88}$$

Expressing (2.88) at site $n' = 1$ and using the relation (2.87) gives

$$g_{S2}(n, 1) = \frac{m}{\beta} \frac{t^n}{t - 1}.$$

We can now combine that relation with (2.87), (2.26), and (2.88) to obtain the function sought

$$g_{S2}(n, n') = \frac{m}{\beta} \frac{t^{|n-n'|+1} + t^{n+n'}}{t^2 - 1}, \quad n, n' \geq 1. \tag{2.89}$$

The procedure used in this section to find the Green's function of the perturbed system can be generalized to obtain the universal equation of the IRT. All matrices in equation (2.82) are defined for, $n' \epsilon [-\infty, \infty]$. We now consider the space D for $n, n' \geq 1$ and rewrite equation (2.82) as

$$\ddot{g}_{S2}(D, D) + \ddot{g}_{S2}(D, M)\vec{A}_{S2}(M, D) = \vec{G}_{S2}(D, D). \qquad (2.90)$$

The index S specifies that all functions are limited to the space of a semi-infinite truncated chain. Equation (2.88) is a particular case of the general equation (2.90) where we have specified the space corresponding to the location of the perturbation by M. In the case of the cleavage of the monoatomic crystal, $M = 1$. A particular form of (2.90) is

$$\ddot{g}_{S2}(D, M) + \ddot{g}_{S2}(D, M)\vec{A}_{S2}(M, M) = \vec{G}_{S2}(D, M). \qquad (2.91)$$

We combine (2.91) and (2.90) to obtain the universal equation of the IRT:

$$\ddot{g}_{S2}(D, D) = \bar{G}_{S2}(D, D) + \bar{G}_{S2}(D, M)\vec{\Delta}^{-1}(M, M)\vec{A}_{S2}(M, D), \qquad (2.92)$$

where

$$\vec{\Delta}(M, M) = \vec{I}(M, M) + \overleftrightarrow{A_{S2}}(M, M). \qquad (2.93)$$

Equation (2.93) introduces the diffusion matrix $\vec{\Delta}$.
The displacement vector $\vec{u}(D)$ is related to the Green's function $\bar{g}_S$ via the relation

$$\vec{u}(D) = \vec{f}(D)\bar{g}_{S2}(D, D), \qquad (2.94)$$

where $\vec{f}$ is some force distribution applied in the space D. Inserting (2.92) into (2.94), we obtain the displacement vector of the perturbed system in terms of the displacement vector of the unperturbed system, U, as

$$\vec{u}(D) = \vec{U}(D) - \vec{U}(M)\vec{\Delta}^{-1}(M, M)\vec{A}_{S2}(M, D). \qquad (2.95)$$

Applying (2.95) to the right side of the cleaved mono-atomic crystal yields

$$u(n') = U(n') - U(1)\Delta^{-1}(1, 1)A_{S2}(1, n') \text{ for } n' \geq 1,$$

with $\Delta^{-1}(1, 1) = \frac{1}{1 + A_{S2}(1.1)} = \frac{t^2 - 1}{t - 1}$ and $A_{S2}(1, n') = \frac{t^{n'} - t^{n'+1}}{t^2 - 1}$. The displacement is therefore

$$u(n') = U(n') + U(1)t^{n'} \quad n' \geq 1.$$

If we choose $U(n') = t^{-n'}$, corresponding to an incident wave coming from $n' = +\infty$, the displacement field in the semi-infinite chain takes the form

$$u(n') = t^{-n'} + t^{n'-1} = e^{-ikn'a} + e^{ik(n'-1)a}.$$

This is a standing wave resulting from the superposition of an incident wave and a reflected wave. We can also obtain this result by writing the equation of motion at site 1 of the cleaved crystal:

$$-m\omega^2 u_1 = \beta(u_2 - u_1).$$

This equation implies that $u_1 - u_0 = 0$, where u_0 is the displacement of the site 0 taken as a fictive site imposing a zero force boundary condition on site 1. We assume that the displacement in the semi-infinite crystal is the sum of a reflected wave and a transmitted wave:

$$u_n = A_i e^{-ikna} + A_r e^{ikna}.$$

Inserting this general solution into the boundary condition leads to the relation between the incident and reflected amplitudes: $A_r = A_i e^{-ika}$ leading to the displacement $u(n) = A_i(e^{-ikna} + e^{ik(n-1)a})$.

2.3.3 Finite Monoatomic Crystal

The finite 1-D monoatomic crystal is formed by cleaving an infinite crystal at two separate locations. This doubly cleaved system is illustrated in Fig. 2.12.

The cleavage operator is a 4x4 matrix expressed in the space of the perturbed sites $(0,1)$ and $(L,L+1)$:

$$\vec{\vec{V}}_0 = \frac{1}{m} \begin{matrix} & \begin{matrix} 0 & \quad 1 & \quad L & \; L+1 \end{matrix} & \\ \begin{bmatrix} \beta & -\beta & 0 & 0 \\ -\beta & \beta & 0 & 0 \\ 0 & 0 & \beta & -\beta \\ 0 & 0 & -\beta & \beta \end{bmatrix} & \begin{matrix} 0 \\ 1 \\ L \\ L+1 \end{matrix} \end{matrix}. \qquad (2.96)$$

The dynamical matrix is composed of three separate blocks corresponding to the three uncoupled regions of the cleaved system of Fig. 2.12, namely regions "1," "2," and "3." Similarly, the Green's function and the surface operators are also block diagonal matrices. Using (2.83), the nonzero components of the surface operator matrix corresponding to the block of the finite segment of crystal "2" are

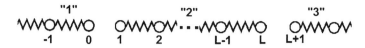

Fig. 2.12 Mono-atomic crystal cleaved between sites $(0,1)$ and $(L,L+1)$ to obtain a finite crystal composed of atoms $[1,L]$

$$
\begin{aligned}
A_{S2}(1,n') &= V_0(1,0)G_0(0,n') + V_0(1,1)G_0(1,n')\\
A_{S2}(L,n') &= V_0(L,L)G_0(L,n') + V_0(L,L+1)G_0(L+1,n')
\end{aligned}
\qquad n' \in [1,L]. \quad (2.97)
$$

The Green's function of the infinite crystal given by (2.26) is inserted into (2.97) to obtain

$$
\begin{aligned}
A_{S2}(1,n') &= \dfrac{-t^{n'}}{t+1}\\[2mm]
A_{S2}(L,n') &= \dfrac{-t^{L-n'+1}}{t+1}
\end{aligned}
\qquad n' \in [1,L]. \quad (2.98)
$$

To apply the universal equation of the IRT, we need the block "2" of the surface operator matrix in the space of the corresponding perturbed sites $M \in [1,L]$, that is

$$
\overleftrightarrow{A}_{S2}(M,M) = \begin{bmatrix} A_{S2}(1,1) & A_{S2}(1,L) \\ A_{S2}(L,1) & A_{S2}(L,L) \end{bmatrix} = \frac{-1}{t+1}\begin{bmatrix} t & t^L \\ t^L & t \end{bmatrix}. \quad (2.99)
$$

The green's function of the finite segment of crystal takes the form

$$
\begin{aligned}
g_{S2}(n,n') = {}&G_0(n,n') - G_0(n,1)\Delta^{-1}(1,1)A_{S2}(1,n') - G_0(n,1)\Delta^{-1}(1,L)A_{S2}(L,n')\\
&- G_0(n,L)\Delta^{-1}(L,1)A_{S2}(1,n') - G_0(n,L)\Delta^{-1}(L,L)A_{S2}(L,n'), \quad n,n'\in[1,L].
\end{aligned} \quad (2.100)
$$

In (2.100),

$$
\overleftrightarrow{\Delta}^{-1}(M,M) = \frac{1}{W}\frac{1}{t+1}\begin{bmatrix} 1 & t^L \\ t^L & 1 \end{bmatrix}, \quad (2.101)
$$

with $W = \det\overleftrightarrow{\Delta} = \frac{1-t^{2L}}{(t+1)^2}$ and according to (2.93) $\overleftrightarrow{\Delta}(M,M) = \overleftrightarrow{I}(M,M) + \overleftrightarrow{A}_{S2}(M,M)$.

Inserting the expressions given by (2.26), (2.98), and (2.101) into (2.100) yields the Green's function of the finite crystal (for $n, n' \in [1,L]$):

$$
g_{S2}(n,n') = \frac{m}{\beta}\left[\frac{t^{|n-n'|+1} + t^{n+n'}}{t^2-1} + \frac{t^{2L+1}}{(t^2-1)(1-t^{2L})}\left(t^{n'-n} + t^{n-n'} + t^{1-n-n'} + t^{n+n'-1}\right)\right]. \quad (2.102)
$$

According to (2.84), the poles of the Green's function are also those of $\overleftrightarrow{\Delta}^{-1}$. Here, these poles are the zeros of . The Eigen values of the finite crystals are, therefore, given by the condition $1 - t^{2L} = 0$. This condition may be rewritten as $t^L - t^{-L} = 0$. For angular frequencies, $\omega \epsilon [0, \omega_0], t = e^{ika}$ and the modes of the finite crystal are given by $e^{ikLa} - e^{-ikLa} = \sin kLa = 0$. These modes correspond to standing waves with wave number conditioned by $k = \frac{p\pi}{La}$ with p being an integer. The displacement field of these standing waves is obtained from (2.95). In components form, (2.95) becomes

$$u(n') = U(n') - U(1)\Delta^{-1}(1, 1)A_{S2}(1, n') - U(1)\Delta^{-1}(1, L)A_{S2}(L, n')$$
$$- U(L)\Delta^{-1}(L, 1)A_{S2}(1, n') - U(L)\Delta^{-1}(L, L)A_{S2}(L, n'), \quad n, n' \epsilon [1, L].$$
$$(2.103)$$

Employing a reference displacement $U(n') = t^{n'}$, (2.103) gives

$$u(n') = t^{n'} + t^{n'}\frac{t}{1 - t^{2L}} + t^{-n'}\frac{t^{2L+2}}{1 - t^{2L}} + t^{n'}\frac{t^{2L}}{1 - t^{2L}} + t^{-n'}\frac{t^{2L+1}}{1 - t^{2L}}.$$

This expression diverges when $1 - t^{2L} = 0$. It is therefore necessary to obtain a finite displacement by renormalizing the previous expression by W. The renormalized displacement then reduces to $u(n') = t^{n'} + t^{-n'+1}$. This expression is that of the displacement of standing waves in the finite crystal.

2.3.4 One-Dimensional Monoatomic Crystal with One Side Branch

The calculation of the displacement in a system composed of a 1-D monoatomic crystal with a finite crystal branch coupled to its side via a spring with constant, β_I, as illustrated in Fig. 2.10, begins with the block matrix describing the Green's function of the uncoupled system $(\beta_I = 0)$

$$\overleftrightarrow{G}_S = \begin{pmatrix} \overleftrightarrow{G}_0 & \overleftrightarrow{0} \\ \overleftrightarrow{0} & \overleftrightarrow{g}'_{S2} \end{pmatrix}, \qquad (2.104)$$

where $\overleftrightarrow{G}_0$ is the Green's function of the infinite crystal (whose components are given by (2.26)) and where $\overleftrightarrow{g}'_{S2}$ is the Green's function of the finite side crystal given by (2.102). This later Green's function is labeled with a "prime" sign to indicate that the spring constants and masses m' and β' of the finite crystal may be different from those of the infinite crystal m and β. The difference between the dynamic matrix of the coupled systems and of the dynamic matrix of the uncoupled system defines a coupling operator:

$$
\ddot{V}_I = \begin{pmatrix} V_I(0,0) & V_I(0,1') \\ V_I(1',0) & V_I(1',1') \end{pmatrix} = \begin{pmatrix} \dfrac{-\beta_I}{m} & \dfrac{\beta_I}{m} \\ \dfrac{\beta_I}{m'} & \dfrac{-\beta_I}{m'} \end{pmatrix}. \tag{2.105}
$$

We note that if the masses in the finite and infinite crystals were the same, the coupling operator would simply be the opposite of the cleavage operator of (2.81). We now use the fundamental equation of the IRT to derive an expression for the displacement field in the coupled system in terms of the Green's function of the constituent crystals making up the uncoupled system and the perturbation operator of (2.105).

To that effect, we first write expressions for the surface operator:

$$
\ddot{A}_0(MD) = \begin{pmatrix} A(0,n) \\ A(0,n') \\ A(1',n) \\ A(1',n') \end{pmatrix} = \begin{pmatrix} V_I(0,0)G_0(0,n) \\ V_I(0,1')g'_{S2}(1',n') \\ V_I(1',0)G_0(0,n) \\ V_I(1',1')g'_{S2}(1',n') \end{pmatrix}. \tag{2.106}
$$

In (2.106), n and n' refer to sites in the infinite crystals and the finite side branch, respectively. The diffusion matrix then takes the form of a 2x2 matrix in the space of the interface sites M:

$$
\begin{aligned}
\ddot{\Delta}(MM) &= \begin{pmatrix} 1 + A(0,0) & A(0,1') \\ A(1',0) & 1 + A(1,1') \end{pmatrix} \\
&= \begin{pmatrix} 1 + V_I(0,0)G_0(0,0) & V_I(0,1')g'_{S2}(1',1') \\ V_I(1',0)G_0(0,0) & 1 + V_I(1',1')g'_{S2}(1',1') \end{pmatrix}.
\end{aligned} \tag{2.107}
$$

The inverse of the diffusion matrix is then

$$
\ddot{\Delta}^{-1}(MM) = \frac{1}{\det\ddot{\Delta}} \begin{pmatrix} 1 + V_I(1',1')g'_{S2}(1',1') & -V_I(0,1')g'_{S2}(1',1') \\ -V_I(1',0)G_0(0,0) & 1 + V_I(0,0)G_0(0,0) \end{pmatrix}. \tag{2.108}
$$

We use (2.95) to obtain the displacement field. For this we also need to assume a form for the reference displacement $U(D) = t^n$. This displacement corresponds to a wave propagating in the infinite crystal and launched from $n = -\infty$. The displacement inside the side crystal is also assumed to be equal to zero. The displacement in the space of the perturbed sites $[0,1']$ take the form

$$
U(M) = (U(0), U(1')) = (1,0). \tag{2.109}
$$

The displacement field at a site $n \geq 1$ along the infinite crystal (i.e., on the right side of the grafted branch) is therefore determined from the equation:

$$
u_n = t^n - (1,0)\begin{pmatrix} \Delta^{-1}(0,0) & \Delta^{-1}(0,1') \\ \Delta^{-1}(1',0) & \Delta^{-1}(1',1') \end{pmatrix}\begin{pmatrix} A(0,n) \\ A(1',n) \end{pmatrix}. \tag{2.110}
$$

Inserting (2.108) and (2.106) into (2.110) yields

$$u_n = t^n - \frac{1}{\det \overleftrightarrow{\Delta}} \overleftrightarrow{V}_I(0,0) G_0(0,n). \tag{2.111}$$

One then combines (2.26), (2.102), (2.105), and (2.111) to obtain

$$u_n = t^n \left(1 + \frac{\beta_I}{\beta} \frac{1}{\det \overleftrightarrow{\Delta}} \frac{t}{t^2 - 1}\right) = t^n T, \tag{2.112}$$

with

$$\det \overleftrightarrow{\Delta} = 1 - \frac{\beta_I}{\beta'} \frac{t' + t'^{2L'}}{(t'-1)(1-t'^{2L'})} - \frac{\beta_I}{\beta} \frac{t}{t^2 - 1}. \tag{2.113}$$

In (2.112), T is the transmission coefficient. We can rewrite (2.113) in the form

$$\det \overleftrightarrow{\Delta} = -\frac{1}{V} - \frac{\beta_I}{\beta} \frac{1}{2i \sin ka}, \tag{2.114}$$

where $-\frac{1}{V} = 1 - \frac{\beta_I}{\beta'} \frac{t'+t'^{2L'}}{(t'-1)(1-t'^{2L'})}$. To obtain equation (2.114), we also defined $t = e^{ika}$. With $t' = e^{ik'a}$, one can show that the quantity V is that given by equation (2.73).

2.3.5 One-Dimensional Monoatomic Crystal with Multiple Side Branches

We now consider N_c side branches of various lengths grafted along an infinite 1-D monoatomic crystal. The spaces D and M for this system are defined as

$$D = \{-\infty, \ldots, -1, 0, 1, \ldots \infty\}$$
$$\cup \left\{\{1', 2', \ldots L'\}, \{1'', 2'', \ldots, L''\}, \{1^{(3)}, 2^{(3)}, \ldots, L^{(3)}\} \ldots, \left\{1^{(N_c)}, 2^{(N_c)}, \ldots, L^{(N_c)}\right\}\right\}$$

and

$$M = \left\{p_1 = 0, 1', p_2, 1'', p_3, 1^{(3)}, \ldots, p_{N_c}, 1^{(N_c)}\right\}.$$

We have located the first finite crystal at site $p_1 = 0$ of the infinite crystal. The second finite crystal is located at site $p_2 > p_1$ of the infinite crystal. The third finite crystal is located at $p_3 > p_2$, etc. In this case, the coupling operator is a $2N_c \times 2N_c$ matrix, whose form is given by

$$\vec{V}_I = \frac{\beta_I}{m} \begin{pmatrix} -1 & 1 & 0 & 0 & \cdots & 0 & 0 \\ 1 & -1 & 0 & 0 & \cdots & 0 & 0 \\ 0 & 0 & -1 & 1 & \cdots & 0 & 0 \\ 0 & 0 & 1 & -1 & \cdots & 0 & 0 \\ \vdots & \vdots & \vdots & \vdots & \cdots & \vdots & \vdots \\ 0 & 0 & 0 & 0 & 0 & -1 & 1 \\ 0 & 0 & 0 & 0 & 0 & 1 & -1 \end{pmatrix}. \tag{2.115}$$

To calculate $\overleftrightarrow{\Delta}(MM) = \overleftrightarrow{I}(MM) + \overleftrightarrow{V}_I(MM)\overleftrightarrow{G}_S(MM)$, one needs the Green's function of the uncoupled system, $\overleftrightarrow{G}_S(MM)$, which takes the form

$$\overleftrightarrow{G}_S(MM) = \begin{pmatrix} G_0(p_1 p_1) & 0 & G_0(p_1 p_2) & 0 & G_0(p_1 p_3) & 0 & \cdots & G_0(p_1 p_{N_c}) & 0 \\ 0 & g_s(1'1') & 0 & 0 & 0 & 0 & \cdots & 0 & 0 \\ G_0(p_2 p_1) & 0 & G_0(p_2 p_2) & 0 & G_0(p_2 p_3) & 0 & \cdots & G_0(p_2 p_{N_c}) & 0 \\ 0 & 0 & 0 & g_s(1''1'') & 0 & 0 & \cdots & 0 & 0 \\ G_0(p_3 p_1) & 0 & G_0(p_3 p_2) & 0 & G_0(p_3 p_3) & 0 & \cdots & G_0(p_3 p_{N_c}) & 0 \\ 0 & 0 & 0 & 0 & 0 & g_s(1^{(3)}1^{(3)}) & \cdots & 0 & 0 \\ \vdots & \vdots & \vdots & \vdots & \vdots & \vdots & \vdots & \vdots & \vdots \\ G_0(p_{N_c} p_1) & 0 & G_0(p_{N_c} p_2) & 0 & G_0(p_{N_c} p_3) & 0 & \cdots & G_0(p_{N_c} p_{N_c}) & 0 \\ 0 & 0 & 0 & 0 & 0 & 0 & \cdots & 0 & g_s(1^{(N_c)}1^{(N_c)}) \end{pmatrix}. \tag{2.116}$$

In this matrix, the odd entries (rows or columns) correspond to locations along the infinite crystal in M and the even entries correspond to the position of the first atom of the finite crystals (also in the space M). From (2.26) and (2.102), the elements of this matrix are therefore

$$G_0(p_i p_j) = \frac{\beta}{m} \frac{t^{|p_i - p_j| + 1}}{t^2 - 1} \tag{2.117}$$

and

$$g_s(1^{(i)}1^{(i)}) = \frac{\beta'}{m'} \frac{t' + t'^{L'i}}{(t'-1)(1 - t'^{2L'i})}. \tag{2.118}$$

We use (2.95) to obtain the displacement field. For this we also need to assume a form for the reference displacement $U(D) = t^n$. This displacement corresponds to a wave propagating in the infinite crystal and launched from $n = -\infty$. The displacement inside the side crystal is also assumed to be equal to zero. The displacement in the space M takes the form

$$U(M) = (U(0), U(1'), U(p_2), U(1''), \ldots, U(p_{N_c}), U(1^{N_c}))$$
$$= (1, 0, t^{p_2}, 0, \ldots, t^{p_{N_c}}, 0). \tag{2.119}$$

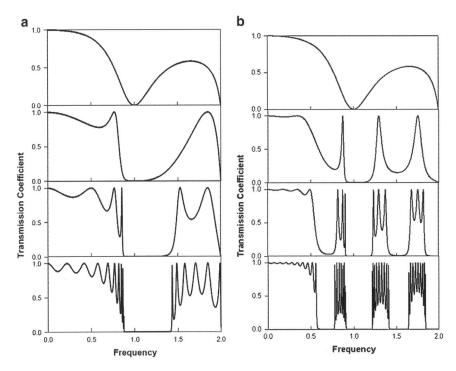

Fig. 2.13 Infinite mono-atomic crystal with (**a**) from top to bottom, a one-atom $(L' = 1)$ side branch located at $p_1 = 0$; two single-atom side branches located at $p_1 = 0$ and at $p_1 = 0$; four single-atom side branches at $p_1 = 0$, $p_2 = 1$, $p_3 = 2$, $p_4 = 3$; and ten one-atom side branches at $p_1 = 0, \ldots, p_{10} = 9$ and (**b**) from top to bottom, a one-atom $(L' = 1)$ side branch located at $p_1 = 0$; two single-atom side branches located at and at $p_1 = 0, p_2 = 4$; four single-atom side branches at $p_1 = 0, p_2 = 4, p_3 = 8, p_4 = 12$; and ten one-atom side branches at $p_1 = 0, \ldots, p_{10} = 36$, $\beta = \beta_I = 1, m = m' = 1$

The displacement field at a site $n \geq p_{N_c}$ along the infinite crystal (i.e., on the right side of the last grafted finite crystal) is therefore determined from (2.95), where we use

$$\vec{A}(M, n) = \begin{pmatrix} V_I(0, 0)G_0(0, n) \\ V_I(1', 0)G_0(0, n) \\ V_I(p_2, p_2)G_0(p_2, n) \\ V_I(1'', p_2)G_0(p_2, n) \\ \vdots \\ V_I(p_{N_c}, p_{N_c})G_0(p_{N_c}, n) \\ V_I(1^{p_{N_c}}, p_{N_c})G_0(p_{N_c}, n) \end{pmatrix} = \frac{\beta_I}{m} \frac{\beta}{m} \frac{t^n}{t^2 - 1} \begin{pmatrix} -t \\ t \\ -t^{1-p_2} \\ t^{1-p_2} \\ \vdots \\ -t^{1-p_{N_c}} \\ t^{1-p_{N_c}} \end{pmatrix}. \quad (2.120)$$

A transmission coefficient is subsequently defined as the ratio $T = u_n/t^n$. For a large number of grafted finite crystals, one has to resort to numerical calculation of the transmission coefficient by inserting (2.115)–(2.120) into (2.95). For the sake of illustration, we have performed such calculations using the limiting case: $\beta = \beta_I$ $= \beta'$ and $m = m'$ (i.e., $t = t'$). The numerical calculation involves the following steps for a series of values of the angular frequency $\omega \leq \omega_0$:

(a) calculating $\xi = 1 - \frac{m\omega^2}{2\beta}$,
(b) calculating $t = \xi + i(1 - \xi^2)^{1/2}$ since $-1 \leq \xi \leq 1$
(c) inserting t into (2.115)–(2.120)
(d) Calculating the transmission coefficient $T(\omega)$

Figure 2.13 illustrates the formation of a band gap by (a) local resonances and (b) band folding effects (Bragg scattering) in the transmission coefficient as a function of frequency for $L' = 1$, $\beta = \beta_I = 1$, $m = m' = 1$. With these conditions, $\omega_0 = 2$. A single one-atom side branch produces one resonant zero of transmission at $\omega = 1$. As one increases the number of side branches, spaced regularly by one interatomic spacing, the periodicity of the infinite chain is conserved and the resonant zero of transmission broadens into a stop band. Two additional dips in transmission on both sides of the resonant stop band form if the side branches are spaced by four atomic spacings. For a large number of side branches spaced by four lattice parameters, these dips broaden and deepen approaching the band gaps that would result from the multiple scattering of waves by a periodic array of side branches.

This example clearly illustrates the contribution of local resonance to wave propagation as well as the contribution of scattering by a periodic array of scatterers. The former mechanism is the foundation of locally resonant structures that determines the properties of acoustic metamaterials. The latter is associated with Bragg's scattering, which is the fundamental mechanism underlying the properties of phononic crystals.

References

1. E.N. Economou, *Green's Functions in Quantum Physics* (Springer, New York, 1990)
2. N.W. Ashcroft, N.D. Mermin, *Solid State Physics* (Brooks Cole, New York, 1976)
3. L. Dobrzynski, Interface response theory of composite systems. Surf. Sci. **200**, 435 (1988)

Chapter 3
One-Dimensional Phononic Crystals

EI Houssaine EI Boudouti and Bahram Djafari-Rouhani

Abstract In this chapter, we discuss the vibrational properties of one-dimensional (1D) phononic crystals of both discrete and continuous media. These properties include the dispersion curves of infinite crystals as well as the confined modes and localized (surface, cavity) modes of finite and semi-infinite crystals. A general rule about the existence of localized surface modes in finite and semi-infinite superlattices with free surfaces is presented. We also present the calculations of reflection and transmission coefficients, particularly in view of selective filtering through localized modes. Most of the results presented in this chapter deal with waves propagating along the axis of the superlattice. However, in the last part of the chapter, we also discuss wave propagation out of the normal incidence and, more particularly, we demonstrate the possibility of omnidirectional transmission gap and selective filtering for any incidence angle. A comparison of the theoretical results with experimental data available in the literature is also presented and the reliability of the theoretical predictions is indicated.

3.1 Introduction

The one-dimensional (1D) phononic crystals called superlattices (SLs) are of great importance in material science. These structures are, in general, composed of two or several layers repeated periodically along the direction of growth. The layers constituting each cell of the SL can be made of a combination of solid–solid or

E.H. EI Boudouti (✉)
LDOM, Département de Physique, Université Mohamed I, 60000 Oujda, Morocco
e-mail: elboudouti@yahoo.fr

B. Djafari-Rouhani
IEMN, UMR-CNRS 8520, UFR de Physique, Université de Lille 1, 59655 Villeneuve d'Ascq, France
e-mail: bahram.djafari-rouhani@univ-lille1.fr

P.A. Deymier (ed.), *Acoustic Metamaterials and Phononic Crystals*,
Springer Series in Solid-State Sciences 173, DOI 10.1007/978-3-642-31232-8_3,
© Springer-Verlag Berlin Heidelberg 2013

solid–fluid-layered media. These materials enter now in the category of so-called phononic crystals (see the first chapter of this book and references therein) [1–3] constituted by inclusions (spheres, cylinders, etc.) arranged in a host matrix along two-dimensional (2D) and three-dimensional (3D) of the space. After the proposal of SLs by Esaki [4], the study of elementary excitations in multilayered systems has been very active. Among these excitations, acoustic phonons have received increased attention after the first observation by Colvard et al. [5] of a doublet associated to folded longitudinal acoustic phonons by means of Raman scattering. The essential property of these structures is the existence of forbidden frequency bands induced by the difference in acoustic properties of the constituents and the periodicity of these systems leading to unusual physical phenomena in these heterostructures in comparison with bulk materials [6–8].

With regard to acoustic waves in solid–solid SLs, a number of theoretical and experimental works have been devoted to the study of the band gap structures of periodic SLs [6–10] composed of crystalline, amorphous semi-conductors, or metallic multilayers at the nanometric scale. The theoretical models used are essentially the transfer matrix [7, 11–13] and the Green's function methods [6, 14–16], whereas the experimental techniques include Raman scattering [5, 17, 18], ultrasonics [19–29], and time-resolved X-ray diffraction [30]. Besides the existence of the band-gap structures in perfect periodic SLs, it was shown theoretically and experimentally that the ideal SL should be modified to take into account the media surrounding the structure as a free surface [14, 15, 31–40], a SL/substrate interface [14, 15, 34, 41, 42], a cavity layer [43–51], etc., which are often used in experiments together with SLs. In addition to the defect modes that can be introduced by such inhomogeneities inside the band gaps, some other works have shown the existence of small peaks in folded longitudinal acoustic phonons and interpreted as confined phonons of the whole finite SL [52–54].

All the above phenomena have been exploited to propose one-dimensional (1D) solid–solid-layered media for several interesting applications as in their 2D and 3D counterparts phononic crystals (see the first chapter of this book and references therein). Among these applications, one can mention (1) omnidirectional band gaps [55–58], (2) the possibility to engineer small-size sonic crystals with locally resonant band gaps in the audible frequency range [59], (3) hypersonic crystals [60–63] with high-frequency band gaps to enhance acousto-optical [49–51] or optomechanical [64, 65] interaction and to realize stimulated emission of acoustic phonons [66], and (4) the possibility to enhance selective transmission through guided modes of a cavity layer inserted in the periodic structure [6, 67] or by interface resonance modes induced by the superlattice/substrate interface [68–70]. The advantage of 1D systems lies in the fact that their design is more feasible and they require only relatively simple analytical and numerical calculations. The analytical calculations enable us to understand deeply different physical properties related to the band gaps in such systems.

In comparison with solid–solid-layered media, the propagation of acoustic waves in the solid–fluid counterparts' structures has received less attention [71]. The first works on these systems have been carried out by Rytov [72] and summarized by Brekhovskikh [71]. Rytov's approach has been used by Schöenberg [73] together with propagator matrix formalism to account for propagation through such a periodic medium in any direction of propagation and at arbitrary frequency. Similar results are also obtained by Rousseau [74]. In the low-frequency limit, it was shown [73] that besides the existence of small gaps, there is one-wave speed for propagation perpendicular to the layering and two-wave speeds for propagation parallel to the layering which are without analogue in solid–solid SLs. The two latter speeds both correspond to compressional waves and their existence is suggestive of Biot's theory [75] of wave propagation in porous media. Alternating solid and viscous fluid layers have been proposed recently [76–78] as an idealized porous medium to evaluate dispersion and attenuation of acoustic waves in porous solids saturated with fluids. The experimental evidence [79] of these waves is carried out using ultrasonic techniques in Al-water and Plexiglas-water SLs. Also, it was shown theoretically and experimentally that finite size layered structures composed of a few cells of solid–fluid layers with one [80, 81] or multiple [82] periodicity may exhibit large gaps and the presence of defect layers in these structures may give rise to well-defined defect modes in these gaps [81]. Recently, solid layers separated by graded fluid layers [83] and piezoelectric composites [84, 85] have shown the possibility of acoustic Bloch oscillations analogous to the Wannier–Stark ladders of electronic states in a biased SL [86].

In this chapter, we discuss the vibrational properties of 1D phononic crystals of both discrete and continuous media. These properties include the dispersion curves of infinite crystals as well as the confined modes and localized (surface, cavity) modes of finite and semi-infinite crystals. We also present the calculations of reflection and transmission coefficients, particularly in view of selective filtering through localized modes. Most of the results presented in this chapter deal with waves propagating along the axis of the SL. However, in the last part of the chapter, we also discuss wave propagation out of the normal incidence and, more particularly, we demonstrate the possibility of omnidirectional transmission gap and selective filtering for any incidence angle. More detailed physical and technical discussions about the band structure, phonon transport, as well as light scattering by acoustic phonons in SLs and multilayered structures are given in the review paper [6].

This chapter is organized as follows: In Sects. 3.2 and 3.3, we give a detailed study on surface and confined longitudinal phonons in infinite, semi-infinite, and finite 1D discrete and continuous phononic crystals. We demonstrated analytically a general rule about the existence of surface modes associated with a SL free-stress surface. This rule predicts the existence of one mode per gap when we consider together two semi-infinite SLs obtained from the cleavage of an infinite SL between two cells. In the case of finite 1D phononic crystals made of N cells, we show the existence of two types of confined modes, namely, there are always $N-1$ modes in the allowed bands, whereas there is one and only one state corresponding to each band gap. In Sect. 3.4, we present briefly the theory of light scattering by longitudinal acoustic phonons by means of the Green's function method. The application of this theory to deduce confined and surface phonons in semiconductor-layered materials is presented. Section 3.5 is devoted to the transmission enhancement assisted by surface resonances when a 1D phononic crystal is inserted between two different substrates. We show that the transmission can reach unity when the number of cells in the phononic crystal is chosen appropriately. The total transmission occurs when the incident acoustic wave interacts with surface resonant phonons localized at the interface between the phononic crystal and one of the substrates. In Sect. 3.6, we show that similarly to the 2D and 3D phononic crystals (see the first chapter of this book and references therein), layered media made of alternating solid–solid and solid–fluid layers may exhibit total reflection of acoustic incident waves in a given frequency range for all incident angles. Also, these structures may be used as acoustic filters that may transmit selectively certain frequencies within the omnidirectional gaps. The transmission filtering can be achieved through the guided modes of a defect layer inserted in the periodic structure.

These investigations are done within the framework of the Green's function method [87, 88] for discrete and continuous composite systems, the so-called *"interface response theory"* associated to such heterostructures. The basic concepts and the fundamental equations of this theory and its application to deduce the necessary ingredients to study acoustic waves in continuous media made of solid and fluid-layered materials are presented in reference [6]. A comparison of the theoretical results with experimental data available in the literature is also presented and the reliability of the theoretical predictions is indicated.

3.2   Surface and Confined Modes in 1D Discrete Phononic Crystals

Even though the problem of vibration modes in an infinite 1D chain of atoms has been the subject of many standard textbooks in solid state physics [89, 90], it is interesting to understand quantum confinement in 1D finite crystal as many fundamental problems are related to low-dimensional physics. In the case of electronic structures, it was shown recently that a finite 1D crystal made of N cells exhibits two types of confined states [91], namely, there are always $N-1$ states in the allowed bands, whereas there is one and only one state corresponding to each band gap [92, 93]. This latter state did not depend on the width of the crystal N. This demonstration has been extended to shear horizontal and sagittal acoustic waves in continuous media made of finite solid–solid [94] and solid–fluid [95] superlattices, respectively. An experimental and theoretical verification of this rule has been given recently for electromagnetic waves in 1D coaxial cables [96]. In this section, we shall give an extension of these results for longitudinal waves in 1D discrete phononic crystals made of two and three different atoms.

3.2.1 Diatomic Chain

In Chap. 2, the band structure of a diatomic crystal made of two different atoms characterized by different masses m_1 and m_2 and coupled by a spring constant β is given (see Fig. 2.7). Now, if we consider a finite structure made of N cells (i.e., $2N$ different atoms), we shall be interested in what follows to the dispersion relation of the discrete (confined) modes associated to standing waves in the finite structure as well as to surface waves induced by the surfaces surrounding the system. The theoretical calculation is based on the interface response theory [87, 88] described in Chap. 2. In the case of an infinite chain of bi-atoms of masses m_1 and m_2 and coupled by a spring constant β (Fig. 3.1a), the dynamical matrix can be written as

$$H(M_m, M_m) = \begin{pmatrix} \ddots & \ddots & 0 & 0 & 0 & 0 \\ \ddots & -\gamma_2 & \beta & 0 & 0 & 0 \\ 0 & \beta & -\gamma_1 & \beta & 0 & 0 \\ 0 & 0 & \beta & -\gamma_2 & \beta & 0 \\ 0 & 0 & 0 & \beta & -\gamma_1 & \ddots \\ 0 & 0 & 0 & 0 & \ddots & \ddots \end{pmatrix} \tag{3.1}$$

where

$$\gamma_i = 2\beta - m_i \omega^2, \quad (i = 1, 2). \tag{3.2}$$

Taking advantage of the periodicity d_1 in the direction of the structure, the Fourier transformed $g^{-1}(k; MM)$ of the above infinite tridiagonal (3.1) matrix within one unit cell has the following form:

$$g^{-1}(k; MM) = \begin{pmatrix} -\gamma_1 & \beta + \beta e^{-jkd_1} \\ \beta + \beta e^{jkd_1} & -\gamma_2 \end{pmatrix} \tag{3.3}$$

where k is the Bloch wave vector in the reciprocal space.

Therefore, one can deduce easily the dispersion relation from $\det(g^{-1}(k; MM)) = 0$ in the following form:

$$\cos(kd_1) = \frac{\gamma_1 \gamma_2}{2\beta^2} - 1 \tag{3.4}$$

where d_1 is the period of the structure. It is also straightforward to Fourier analyze back into real space all the elements of $g(k; MM))$ and obtain all the interface elements of g in the following form:

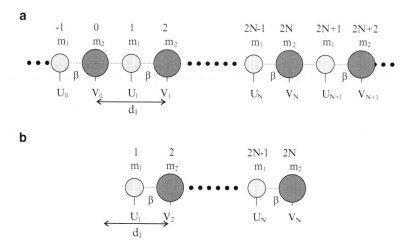

Fig. 3.1 (a) Schematic representation of an infinite structure made of two atoms ($m_1 - m_2$) coupled by a spring constant β. (b) Finite structure made of N bi-atoms. d_1 is the period of the crystal

$$g(n, \ 1; \ n', \ 1) = \left(\frac{\gamma_2}{\beta^2}\right)\frac{t^{|n-n'|+1}}{t^2 - 1}, \tag{3.5}$$

$$g(n, \ 1; n', \ 2) = \frac{1}{\beta}\left(\frac{t^{|n-n'|+1}}{t^2 - 1} + \frac{t^{|n-n'-1|+1}}{t^2 - 1}\right), \tag{3.6}$$

$$g(n, \ 2; n', \ 1) = \frac{1}{\beta}\left(\frac{t^{|n-n'|+1}}{t^2 - 1} + \frac{t^{|n-n'+1|+1}}{t^2 - 1}\right), \tag{3.7}$$

$$g(n, \ 2; n', \ 2) = \left(\frac{\gamma_1}{\beta^2}\right)\frac{t^{|n-n'|+1}}{t^2 - 1}. \tag{3.8}$$

where $t = e^{ikd_1}$ and n and n' indicate the positions of the cells with two atoms labeled 1 and 2.

Now, we consider a finite structure made of $2N$ atoms (i.e., N bi-atoms) obtained from the cleavage of an infinite structure in the interface space $M = \{0, 1, 2N, 2N + 1\}$ (Fig. 3.1b), which starts with atom 1 of mass m_1 and terminates with atom $2N$ of mass m_2. The cleavage operator is given by

$$V_{\text{cl}}(MM) = h(MM) - H(MM) \tag{3.9}$$

where h is the dynamical matrix of the whole system decoupled to three subsystems after eliminating the springs connecting atoms 0 and 1 from one side and $2N$ and

$2N + 1$ from the other side. Therefore, $V_{cl}(MM)$ is a 4×4 matrix, defined in the interface space M and can be written as follows:

$$V_{cl}(MM) = \begin{pmatrix} \beta & -\beta & 0 & 0 \\ -\beta & \beta & 0 & 0 \\ 0 & 0 & \beta & -\beta \\ 0 & 0 & -\beta & \beta \end{pmatrix}. \tag{3.10}$$

The operator $\Delta(MM)$ is defined by the relation [87, 88]:

$$\Delta(MM) = I(MM) + V_{cl}(MM)g(MM) \tag{3.11}$$

where $g(MM)$ can be obtained from (3.5) in the interface space $M = \{0, 1, 2N, 2N + 1\}$:

$$g(MM) = \frac{t}{\beta(t-1)} \begin{pmatrix} \dfrac{\gamma_1}{\beta}\dfrac{1}{t+1} & 1 & \dfrac{\gamma_1}{\beta}\dfrac{t^{2N}}{t+1} & t^{2N} \\ 1 & \dfrac{\gamma_2}{\beta}\dfrac{1}{t+1} & t^{2N-1} & \dfrac{\gamma_2}{\beta}\dfrac{t^{2N}}{t+1} \\ \dfrac{\gamma_1}{\beta}\dfrac{t^{2N}}{t+1} & t^{2N-1} & \dfrac{\gamma_1}{\beta}\dfrac{1}{t+1} & 1 \\ t^{2N} & \dfrac{\gamma_2}{\beta}\dfrac{t^{2N}}{t+1} & 1 & \dfrac{\gamma_2}{\beta}\dfrac{1}{t+1} \end{pmatrix}. \tag{3.12}$$

Therefore, one can deduce the expression of $\Delta_s(M_sM_s)$ of the finale structure in the space of the terminated surfaces $M_s = \{0, 2N\}$ (Fig. 3.1b) as follows:

$$\Delta_s(M_sM_s) = \begin{pmatrix} -\dfrac{1}{t-1} + \dfrac{\gamma_2}{\beta}\dfrac{1}{t^2-1} & \left(-\dfrac{\gamma_1}{\beta}\dfrac{1}{t^2-1} + \dfrac{1}{t-1}\right)t^{2N} \\ \left(\dfrac{1}{t-1} - \dfrac{\gamma_2}{\beta}\dfrac{1}{t^2-1} +\right)t^{2N} & -\dfrac{1}{t-1} + \dfrac{\gamma_1}{\beta}\dfrac{1}{t^2-1} \end{pmatrix}. \tag{3.13}$$

The discrete modes are given by the following equation [87, 88]:

$$det(\Delta_s) = 0. \tag{3.14}$$

After some algebraic calculation, (3.14) can be written in the following form:

$$\left(t + \frac{\beta}{\beta - \gamma_1}\right)\left(t + \frac{\beta}{\beta - \gamma_2}\right)(t^{2N} - 1) = 0. \tag{3.15}$$

When k is real, which corresponds to allowed bands, the eigenmodes are given by vanishing the third term in (3.15), namely:

$$sin(Nkd_1) = 0, \tag{3.16}$$

which gives

$$kd_1 = \frac{m\pi}{N}, \quad m = 1, 2, \ldots, N - 1. \tag{3.17}$$

When k is imaginary, which corresponds to band gaps, the eigenmodes are given by vanishing the first two terms in (3.15), namely:

$$t = \frac{\beta}{\gamma_1 - \beta} \tag{3.18}$$

and

$$t = \frac{\beta}{\gamma_2 - \beta} \tag{3.19}$$

with the condition:

$$|t| < 1 \tag{3.20}$$

that ensures the decaying of the waves far from the surface. By using (3.4), (3.18) and (3.19) can be written in a compact form:

$$\beta(\gamma_1 + \gamma_2) - \gamma_1\gamma_2 = 0 \tag{3.21}$$

Equations (3.18) [respectively (3.19)] and (3.21) can be restricted to the case $m_2 < m_1$ (respectively $m_1 < m_2$). These results show that localized surface modes appear only when the light atom is at the surface. In addition, (3.15) clearly shows that surface modes do not depend on N and therefore (3.21) gives the surface modes associated to two semi-infinite chains obtained from the cleavage of an infinite chain between atoms 1 and 2.

Figure 3.2a gives the variation of the frequencies as function of kd_1/π inside the first Brillouin zone for a diatomic chain such that $m_1 = 2.09$ g, $m_2 = 4.08$ g and $\beta = 6.3 \; 10^6$ N/m. These parameters are taken from the experimental work by Hladky et al. [97] on welded spheres, i.e., two alternating steel spheres of different diameters. The band structure of the infinite system is given by dots, whereas the confined modes of a finite structure made of $N = 4$ bi-atoms are shown by solid circles and surface modes are sketched by open circles.

As mentioned in Chap. 2, the first branch starts at the origin for $\frac{kd_1}{\pi} = 0$ and increases progressively until the frequency $\sqrt{\frac{2\beta}{m_2}}$ for $\frac{kd_1}{\pi} = 1$. The higher branch starts at the frequency $\sqrt{\frac{2\beta(m_1m_2)}{(m_1+m_2)}}$ for $\frac{kd_1}{\pi} = 0$ and decreases progressively until the frequency $\sqrt{\frac{2\beta}{m_1}}$ for $\frac{kd_1}{\pi} = 1$.

One can see clearly that each branch contains $N - 1$ confined modes. The gap width is $\sqrt{\frac{2\beta}{m_2}} - \sqrt{\frac{2\beta}{m_1}}$. Inside the gap, there is a localized mode induced by the surface terminated with the light atom. This mode is independent of N as it is

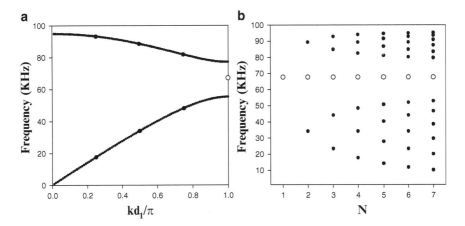

Fig. 3.2 (a) Variation of the frequencies versus $\frac{kd_1}{\pi}$ inside the first Brillouin zone for a diatomic linear chain. The masses are fixed to $m_1 = 2.09$ g, $m_2 = 4.08$ g and the spring constant $\beta = 6.3$ 10^6 N/m. The *dots* represent the band structure of the infinite chain, the *solid circles* give the confined modes of the finite structure made of $N = 4$ bi-atoms, whereas the open circles show the surface modes inside the band gaps. (**b**) Variation of the frequencies versus the number N of bi-atoms. *Solid* and *open circles* are confined and surface modes, respectively

illustrated in Fig. 3.2b where we have represented the variation of the frequency as function of the number of bi-atoms N in the finite crystal. It is worth noting that this simple analytical model gives a clear explanation to the numerical and experimental results by Hladky et al. [97] about the independence of the surface modes on the width of the finite crystal.

3.2.2 Tri-atomic Chain

Figure 3.3a gives a schematic representation of an infinite linear chain made of three atoms of masses m_1, m_2, and m_3 repeated periodically. The spring constant β is assumed to be the same for all atoms. Figure 3.3b shows a finite linear chain obtained from the cleavage of the infinite structure between the sites $(0, 1)$ on one side and $(3N, 3N + 1)$ on the other side. By using the same procedure as above for a bi-atomic chain, one obtains the dispersion relation of the infinite chain (Fig. 3.3a):

$$\cos(kd_2) = \frac{\gamma_1\gamma_2\gamma_3 - \beta^2(\gamma_1 + \gamma_2 + \gamma_3)}{2\beta^3} \qquad (3.22)$$

where d_2 is the period and γ_1, γ_2, and γ_3 are given by (3.2). Also, the eigenmodes of the finite chain (Fig. 3.3b) are given by

$$\left(t - \frac{\beta^2}{\gamma_1\gamma_2 - \beta^2 - \beta\gamma_2}\right)\left(t - \frac{\beta^2}{\gamma_2\gamma_3 - \beta^2 - \beta\gamma_2}\right)(t^{2N} - 1) = 0 \qquad (3.23)$$

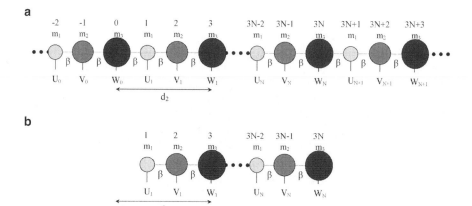

Fig. 3.3 (**a**) Schematic representation of the infinite structure made of three atoms ($m_1 - m_2 - m_3$) coupled by the same spring constant β. (**b**) Schematic representation of a finite structure composed of N tri-atoms

For k real, the third term in (3.23) gives the same expression as in (3.17). However, if k is imaginary, the first two terms in (3.23) give the localized surface modes, namely:

$$t = \frac{\beta^2}{\gamma_1\gamma_2 - \beta^2 - \beta\gamma_2} \tag{3.24}$$

and

$$t = \frac{\beta^2}{\gamma_2\gamma_3 - \beta^2 - \beta\gamma_2} \tag{3.25}$$

with the condition:

$$|t| < 1 \tag{3.26}$$

From (3.22), (3.24), and (3.25), the dispersion relation of surface modes can be written in a compact form:

$$\beta\gamma_2(\gamma_1 + \gamma_3) + \beta^2(\gamma_1 - \gamma_2 + \gamma_3) - 2\beta^3 - \gamma_1\gamma_2\gamma_3 = 0. \tag{3.27}$$

In the particular case where the masses m_1 and m_3 are identical, (3.24) and (3.25) become:

$$t = \pm 1. \tag{3.28}$$

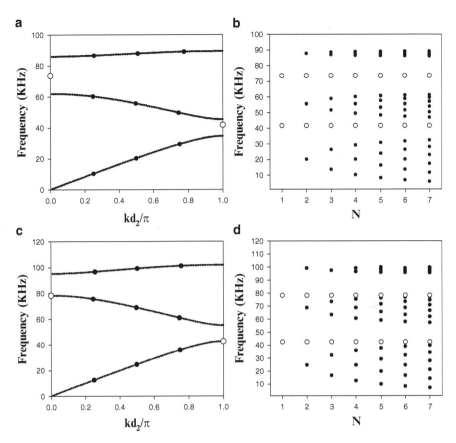

Fig. 3.4 (**a**) Variation of the frequencies versus $\frac{kd_2}{\pi}$ inside the first Brillouin zone for a tri-atomic linear chain constituted of three atoms of masses $m_1 - m_2 - m_3$. The masses are fixed to $m_1 = 2.09$ g, $m_2 = 4.08$ g, $m_3 = 6.07$ g and the spring constant is taken such that $\beta = 6.3$ 10^6 N/m. The *dots* represent the band structure of the infinite chain, the *solid circles* give the confined modes of the finite structure made of $N = 4$ tri-atoms, whereas the *open circles* show the surface modes inside the band gaps. (**b**) Variation of the frequencies versus the number N of tri-atoms. *Solid* and *open circles* are confined and surface modes respectively. (**c, d**) The same as in (**a**) and (**b**), but for a symmetric tri-atomic chain such that $m_1 = m_3 = 2.09$ g, $m_2 = 4.08$ g, and $\beta = 6.3 \times 10^6$ N/m

This result clearly shows that surface modes fall at the edge of the band gaps if the unit cell is symmetric.

Similarly to the case of a bi-atomic chain, (3.24) [respectively (3.25)] and (3.26) can be restricted to the case $m_3 < m_1$ (respectively $m_1 < m_3$). These results show that localized surface modes appear only when the structure terminates with the light atom at the surface.

Figure 3.4a shows the variation of the frequencies as function of kd_2/π for a tri-atomic chain such that $m_1 = 2.09$ g, $m_2 = 4.08$ g, $m_3 = 6.07$ g, and $\beta = 6.3$ 10^6 N/m. The band structure of the infinite system is given by dots, whereas the confined modes of a finite structure made of $N = 4$ tri-atoms are shown by solid circles and surface modes are sketched by open circles.

As predicted, because of three atoms constituting each cell, the band structure is composed of three branches separated by two gaps. On can notice the formation of $N - 1$ confined modes inside each band and one mode per gap induced by the weak atom at the surface. These modes are independent of N as it is illustrated in Fig. 3.4b).

Figure 3.4c, d give the same results as in Fig. 3.4a, b, respectively, but for a symmetric tri-atomic chain, namely, $m_1 = m_3 = 2.09$ g, $m_2 = 4.08$ g, and $\beta = 6.310^6$ N/m. Contrary to the previous case, the surface modes fall exactly at the edges of the band gaps in accordance with the analytical results [(3.28)]. These results are similar to those obtained recently for photonic crystals made of coaxial cables [96].

3.3 Surface and Confined Modes in 1D Continuous Phononic Crystals

3.3.1 Interface Response Theory of Continuous Media

In Chap. 2, Deymier and Dobrzynski exposed the interface response theory for 1D discrete media, which allows calculating the Green's function of any composite material. In what follows, we present the basic concept and the fundamental equations of this theory for continuous 1D media [87, 88]. Let us consider any composite material contained in its space of definition D and formed out of N different homogeneous pieces located in their domains D_i. Each piece is bounded by an interface M_i, adjacent in general to j ($1 \leq j \leq J$) other pieces through subinterfaces domains M_{ij}. The ensemble of all these interface spaces M_i will be called the interface space M of the composite material. The elements of the Green's function $g(DD)$ of any composite material can be obtained from [87, 88]

$$g(DD) = G(DD) - G(DM)G^{-1}(MM)G(MD)$$
$$+ G(DM)G^{-1}(MM)g(MM)G^{-1}(MM)G(MD), \qquad (3.29)$$

where $G(DD)$ is the reference Green's function formed out of truncated pieces in D_i of the bulk Green's functions of the infinite continuous media and $g(MM)$ the interface element of the Green's function of the composite system. The knowledge of the inverse of $g(MM)$ is sufficient to calculate the interface states of a composite system through the relation [87, 88]

$$det[g^{-1}(MM)] = 0 \qquad (3.30)$$

Moreover, if $U(D)$ represents an eigenvector of the reference system, (3.29) enables the calculation of the eigenvectors $u(D)$ of the composite material [87, 88]

$$u(D) = U(D) - U(M)G^{-1}(MM)G(MD) + U(M)G^{-1}(MM)g(MM)G^{-1}(MM)G(MD).$$
(3.31)

In (3.31), $U(D)$, $U(M)$, and $u(D)$ are row vectors. Equation (3.31) provides a description of all the waves reflected and transmitted by the interfaces, as well as the reflection and transmission coefficients of the composite system. In this case, $U(D)$ is a bulk wave launched in one homogeneous piece of the composite material [6].

3.3.2 Inverse Surface Green Functions of the Elementary Constituents

We consider an infinite homogeneous isotropic material i characterized by its characteristic impedance $Z_i = \rho_i v_i$ where ρ_i is the mass density, v_i the longitudinal velocity of sound. We limit ourselves to the simplest case of longitudinal vibrations in isotropic crystals with (001) interfaces where the field displacement $u(z)$ is along the axis z (perpendicular to the layers). The corresponding bulk equation of motion for medium i is given by

$$\left(\rho^{(i)}\omega^2 + C_{11}^{(i)} \frac{d^2}{dz^2} \right) u(z) = 0,$$
(3.32)

where $\rho^{(i)}$ and $C_{11}^{(i)}$ are, respectively, the mass density and the elastic constant and ω is the frequency of the vibrations.

Equation (3.32) can be written as

$$C_{11}^{(i)} \left(\frac{d^2}{dz^2} - \alpha_i^2 \right) u(z) = 0,$$
(3.33)

where

$$\alpha_i = -j\frac{\omega}{v_i}, v_i = \sqrt{\frac{C_{11}^{(i)}}{\rho^{(i)}}} \quad \text{and} \quad j = \sqrt{-1}.$$
(3.34)

The corresponding bulk Green's function for medium i is given by the equation:

$$C_{11}^{(i)} \left(\frac{d^2}{dz^2} - \alpha_i^2 \right) G_i(z - z') = \delta(z - z'),$$
(3.35)

whose solution can be written as [6]

$$G_i(z, z') = \frac{-j}{2\omega Z_i} e^{-\alpha_i|z-z'|}, \tag{3.36}$$

Before addressing the problem of periodic lamellar structures, it is helpful to know the surface elements of its elementary constituents, namely, the Green function of a finite slab of length d_i bounded by two free surfaces located at $z = -d_i/2$ and $+d_i/2$. These surface elements can be written in the form of a (2×2) matrix $g(MM)$, within the interface space $M_i = \{-\frac{d_i}{2}, +\frac{d_i}{2}\}$. The inverse of this matrix takes the following form [6]

$$[g(MM)]^{-1} = \begin{pmatrix} \dfrac{-\omega Z_i C_i}{S_i} & \dfrac{\omega Z_i}{S_i} \\ \dfrac{\omega Z_i}{S_i} & \dfrac{-\omega Z_i C_i}{S_i} \end{pmatrix}. \tag{3.37}$$

where $C_i = \cos(\omega d_i/v_i)$ and $S_i = \sin(\omega d_i/v_i)$ in equation (3.37).

The inverse of the surface element of a semi-infinite substrate s characterized by its impedance Z_s and bounded by its surface $z = 0$ is given by [6]

$$[g_s(0, \ 0)]^{-1} = j\omega Z_s. \tag{3.38}$$

3.3.3   Dispersion Relations of Finite and Semi-infinite Periodic 1D Structures

Consider an infinite superlattice made of a periodic repetition of a given 1D cell (Fig. 3.5c). The cell could be a multilayer structure, a multiwaveguide system, etc. Using the Green's function formalism, each cell is characterized by a 2×2 matrix constituted by the Green's function elements on the surface bounding the cell (Fig. 3.5a). The inverse of this matrix can be written explicitly as

$$[g(MM)]^{-1} = \begin{pmatrix} a & b \\ b & c \end{pmatrix}, \tag{3.39}$$

where $M = \{0, 1\}$ (see Fig. 3.5a). The four matrix elements are real quantities functions of the different parameters of the constituents inside the unit cell. The elements of a, b, and c for a unit cell made of j layers can be obtained by a linear juxtaposition of the 2×2 matrices of each layer i [(3.37)]. Then by inverting the whole $(j + 1) \times (j + 1)$ matrix and keeping only the elements at the extremities of this matrix, one obtains the 2×2 matrix of the unit cell, which we invert once again to obtain (3.39). It is worth noting that in general $a \neq c$; however, if the cell

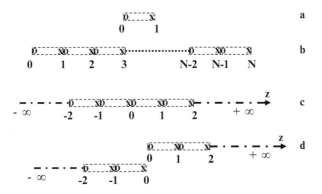

Fig. 3.5 Schematic representation of (**a**) a finite cell bounded by the space of interfaces M = {0, 1}, the circle and the cross indicate the left and the right surfaces of the cell, respectively. (**b**) A finite SL constituted of N cells. (**c**) An infinite SL. (**d**) Two semi-infinite SLs obtained from the cleavage of the infinite SL (**c**) between two cells. Notice the similarities between the surfaces ending the two complementary SLs [(**d**)] and those corresponding to a finite SL [(**a**), (**b**)]

is symmetric then $a = c$. The eigenmodes of the elementary cell are given by (3.30), namely

$$ac - b^2 = 0. \tag{3.40}$$

Now, the Green's function of the infinite SL made of a periodic repetition of a given cell (Fig. 3.5c) is obtained by a linear juxtaposition of the 2×2 matrix [Eq. (3.39)] in the interface domain of all the sites n. We obtain a tridiagonal matrix where the diagonal and off-diagonal elements of this matrix are given, respectively, by $a + c$ and b.

Taking advantage of the translational periodicity of this system along the z axis, this matrix can be Fourier transformed as [6]

$$[g(k, MM)]^{-1} = 2b[cos(kD) - \xi] \tag{3.41}$$

where k is the modulus of the one-dimensional reciprocal vector (Bloch wave vector), D is the period of the SL, and $\eta = -(a + c)/2b$.

The dispersion relation of the infinite periodic SL (Fig. 3.5c) is given by (3.30) and (3.41), namely

$$cos(kD) = -(a + c)/2b. \tag{3.42}$$

On the other hand, in the k space, the surface Green's function is

$$[g(k, MM)] = \frac{1}{2b[cos(kD) - \xi]} \tag{3.43}$$

After inverse Fourier transformation, (3.43) gives

$$g(n, n') = \frac{1}{b} \frac{t^{|n-n'|+1}}{t^2 - 1}$$ (3.44)

where n and n' denote the positions of the different interfaces between the cells and $t = e^{ikD}$.

Consider now a finite SL bounded by the two surfaces $n = 0$ and $n = N$ (Fig. 3.5b). The 2×2 Green's function matrix in the space of interface $M' = \{0, N\}$ of the finite SL can be written as [6]

$$g(M'M') = \frac{1}{\Delta} \begin{pmatrix} a + \frac{b}{t} - t^{2N}(a + bt) & -bt^N \left(t - \frac{1}{t}\right) \\ -bt^N \left(t - \frac{1}{t}\right) & -a - bt + t^{2N}\left(a + \frac{b}{t}\right) \end{pmatrix},$$ (3.45)

where

$$\Delta = (a + bt)\left(a + \frac{b}{t}\right)(1 - t^{2N})$$ (3.46)

The eigenmodes of the finite SL are given by the poles of the Green's function, namely $\Delta = 0$, or equivalently

$$\left(\frac{1}{t} + \frac{a}{b}\right)\left(t + \frac{a}{b}\right)(1 - t^{2N}) = 0.$$ (3.47)

This expression shows that there are two types of eigenmodes in the finite structure:

1. If the wave vector k is imaginary (modulo π) which corresponds to a forbidden band (gap), then the eigenmodes are given by the two first terms of (3.47), namely

$$t = -\frac{a}{b}$$ (3.48)

and

$$t = -\frac{b}{a},$$ (3.49)

whereas the third term in (3.47) cannot vanish inside the gap since t should satisfy the condition

$$|t| < 1$$ (3.50)

to ensure the decaying of surface states from the surface.

In addition, we remark that if $N \to \infty$ the term t^{2N} vanishes and therefore the two expressions [Eqs. (3.48) and (3.49)] give the surface modes for two semi-infinite SLs with complementary surfaces (Fig. 3.5d).

Equations (3.48) and (3.49) can be written in a unique explicit form by replacing them in (3.42) and factorizing by the factor $\frac{1}{b}$, one obtains

$$ac - b^2 = 0. \tag{3.51}$$

Therefore, the surface modes of one semi-infinite SL are given by (3.51) together with the condition $\left|\frac{a}{b}\right| < 1$, whereas the surface modes of the complementary SL are given by (3.51) but with the condition $\left|\frac{b}{a}\right| < 1$. This result shows that if a surface mode appears on the surface of one SL, it does not appear on the other surface of the complementary SL. Moreover, (3.51) shows that the expression giving the surface modes for two complementary SLs is exactly the same expression giving the eigenmodes of one cell [Eq. (3.40)]. In the particular case of a symmetric cell (i.e., $a = c$), then (3.51) reduces to $a = \pm b$. This expression corresponds to a band gap edge as $cos(kD) = \pm 1$ [Eq. (3.42)].

In addition to these results, let us recall briefly another result concerning the existence of surface modes associated to two semi-infinite SLs obtained by the cleavage of an infinite SL, namely [14, 15] there exists as many surface modes as minigaps. These modes are associated with either one or the other surface of the two SLs. These results concern especially the transverse elastic waves in layered media [14, 15] and electromagnetic waves in quasi-one-dimensional waveguides [98, 99].

2. If the wave vector k is real which corresponds to an allowed band, then the eigenmodes of the finite SL are given by the third term in (3.47), namely

$$sin(NkD) = 0, \tag{3.52}$$

which gives

$$kD = \frac{m\pi}{N}, \quad m = 1, 2, \ldots, N - 1, \tag{3.53}$$

whereas the first and second terms in (3.47) cannot vanish in the bulk bands.

From the above results, one can deduce that a finite SL constituted of N cells gives rise to $N - 1$ modes inside the bulk bands of the SL and one surface mode in each gap of the SL that may be attributed to one of the two surfaces surrounding the finite SL. The surface modes are independent of N and coincide with those of two semi-infinite SLs obtained by the cleavage of an infinite SL between two cells. In what follows, we shall give some numerical examples of these results in the case of longitudinal acoustic wave propagation in layered media.

3.3.4 Transmission and Reflection Coefficients

We shall consider a finite multilayer system sandwiched between two different homogeneous semi-infinite media having indexes s and s', respectively. We have two interfaces bounding the multilayer system, we shall call them l (left) and r (right) respectively. The inverse of the Green's-function projected in the space of the interfaces can be expressed as

$$g_S^{-1} = \begin{bmatrix} g_S^{-1}(l,\,l) = a + j\omega Z_s & g_S^{-1}(l,\,r) = b \\ g_S^{-1}(r,\,l) = b & g_S^{-1}(r,\,r) = c + j\omega Z_{s'} \end{bmatrix}$$

Let us then consider an incident wave in the semi-infinite medium s

$$u(z) = \exp(-i\alpha_s z), \tag{3.54}$$

where $\alpha_s = \omega/v_s$. Following the expressions detailed in [100], it can be found that the transmitted wave in medium s' has the form

$$u_T(z) = -2i\alpha_s g_S(l,r)\exp(-i\alpha_{s'} z), \tag{3.55}$$

whereas the reflected wave has the form

$$u_R(z) = -[1 + 2i\alpha_s g_S(l,l)]\exp(i\alpha_s z). \tag{3.56}$$

It is then clear that (3.55) and (3.56) can be written as

$$u_T(z) = C_T \exp(-i\alpha_{s'} z),$$
$$u_R(z) = C_R \exp(i\alpha_s z), \tag{3.57}$$

where C_T and C_R are the transmission and reflection amplitudes given by

$$C_T = -2i\alpha_s g_S^{-1}(l,r)\det|g_S|,$$
$$C_R = -[1 + 2i\alpha_s g_S^{-1}(l,l)\det|g_S|]. \tag{3.58}$$

3.3.5   Numerical Results

3.3.5.1   Case of a Finite Periodic Structure Made of Asymmetric Cells

In what follows, we consider a finite periodic phononic crystal where each cell is made of two layers characterized by lengths d_A and d_B, longitudinal wave velocities v_A and v_B and impedances Z_A and Z_B, respectively. When applied to a SL made of a periodic repetition of layers A and B, the general dispersion relation [Eq. (3.42)] gives the well-known relation

$$\cos(kD) = C_A C_B - 0.5(Z_A/Z_B + Z_B/Z_A)S_A S_B \tag{3.59}$$

where $C_{A,B} = cos(\omega d_{A,B}/\upsilon_{A,B})$, $S_A = sin(\omega d_{A,B}/\upsilon_{A,B})$, and $D = d_A + d_B$ is the period.

In the particular case where $d_A/\upsilon_A = d_B/\upsilon_B$, one can show easily that the limits of the first band gap lying at the Brillouin zone edge (i.e., $kD = \pi$) are given by

$$\Omega_\pm = \Omega_0 \pm sin^{-1}(|Z_A - Z_B|/(Z_A + Z_B)) \tag{3.60}$$

where $\Omega = \omega d_{A,B}/\upsilon_{A,B}$ is the reduced frequency and $\Omega_0 = \omega d_{A,B}/\upsilon_{A,B} = \pi/2$ is the central gap frequency corresponding to quarter wavelength layers.

If the impedances Z_A and Z_B are very close, (3.60) shows that the band gap width $\Delta\Omega = \Omega_+ - \Omega_-$ is proportional to the difference between impedances Z_A and Z_B. However, if the mismatch between impedances is higher such that for example $Z_A/Z_B = 2$, then (3.59) becomes

$$cos(kD) = 1 - \frac{9}{4} sin^2(\Omega) \tag{3.61}$$

The limits of the band gaps are given by the successive sequences $kD = 0, \pi, \pi,$ $0, 0, \pi, \pi, 0 \ldots$ and therefore $\Omega \simeq 0, 0.39\pi, 0.61\pi, \pi, \pi, 1.39\pi, 1.61\pi, 2\pi, \ldots$. These results show that the band gap structure is periodic every $\Omega = \pi$. Therefore, we limited ourselves to the reduced frequency region $0 \leq \Omega \leq \pi$. The width of the successive bands is about 0.4π, the width of the gaps at the edge of the Brillouin zone is about 0.2π, whereas the width of the gaps at the center of the Brillouin zone vanishes. These results are confirmed in Fig. 3.6 where we have plotted the dispersion curves (frequency versus kD) for the periodic structure depicted above.

Inside the first gap $kD = \pi \pm j\kappa$, the dispersion relation [Eq. (3.61)] becomes

$$cosh(\kappa/2) = \frac{3}{2\sqrt{2}} sin(\Omega). \tag{3.62}$$

Equation (3.62) gives the imaginary part κ of the reduced wave vector kD inside the gaps which is responsible for the attenuation of the modes that may lie inside these gaps when a defect is inserted in the structure such as the surface[6]. From the above results, one can deduce that the center of the first gap is given by $\Omega = \pi/2$ and the value of κ at this frequency is $\kappa \simeq 0.69$ [Eq. (3.62)][see the dashed curves in Fig. 3.6].

As concerns the eigenmodes of a finite SL (illustrated in Fig. 3.6 for $N = 4$ cells), one can distinguish, as described in Sect. 3.3.3, the surface modes [Eq. (3.51)] lying inside the forbidden bands and the bulk modes [Eq. (3.53)] lying inside the allowed bands. The expression giving the surface modes [Eq. (3.51)] can be written as [14, 15]

$$Z_A C_A S_B + Z_B C_B S_A = 0, \tag{3.63}$$

Fig. 3.6 Band structure of an
infinite SL. Each cell is made
of two layers characterized by
lengths d_A and d_B,
longitudinal wave velocities
v_A and v_B, and impedances Z_A
and Z_B, respectively. *Solid*
and *open circles* correspond
to the eigenmodes of a finite
structure made of $N = 4$ cells
with $d_A/d_B = v_A/v_B$ and
$Z_A/Z_B = 2$

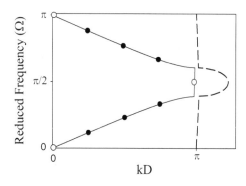

together with the condition [Eq. (3.50)]

$$\left| C_A C_B - \frac{Z_B}{Z_A} S_A S_B \right| < 1 \tag{3.64}$$

when the structure is terminated by layer A, and

$$\left| C_A C_B - \frac{Z_A}{Z_B} S_A S_B \right| < 1 \tag{3.65}$$

when the structure is terminated by layer B.

In the particular case considered here, $C_A = C_B = cos(\Omega)$ and $S_A = S_B = \sin(\Omega)$. Therefore, (3.63) becomes simply

$$sin(2\Omega) = 0, \quad \text{i.e.,} \, \Omega = m\pi/2 \tag{3.66}$$

where m is an integer. If m is even (i.e., $\Omega = 0, \pi, 2\pi, \ldots$), then neither (3.64) nor (3.65) are fulfilled since the left-hand term in these equations is unity. As mentioned above, this situation corresponds to the center of the Brillouin zone ($kD = 0$) where the band gaps close. However, if m is odd (i.e., $\Omega = \pi/2, 3\pi/2, \ldots$), then only Eq. (3.64) is fulfilled since $Z_B < Z_A$, which means that all the surface modes appear on the surface of the structure terminated by layer A and no surface modes appear when the structure terminates with layer B. In Fig. 3.6 we have plotted by open circles the surface mode lying in the first gap at $\Omega = \pi/2$ as well as the frequencies lying at the band gap edges (i.e., $\Omega = 0, \pi, \ldots$). Apart from these modes, there exists $N-1 = 3$ modes in each band given by Eq. (3.53).

Figure 3.7 shows the variation of the eigenmodes of a finite SL as a function of the number of cells N. For $N = 1$ (one cell), the eigenmodes are given by (3.66) and we can distinguish the modes lying at the closing of the band gaps (i.e., $\Omega = 0$ and $\Omega = \pi$) and the surface mode lying at the center of the band gap (i.e., $\Omega = \pi/2$). When N increases, the above modes remain constant, whereas there exist $N-1$ modes in each band for every value of N in accordance with the analytical results in Sect. 3.3.3.

Fig. 3.7 Variation of the eigenmodes of the finite SL as a function of the number of cells N. *Open* and *solid circles* have the same meaning as in Fig. 3.6

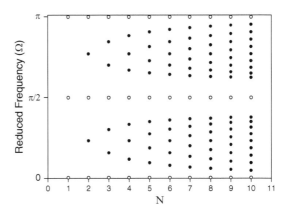

An analysis of the local density of states (LDOS) (Fig. 3.8) shows that the surface modes lying in the first gap at $\Omega = \pi/2$ exhibit a strong localization at the surface with almost the same localization length as far as N exceeds 3. It is worth noting that the LDOS reflects the behavior of the square modulus of the displacement field inside the structure.

3.3.5.2 Case of a Finite Structure with Symmetric Cells

In what follows, we consider a finite periodic structure made of symmetric cells. Each cell is composed of a layer of type B inserted between two layers of type A. Therefore, each cell becomes equivalent to a $A/B/A$ tri-layer. In this case, the dispersion relation [Eq. (3.42)] becomes

$$cos(kD) = C_A^2 C_B - C_B^2 C_A - C_A S_A S_B (Z_A/Z_B + Z_B/Z_A), \qquad (3.67)$$

where $D = 2d_A + d_B$. In the particular case where $d_A/v_A = d_B/v_B$ and $Z_A/Z_B = 2$, the above equation becomes simply

$$cos(kD) = \frac{cos(\Omega)}{2}(9\,cos^2(\Omega) - 7). \qquad (3.68)$$

The band gap edges are given by $cos(kD) = \pm 1$, namely, $cos(\Omega) = \pm 1, \pm 1/3$ and $\pm 2/3$. Therefore (see Fig. 3.9),

$$\Omega = 0, 0.27\pi, 0.39\pi, 0.61\pi, 0.73\pi, \pi, \ldots \qquad (3.69)$$

Inside the first two gaps, $kD = \pi \pm j\kappa$ and $kD = \pm j\kappa$, respectively. Then the attenuation coefficient κ [Eq. (3.68)] satisfies the equation

$$cosh(\kappa) = \left| \frac{cos(\Omega)}{2}(9\,cos^2(\Omega) - 7) \right|. \qquad (3.70)$$

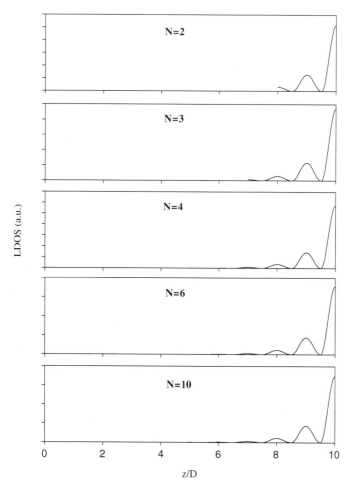

Fig. 3.8 The local density of states (LDOS) [in arbitrary units] as a function of the space position z for the mode lying at the central gap frequency $\Omega = \pi/2$ (Fig. 3.7) for $N = 2$ (**a**), 3 (**b**), 4 (**c**), 6 (**d**), and 10 (**e**). The finite SL is terminated by layers A and B at the left and the right of the structure, respectively

Thus, one can deduce the reduced frequencies at the center of the first two gaps, namely

$$cos(\Omega) = \frac{7\sqrt{7}}{9\sqrt{3}}, \quad \text{i.e.,} \quad \Omega \simeq 0.33\pi, \quad \text{and} \quad \Omega \simeq 0.67\pi \qquad (3.71)$$

as well as the corresponding values of κ ($\kappa \simeq 0.59$, see the dashed curves in Fig. 3.9).

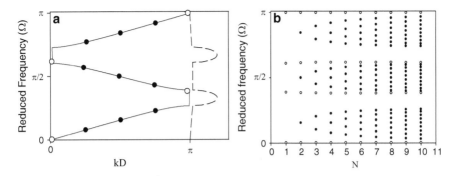

Fig. 3.9 (a) Same as in Fig. 3.6 but here the cell is taken symmetrical, i.e., an A/B/A tri-layer. (b) Variation of the eigenmodes of the finite SL as a function of the number of cells N. *Open* and *solid circles* have the same meaning as in Fig. 3.7

The surface modes [Eq. (3.51)] for a tri-layer SL are given in general by [15]

$$2C_A S_A C_B + S_B \left(C_A^2 \frac{Z_A}{Z_B} + S_A^2 \frac{Z_B}{Z_A} \right) = 0 \tag{3.72}$$

In the particular case considered here, $C_A = C_B = cos(\Omega)$, $S_A = S_B = sin(\Omega)$, and $Z_A/Z_B = 2$. Thus, (3.72) becomes

$$sin(\Omega)(9\,cos^2(\Omega) - 1) = 0, \tag{3.73}$$

which leads to

$$sin(\Omega) = 0, \quad \text{i.e.,} \quad \Omega = 0, \pi, 2\pi. \tag{3.74}$$

or

$$cos(\Omega) = \pm 1/3, \quad \text{i.e.,} \quad \Omega = 0.39\pi,\ 0.61\pi,\ 1.39\pi,\ 1.61\pi,\ \ldots \tag{3.75}$$

However, the two latter equations give $cos(kD) = \pm 1$ [Eq. (3.68)]. Consequently, as mentioned in Sect. 3.3.3, the finite periodic SL with symmetric cells do not exhibit surface modes inside the band gaps, but leads only to a constant frequency band edge modes. These results are similar to those found by Ren [94] for transverse waves in finite one-dimensional systems using another method of calculation. Of course, in addition to the band edge modes, one can expect $N-1$ modes in each band given by Eq. (3.53).

Figure 3.9a, b resume the numerical results corresponding to the analytical results detailed above. Among the different modes, one can distinguish the band-edge modes plotted by open circles, these modes fall at a constant frequency independent of the number of cells N (see Fig. 3.9b) and the bulk band modes ($N-1 = 3$) lying inside each allowed band (solid circles).

3.4   Light Scattering by Longitudinal Acoustic Phonons

In this section, we shall give some experimental results related to the determination
of bulk and surface modes in finite superlattices. In particular, we shall concentrate
essentially on experimental measurements based on light scattering by longitudinal
acoustic phonons.

The principle of this scattering can be summarized as follows: the propagation of
an acoustic wave in the superlattice excites periodic variations of strain which in
turn induce a modulation of the dielectric tensor ε_{ij} from the photo-elastic coupling
to elastic fluctuations,

$$\delta\varepsilon_{ij} = \varepsilon_{ii}\varepsilon_{jj} \sum_{kl} P_{ijkl} \frac{1}{2}\left(\frac{\partial u_k}{\partial x_l} + \frac{\partial u_l}{\partial x_k}\right) \tag{3.76}$$

P_{ijkl} are the elements of the photoelastic tensor and can be considered as
functions of z. The coupling of incident light to phonons gives rise to a polarization
in the superlattice which creates a scattered field. We are interested in pure
longitudinal phonons along the axis z of a multilayer structure composed of cubic
materials with (001) interfaces. In this case, we can assume that all the electromagnetic fields (incident, scattered, and polarization waves) are polarized parallel to the
x axis and propagates along z. Then each medium α in the structure can be
characterized by an elastic constant C_α (which means C_{11}), the mass density ρ_α,
the dielectric constant $\varepsilon_\alpha = n_\alpha^2$ (where n is the index of refraction in the medium α),
and one photoelastic constant $p_\alpha = -\varepsilon_\alpha^2 P_{1133}^\alpha$.

Equation (3.76) becomes for each medium α

$$\delta\varepsilon_\alpha = p_\alpha \frac{\partial u_\alpha(z)}{\partial z}. \tag{3.77}$$

The calculation of the emitted electric field $E_s(z,t)$ when the superlattice is
submitted to an incident electromagnetic field can be done following the Green's
function method [101, 102]

$$E_s(z,\ t) = -\frac{\omega_i^2}{\varepsilon_0 c^2} \sum_\alpha \int p_\alpha G(z,z') \frac{\partial u_\alpha(z')}{\partial z'} E_i^0(z',t)\,\mathrm{d}z'. \tag{3.78}$$

Here ω_i is the angular frequency of the incident wave, ε_0 and c are the permittivity and the speed of light in vacuum, respectively, $E_i^0(z',\ t)$ is the electric field in
the SL, and $G(z, z')$ is the Green's function associated with the propagation of an
electromagnetic field along z in the vacuum/superlattice system in the absence of
acoustic deformation.

In the particular case where the dielectric modulation of the multilayer structure
can be neglected (which happens when the layers are thin as compared to the optical

wavelengths), the system can be considered as an homogeneous effective medium from the optical point of view, then $E_i^0(z') = E_i^0 e^{ik_i z'}$ is a plane wave (instead of being a Bloch wave) and $G(z, z') \propto e^{ik_s(z-z')}$. k_i and k_s are the wave vectors of the incident and scattered waves. Therefore, (3.78) becomes

$$E_s(z) \propto -\frac{\omega_i^2}{\varepsilon_0 c^2} \sum_\alpha \int p_\alpha e^{iqz'} \frac{\partial u_\alpha(z')}{\partial z'} dz' \qquad (3.79)$$

where

$$q = k_i - k_s = 2k_i = 4\pi n_{\text{eff}}/\lambda \qquad (3.80)$$

is the wave vector of the phonon in the backscattering geometry and n_{eff} is the effective index of refraction of the effective medium.

A great deal of work has been devoted to light scattering from acoustic phonons in multilayered structures, since the first observation of folded longitudinal-acoustic modes by Colvard et al. [5]. Several experimental studies have been reported on GaAs-Ga$_x$ Al$_{1-x}$ As and Si-Ge$_x$ Si$_{1-x}$ systems. As mentioned before, in an ideal periodic structure (superlattice) consisting of an infinite sequence of building blocks AB made of different semiconductors A and B, the branches of the acoustic-phonon dispersion are back-folded inside the Brillouin zone due to the periodicity of the system. In the Raman process involving longitudinal acoustic phonons in backscattering geometry along the growth direction (z), crystal momentum is conserved, i.e., the wave vector transfer to the phonon corresponds to the sum of the magnitudes of the wave vectors of incident and scattered photons k_i and k_s, respectively. Characteristic doublets are observed in the spectrum which reflects the folding of the superlattice dispersion curves in the first Brillouin zone. Crystal momentum conservation at the doublet frequencies implies that all partial waves are coherently scattered, i.e., all layers of the superlattice contribute constructively to the total intensity. Therefore, the doublets are very sharp and pronounced.

In a real superlattice, the coherence of the scattering contributions from the individual layers is partly removed due to interface roughness and layer thickness fluctuations, finite-size effect of the superlattice as well as the effect of different defects that may be introduced inside these systems such as surfaces, interfaces, and defect layers (cavities, buffers, etc.).

In view of the relatively small thicknesses of the layers in the superlattice, the acoustic phonon Raman scattering can be obtained from (3.79) as

$$I(\omega) \propto \left| \sum_\alpha \int p_\alpha e^{iqz'} \frac{\partial u_\alpha(z')}{\partial z'} dz' \right|^2. \qquad (3.81)$$

Here we assume that the light propagates like in a homogeneous medium and u (z) is the normalized lattice displacement. Figure 3.10a gives the experimental

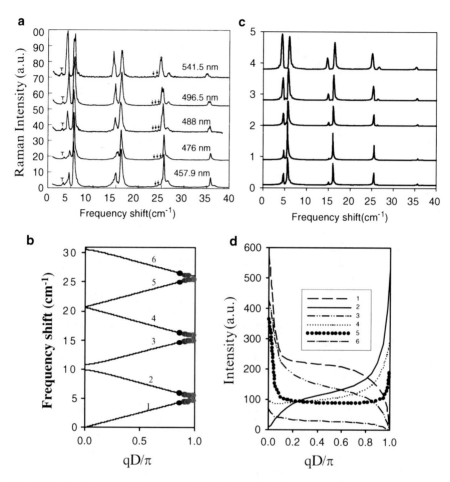

Fig. 3.10 (**a**) Room-temperature acoustic phonon Raman spectra in the Si/Si$_{0.52}$ Ge$_{0.48}$ superlattice, excited with the five studied laser lines (After [53]). (**b**) Calculated dispersion curves of the infinite SL (*solid lines*). The *filled circles* are obtained from the doublets of the theoretical spectra sketched in (**c**) using our theoretical model. (**d**) Variation of the intensities of the six first folded branches as functions of the diffusion wave vector D (qD/π)

results of Raman intensity obtained by Zhang et al. [52, 53] for a superlattice composed of 15 periods of 20.5 nm of Si and 4.9 nm of Si$_{0.52}$ Ge$_{0.48}$ epitaxially grown on a [101] oriented Si substrate. The different curves in Fig. 3.10a correspond to different laser wavelengths (i.e., different phonon wave-vectors). By reporting the frequency positions of the doublets within the band gap structure (Fig. 3.10b), good agreement between the dispersion curves of the infinite superlattice (full curves) and the experimental results (dots) has been obtained. These results enable one to deduce a precise measurement of the width of the first three gaps. Besides the description of the band gap structure, the Raman spectra show also small features (indicated by small vertical arrows in Fig. 3.10a) which are interpreted as confined modes (discrete modes) due to the finite-size structure of the superlattice. By using our theoretical model [El Boudouti et al. (unpublished)], we have reproduced theoretically in Fig. 3.10c the different Raman spectra of Fig. 3.10a and the agreement between theoretical and experimental results is quite good. In Fig. 3.10d we have calculated the intensity variation of different phonon branches labeled 1–6 in Fig. 3.10b within the reduced Brillouin zone. The intensities show drastic variations, especially for q close to the Brillouin zone edges. The Brillouin line (branch labeled 1) is the most intense mode for a large range of q values except near the zone boundary. These behaviors are similar to the theoretical predictions obtained by He et al. [103] on GaAs-Ga$_x$ Al$_{1-x}$ As superlattices.

Besides the doublets associated to folded longitudinal acoustic phonons, Lemos et al. [104] have shown the existence of additional modes between the doublets which are induced by a cap layer deposited at the surface of the superlattice. These modes fall inside the gap located at ~ 15 cm^{-1}. The superlattice is composed of 20 periods of 21.5 nm of Si and 5.0 nm of Ge$_{0.44}$Si$_{0.56}$ terminated by a cap layer made of Ge$_{0.44}$Si$_{0.56}$ with a thickness $d_c = 1.5$ nm. The top and bottom curves in Fig. 3.11a are drawn for two different wavelengths 514.5 nm and 496.5 nm, respectively. By using our theoretical model [El Boudouti et al. (unpublished)], we have reproduced correctly (Fig. 3.11b) the main features of these results, except that the intensity of the gap-mode greatly exceeds the observed value. To confirm that the gap mode is induced by the cap layer, we have calculated the local DOS as a function of the space position for the mode lying at ~ 15 cm^{-1}. The spatial localization of this mode (see Fig. 3.11c) shows clearly that it is localized in the cap layer and decreases inside the SL.

3.5 Transmission Enhancement Assisted by Surface Resonance

The possibility of the enhanced transmission from a semi-infinite solid to a semi-infinite fluid, in spite of a large mismatch of their acoustic impedances, has been shown theoretically and experimentally [68–70]. The transmission occurs through the surface resonances induced by a 1D solid–solid-layered structure inserted between these two media. These resonances are attributed to the SL/fluid interface [69] and coincide with the surface modes of the semi-infinite SL terminated with the layer having the lower acoustic impedance [31]. Recently [105], the possibility of the so-called extraordinary acoustic transmission assisted by surface resonances between two fluids has been shown. The structure consists in separating the two fluids by a rigid film flanked on both sides by finite arrays of grooves. The transmission followed by a strong collimation of sound arises through a single hole perforated in the film.

By analogy with the previous works on this subject [69], we show the possibility of enhanced transmission between two fluids by inserting a solid–fluid-layered material between these two fluids. Besides the possibility of selective transmission,

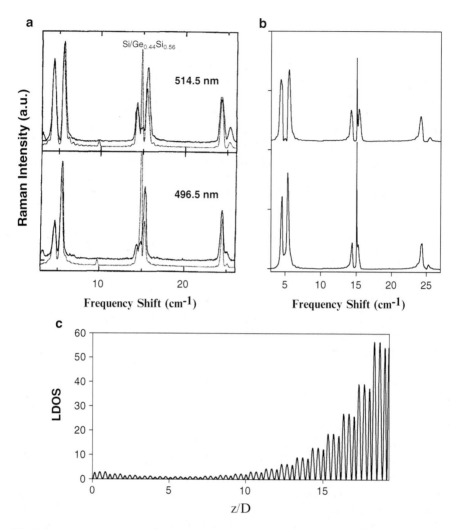

Fig. 3.11 (a) Raman spectra of a SL composed of 20 periods of 21.5 nm of Si and 5.0 nm of $Ge_{0.44}Si_{0.56}$ terminated by a cap layer made of $Ge_{0.44}Si_{0.56}$ with a thickness $d_c = 1.5$ nm. The *top* and *bottom curves* are drawn for two different wavelengths 514.5 nm and 496.5 nm, respectively (After [104]). (b) Theoretical results obtained from our theoretical model. (c) LDOS as function of the space position z for the surface mode located at ~15 cm^{-1}

this structure enables from a practical point of view to separate the two fluids which are in general miscible. We give a simple analytical expression of the effective acoustic impedance of the finite SL that enables to deduce easily the optimal value N of layers in the SL to reach total transmission. In addition to the amplitude analysis, we study also the behavior of the phase time around the surface resonances as a function of N.

As in the previous work [69], we consider a structure formed by a finite solid–fluid SL composed of N solid layers of impedance Z_s separated by N–1 fluid layers of impedance Z_f and inserted between two fluids of impedances Z_{f1} and Z_{f2}. In the particular case of normal incidence ($k_\parallel = 0$) and assuming quarter wavelength layers, i.e., $\frac{\omega}{v_l}d_s = \frac{\omega}{v_f}d_f = \frac{\pi}{2}$, the inverse of the Green's function of the finite SL with free surfaces becomes [95]

$$g(MM)^{-1} = \begin{pmatrix} 0 & Z_f\left(\frac{Z_s}{Z_f}\right)^N \\ Z_f\left(\frac{Z_s}{Z_f}\right)^N & 0 \end{pmatrix}. \tag{3.82}$$

which is equivalent to the inverse Green's function of a quarter wavelength layer with an effective acoustic impedance $Z_e = Z_f\left(\frac{Z_s}{Z_f}\right)^N$. Then we can use the well-known relation [106] that enables to use an intermediate layer to form an antireflection coating between two different semi-infinite media, namely, $Z_{f1}\,Z_{f2} = Z_e^2$. Then we get easily

$$N = \frac{1}{2}\frac{\ln\left(\frac{Z_{f1}Z_{f2}}{Z_f^2}\right)}{\ln\left(\frac{Z_s}{Z_f}\right)}. \tag{3.83}$$

This relation requires a suitable choice of the materials in order to get a positive value of N greater than unity. In particular, the solid and fluid media constituting the SL should have close impedances.

An example is illustrated in Fig. 3.12 for a SL composed of Al and Hg and sandwiched between water (incident medium) and Hg (detector medium). The elastic parameters of the materials are given in Table 3.1. The thicknesses of the layers in the SL are chosen such that $\frac{d_s}{v_l} = \frac{d_f}{v_f}$. One can see clearly that selective transmission occurs around the reduced frequency $\Omega_0 = \frac{\omega d_s}{v_l} = \frac{\omega d_f}{v_f} = (2n + 1)\frac{\pi}{2}$ for a number of cells such that $N = 11$ according to (3.83). Far from $N = 11$, the transmission decreases significantly as it is illustrated in the inset of Fig. 3.12. As a matter of comparison, we have also sketched by horizontal line the transmission rate between water and Hg in the absence of the finite SL. The resonances in Fig. 3.12 are of Breit-Wigner type [69] with a lorentzian shape because of the absence of transmission zeros at normal incidence. Zhao et al. [70] have attributed the resonances lying in the middle of the gaps of the SL to the interference effect of acoustic waves reflected from all periodically aligned interfaces. This explanation is of course correct but a physical interpretation is still needed. We show that the resonances are actually surface resonances induced by the interface between the SL and water. Indeed, the dispersion relation giving the surface modes of a SL ended with a solid layer in contact with vacuum are given by (see Sect. 3.3.5.1 and Fig. 3.7)

$$\Omega_0 = \frac{\omega d_s}{v_l} = \frac{\omega d_f}{v_f} = (2n + 1)\frac{\pi}{2} \tag{3.84}$$

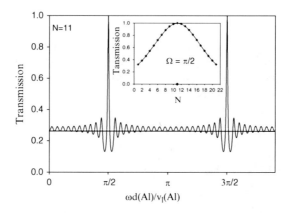

Fig. 3.12 Transmission rate for a finite SL composed of $N = 11$ layers of Al separated by $N - 1 = 10$ layers of Hg. The structure is inserted between water (incident medium) and Hg (detector medium). The *inset* shows the variation of the maxima of the transmission as a function of the number of unit cells N for the mode situated at $\frac{\omega d_{Al}}{v_1(Al)} = \frac{\pi}{2}$. The *straight horizontal line* correspond to the transmission rate between water and Hg (i.e., without the finite SL)

Table 3.1 Elastic parameters of aluminum, glass, water, and mercury

Materials	Mass density $\rho(kg/m^3)$	Longitudinal velocity (m/s)	Wave impedance Z (kg/m^3s)
Aluminum	2.716×10^3	6.17×10^3	16.758×10^6
Glass	2.427×10^3	5.40×10^3	13.106×10^6
Water	1.00×10^3	1.479×10^3	1.479×10^6
Mercury	13.500×10^3	1.450×10^3	19.575×10^6

In addition to (3.84), the supplementary condition (3.64) that ensures the decaying of surface modes from the surface becomes

$$Z_s < Z_f \tag{3.85}$$

This condition is fulfilled in the case of a SL made of Al-Hg. Now, when the Al layer of the SL is in contact with water (instead of vacuum), this latter medium does not affect considerably the position of the surface resonances as the impedance of water is much smaller than Al. In order to confirm the above analysis, we have also sketched the local density of states (LDOS) as a function of the space position z (Fig. 3.13) for the mode lying at $\Omega_0 = \pi/2$. This figure clearly shows that this resonance is localized at the surface of the SL and decreases inside its bulk. Let us notice that the LDOS reflects the square modulus of the displacement field. Therefore, these results show without ambiguity that the transmission is enhanced by surface resonances.

Besides the amplitude of the transmission, we have also analyzed the behavior of the phase time (Fig. 3.14). One can notice a strong delay time at the frequencies corresponding to surface resonances, reflecting the time spent by the phonon at the SL/water interface before its transmission. Contrary to the amplitude (see the inset

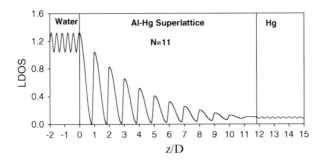

Fig. 3.13 Variation of the local density of states (LDOS) (in arbitrary units) as a function of the space position z/D for the surface resonance situated at $\dfrac{\omega d_{Al}}{v_l(Al)} = \dfrac{\pi}{2}$ in Fig. 3.12 and for $N = 11$

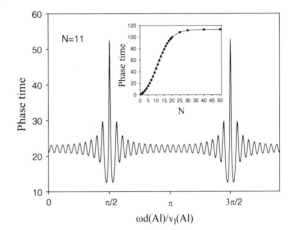

Fig. 3.14 Same as in Fig. 3.12, but for the transmission phase time (in units of $d_{Al}/v_l(Al)$)

of Fig. 3.12), the phase time at the surface resonance goes asymptotically to a limiting value (~110) (in units of $d_s/v_l(Al)$) when N increases. This result known as the Hartman effect [107] arises for classical waves tunneling through a barrier where the phase time saturates to a constant value for a sufficiently barrier's thickness. This phenomenon has been observed experimentally [108] and explained theoretically [109–111] in 1D photonic crystals. For a frequency lying in the allowed bands, the phase time (not shown here) increases linearly as a function of N.

The above results can be explained in terms of the DOS. Indeed, due to the similarity between the DOS and the phase time (for more details, see [100]), Fig. 3.14 reflects also the DOS where the different resonant modes are enlarged because of their interaction with the bulk waves of the surrounding media. When N increases, the number of oscillations in the bulk bands (which is related to the number of cells in the system) and the corresponding DOS increase. However, the behavior is different for the peak associated to the surface resonance. Indeed, for low values of N, the localization of this mode increases as a function of N because the mode interacts less with the second substrate. So, its width decreases and its maximum increases to ensure an area equal to unity under the resonance peak. However, the peak width cannot decrease indefinitely and reaches a threshold because of its interaction with the first substrate. Therefore, the DOS (or the phase time) saturates to a constant value. We have also examined the group velocity v_g which is inversely proportional to the phase time. We have found that v_g oscillates around the mean velocity $v_m = D(d_f/v_f + d_s/v_l)^{-1}$ inside the bands, whereas this quantity is strongly reduced around the surface resonance. Therefore, such structures can be used as a tool to reduce the speed of wave propagation.

As a matter of completeness we have also checked two other cases: (1) the case where there is no surface resonance in the gap of the SL. This can be obtained by using Hg on both sides of the structure. In this case, even if (3.84) and (3.85) are satisfied, (3.83) gives inacceptable value of N ($N < 0$). In spite of the absence of surface resonances, the phase time saturates to a constant value (~ 17) (in units of $d_s/v_l(Al)$) at the mid-gap frequencies, because of the Hartman effect [107, 109]. This value is much smaller than in the presence of a surface resonance. (2) The case where there is two surface resonances in the gap of the SL. This can be obtained by using water on both sides of the structure. In this case, (3.84) and (3.85) are satisfied and (3.83) gives $N \simeq 22$. Because of the existence of two symmetrical surfaces that can support surface modes, one obtains a large surface resonance at $\Omega_0 = \pi/2, 3\pi/2,$ ... for $N = 22$. For smaller values of N, this resonance splits into two distinguished resonances around Ω_0 because of the interaction between the two surfaces. A total transmission is still obtained at each resonance. On the contrary, for higher values of N ($N > 22$), there is a single peak in the transmission because the two surface resonances become decoupled, although being enlarged due to their interaction with the substrates. In this case, the transmission peak decreases as far as N increases.

A recent experiment has been realized by Zhao et al. [70] on a layered structure that consists of an alternative stacking of aluminum and glass planar sheets, which have the same dimensions: $12 \times 12\ cm$ section and $3\ cm$ thickness. The experimental setup is based on the ultrasonic transmission technique. Figure 3.15a gives a schematic diagram of the sample and the experimental setup, showing that the emitter contact transducer is coupled to substrate using a coupling gel and the last layer is immersed in water. The receiver transducer is placed at a distance away from the interface of the last layer B and water. A pulse generator produces a short duration pulse. The pulses transmitted through the sample were detected by an immersion transducer, which has a central frequency of 0.5 MHz and a diameter of 12.5 mm. Because of the limit of central frequency of the transducers, only the first peculiar transmission peak was studied [70].

Figure 3.15b–d give the transmission coefficient versus the frequency for different structures as depicted in the insets where materials A, B, and C denote aluminum, glass, and water, respectively. The elastic parameters of these materials are given in Table 3.1. The layers in the SL are chosen such that $\omega d_A/v_A = \omega d_B/v_B = \pi/2$ (i.e., quarter wavelength layers). In the case where the SL starts with layer A and terminates with layer B, Fig. 3.15b shows a peculiar peak in the first gap around $f = 0.5\ MHz$ for a finite SL composed of $N \simeq 4$ periods. An analytical expression giving the number of bi-layers necessary to attain the transmission unity has been

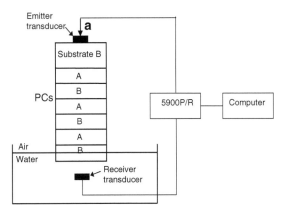

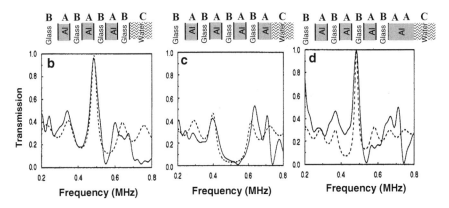

Fig. 3.15 (a) Configuration of a finite SL and the liquid detector. The free surface of the sample is immersed in liquid. Two kinds of solid layers (A and B) are alternately stacked in the sample. (b), (c), (d) Transmission rate of acoustic waves in water (medium C) for three different samples as described in the *insets*. *Dashed* and *solid curves* represent theoretical and experimental results respectively. (After [70])

derived and shown to be $N = \frac{1}{2} \frac{ln(Z_B/Z_C)}{ln(Z_A/Z_B)}$ (i.e., $N \simeq 4$ in the present case). The experimental measurements (solid curves) agree well with the theoretical results (dashed curves). Inside the bands, the transmission coefficient oscillates around the transmission value $T \simeq 0.36$ when the wave is transmitted directly from the substrate B (i.e., glass) to water without the presence of the finite SL.

In addition to the expression giving the optimized value of N to get the total transmission, the authors have attributed this enhancement to interference of the acoustic waves reflected from all the interfaces. Whereas Mizuno [69] has associated these peculiar transmission phenomenon to surface (or interface) waves that can exist between the last layer and the receiver medium (i.e., water). In addition to this structure (considered as the reference system), Zhao et al. [70] have studied also experimentally two other structures: the first one consists on a SL ending by layer A layer (i.e., Al) on both sides (Fig. l5c) and the second structure consists of a SL starting with A layer and ending with two layers A (i.e., layer A with thickness $2\,d_A$) (Fig. 15d). Figure 3.15c did not show any selective transmission around 0.5 MHz, whereas Fig. 15d exhibits the reappearance of the transmission mode. The authors have explained in terms of interference phenomenon between the different materials how the selective transmission can appear and disappear depending on the nature of the layer in contact with the last substrate (i.e., water).

In order to give another insight and explanation to these results, we have taken the same structure as in Fig. 15a (i.e., composed of N periods A-B, but ended with a cap layer (e.g., D) in contact with the substrate C as follows: $B|A|B|A|B|A|B|....|A|B|A|B|D|C$. We suppose that A and B are quarter wavelength layers (as in Zhao's work), whereas the cap layer D can take any thickness d_D, velocity v_D, density ρ_D, and impedance Z_D. One can show that the transmission coefficient reaches unity in three situations, namely [95]

$$\text{(i)}\quad N = \frac{1}{2}\frac{ln(Z_B/Z_C)}{ln(Z_A/Z_B)}\quad\text{and}\quad Z_C = Z_D$$

$$\text{(ii)}\quad N = \frac{1}{2}\frac{ln\left(Z_D^2/Z_BZ_C\right)}{ln(Z_A/Z_B)}\quad\text{and}\quad cos(\omega d_D/v_D) = 0$$

$$\text{(iii)}\quad N = \frac{1}{2}\frac{ln(Z_B/Z_C)}{ln(Z_A/Z_B)}\quad\text{and}\quad sin(\omega d_D/v_D) = 0 \qquad (3.86)$$

These three conditions can explain easily the three spectra in Fig. 3.15. One can see that Fig. 15b corresponds to the first situation (i) and gives $N \simeq 4$ as in Zhao's work [70]. Figure 3.15c corresponds to the second situation (ii) where the cap layer $D = A$; in this case one can check easily that N becomes negative which means that this condition cannot be fulfilled and therefore the transmission cannot reach unity around $f \simeq 0.5$ MHz. Figure 3.15d corresponds to the third situation (iii) where the cap layer $D = 2A$ (i.e., the double layer $d_D = 2d_A$). In this case, D becomes a half wavelength layer (i.e., $sin(\omega d_D/v_D) = 0$ or $\omega d_D/v_D = m\pi$) and $N \simeq 4$.

It is worth mentioning that Zhao et al. [70] have also studied theoretically the situation where the receiver substrate presents a high impedance like tungsten for example. In this case, it was found that contrary to the situation where the system is in contact with water, the selective transmission arises when the SL terminates with A layer (i.e., aluminum) or B layer (i.e., glass) but with double thickness. The optimized number of periods to reach the maximum transmission when $\omega d_A/v_A \neq \omega d_B/v_B$ has been also examined numerically.

3.6   Omnidirectional Reflection and Selective Transmission

3.6.1   Case of Solid–Solid-Layered Media

In the field of photonic band gap materials, it has been argued during the last years [112–114] that one-dimensional structures such as superlattices can also exhibit the property of omnidirectional reflection, i.e., the existence of a transmission band gap for any incident wave independent of the incidence angle and polarization. However, because the photonic band structure of a superlattice does not display any absolute band gap (i.e., a gap for any value of the wave vector), the property of omnidirectional reflection holds in general when the incident light is launched from vacuum, or from a medium with relatively low index of refraction (or high velocity of light). To overcome this difficulty, when the incident light is generated in a high refraction index medium, a solution [115] that consists to associate with the superlattice a cladding layer with a low index of refraction has been proposed. This layer acts like a barrier for the propagation of light.

The object of this section is to examine the possibility of realizing one-dimensional structures that exhibit the property of omnidirectional reflection for acoustic waves. In the frequency range of the omnidirectional reflection, the structure will behave analogously to the case of 2D and 3D phononic crystals, i.e., it reflects any acoustic wave independent of its polarization and incidence angle. We shall show that a simple superlattice can fulfill this property, provided the substrate from which the incident waves are launched is made of a material with relatively high acoustic velocities of sound. However, the substrate may have relatively low acoustic velocities, according to the large varieties in the elastic properties of materials. Then, we propose two alternative solutions to overcome the difficulty related to the choice of the substrate, in order to obtain a frequency domain in which the transmission of sound waves is inhibited even for a substrate with low velocities of sound. As mentioned in the case of photonic band gap materials, one solution would be to associate the superlattice with a cladding layer having high velocities of sound in order to create a barrier for the propagation of acoustic waves. Another solution will consist of associating two superlattices chosen appropriately in such a way that the superposition of their band structures displays a complete acoustic band gap [56, 57].

First, we emphasize that a single superlattice can display an omnidirectional reflection band, provided the substrate is made of a material with relatively high velocities of sound. The expressions of the transmission and reflection coefficients and densities of states are cumbersome. We shall avoid the details of these calculations which are given in [67].

Let us first examine the so-called projected band structure of a superlattice, i.e., the frequency ω versus the wave vector $k_\parallel$ (parallel to the layers). Figure 3.16 displays the phononic band structure of an infinite superlattice composed of Al and W materials with thicknesses d_1 and d_2, such as $d_1 = d_2 = 0.5D$, D being the period of the superlattice. We have used a dimensionless frequency $\Omega = \omega D / C_t(\text{Al})$, where $C_t(\text{Al})$ is the transverse velocity of sound in Al (the elastic parameters of the materials are listed in Table 3.2). The left and right panels, respectively, give the band structure for transverse and sagittal acoustic waves. For every value of $k_{//}$, the shaded and white areas in the projected band structure, respectively, correspond to the minibands and to the minigaps of the superlattice, where the propagation of acoustic waves is allowed or forbidden. Due to the large contrast between the elastic parameters of Al and W, the minigaps of the superlattice are rather large in contrast

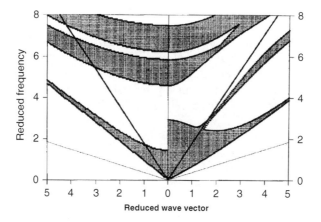

Fig. 3.16 Projected band structure of sagittal (*right panel*) and transverse (*left panel*) elastic waves in a W/AL superlattice. The reduced frequency $\Omega = \omega D/C_t(\text{Al})$ is presented as a function of the reduced wave vector $k_\parallel D$. The *shaded* and *white areas*, respectively, correspond to the minibands and minigaps of the superlattice. The *heavy* and *thin straight lines* correspond, respectively, to sound velocities equal to transverse and longitudinal velocities of sound in epoxy

Table 3.2 Elastic parameters of the materials involved in the calculations

Materials	Mass density (kg m^{-3})	C_t (m/s)	C_l(m/s)
W	19,300	2,860	5,231
Al	2,700	3,110	6,422
Si	2,330	5,845	8,440
Epoxy	1,200	1,160	2,830
Pb	10,760	850	1,960
Nylon	1,110	1,100	2,600
Plexiglas	1,200	1,380	2,700
Water	1,000	–	1,490

to the case of other systems such as GaAs-AlAs superlattices. Nevertheless, it can be easily noticed that the band structure shown in Fig. 3.16 does not display any absolute gap, this means a gap existing for every value of the wave vector $k_\parallel$. However, the superlattice can display an omnidirectional reflection band in the frequency range of the minigap ($2.952 < \Omega < 4.585$) if the velocities of sound in the substrate are high enough. More precisely, let us assume that the transverse velocity of sound in the substrate $C_t(s)$ is greater than 5,543 m/s, (the heavy line in Fig. 3.16 indicates the sound line with the velocity 5,543 m/s). For any wave launched from this substrate, the frequency will be situated above the sound line $\omega = C_t(s)k_\parallel$, i.e., above the heavy line in Fig. 3.16. When the frequency falls in the range $2.952 < \Omega < 4.585$ (corresponding to the minigap of the superlattice at $k_\parallel = 0$), the wave cannot propagate inside the superlattice and will be reflected back. Thus, the frequency range $2.952 < \Omega < 4.585$ corresponds to an omnidirectional reflection band for the chosen substrate. Generally speaking, the above condition expresses that the cone defined by the transverse velocity of sound in the substrate contains a minigap of the superlattice. With the Al/W superlattice, this condition is, for instance, fulfilled if the substrate is made of Si [55–57]. Of course, in practice, due to the finiteness of the omnidirectional mirror, one can only impose that the transmittance remains below a given threshold (for instance, 10^{-3} or 10^{-2}). A recent experiment [58] has been performed by Manzanares-Martinez on Pb/Epoxy SLs to show the occurrence of such omnidirectional band gaps.

There exist different ways to realize selective transmission through layered solid–solid structures. One way consists to insert a defect layer (cavity) within the structure. The filtering is carried out through the resonant modes of the cavity. An example is shown in Fig. 3.17a for a SL composed of five layers of Al and four layers of W. The cavity is made of epoxy and inserted in the middle of the Al-W SL. The whole system is embedded between two Si substrates.

Figure 3.17a gives the dispersion curves associated to defect modes in the first gap of the SL. Because of the low velocities of sound in epoxy as compared to Si, the defect branch is almost flat and falls around $\Omega \simeq 3.48$, which means that the transmission filtering arises around almost the same frequency for all incident angles and polarizations of the waves. Figure 3.17b shows the evolution of the maximum of the transmission coefficient as function of the incident angle along the defect branch. Depending on the polarization of the incident wave, one can have two possibilities (1) the incident wave with shear-horizontal polarization is completely transmitted (straight horizontal line), (2) an incident wave with shear-vertical polarization gives rise to two transmitted waves, one longitudinal (dashed dotted curve) and the other shear-vertical (dashed curve). The two latter curves present a noticeable variation for the incident angles $0° < \theta < 45°$ with an important conversion of modes from transverse-vertical to longitudinal around $\theta \simeq 19°$. For $45° < \theta < 90°$ (i.e., $C_t(Si) < C < C_l(Si)$), the longitudinal component of the transmitted wave vanishes, whereas the transverse component continues to exist. The number of defect branches inside the omnidirectional gap depends on the size of the defect layer, this number increases as function of the thickness of the defect layer. Let us mention that the existence and the behavior of localized sagittal modes induced by defect layers within SLs have been the subject of recent studies [116,117]. Resonances and mode conversions of phonons scattered by SLs with and without inhomogeneities have been discussed [117–119]. In addition, group velocities in the infinite and finite SLs have been calculated [120, 121]. In a frequency gap, their magnitude in the finite SL becomes much larger than that in the band region, and increases as the periodicity N increases [121]. This N dependence is qualitatively different depending on whether the gap in the corresponding infinite SL is due to the intramode or intermode Bragg reflection. The frequency gaps associated with intramode and intermode reflections lay, respectively, at the edges and within the Brillouin zone. The latter modes are strongly related to the conversion mode effect.

Some years ago, Manzanares-Martinez et al. [58] have demonstrated experimentally and theoretically the occurrence of omnidirectional reflection in a finite SL made of a few periods of Pb/epoxy and sandwiched between substrates made of Nylon. The parameters of the materials are given in Table 3.2. The thicknesses of

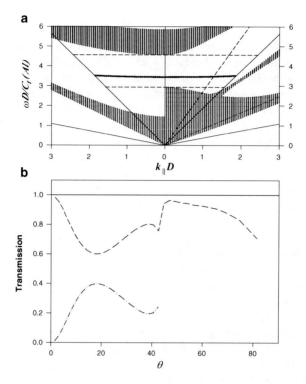

Fig. 3.17 (a) Band gap structure of transverse and sagittal modes as described in Fig. 3.16. The *bold line* inside the omnidirectional gap represents the defect branch induced by a cavity layer made of epoxy inserted in the middle of the finite Al/W SL embedded between two Si substrates. The *thick* (*thin*) *straight* and *dashed lines* gives respectively the transverse and longitudinal velocities of sound in Si (epoxy). (b) Amplitudes of the transmitted waves along the defect branch in (a) as a function of the incident angle θ. The horizontal line with total transmission corresponds to shear-horizontal wave, whereas *dashed* and *dashed-dotted curves* correspond to shear-vertical and longitudinal transmitted waves respectively

the layers were chosen so that the structure has its omnidirectional gap in the working frequencies of the transducers. They took layers of the same thickness 1 mm so as to generate a gap centered at around 300 kHz.

Figure 3.18a displays the band gap structure for transverse and sagittal acoustic waves. An omnidirectional gap is predicted in the frequency region 273 kHz $\leq f$ $\leq$ 371 kHz, which corresponds to the normal incidence band edges of the sagittal modes. However, it is worth noting that the proposed structure does not have the property of omnidirectional reflection for transverse waves for which the velocity is about half of the longitudinal waves. The transmission measurements (Fig. 3.18b) have been performed for longitudinal incident waves and the waves detected after travelling the system, consist of the projection in the radial direction of the transmitted waves (longitudinal and transverse). Figure 3.18b shows the transmission amplitude measured at different angles of incidence for the samples analyzed.

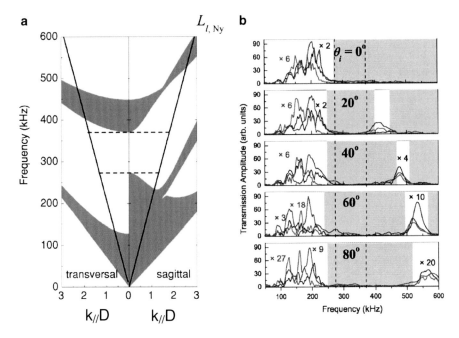

Fig. 3.18 (a) Projected band structure of sagittal and transverse elastic waves in a Pb/epoxy superlattice with layer thickness $d_1 = d_2 = 1$ mm. The frequencies (in kHz) are represented as function of the reduced parallel wave vector $k_{//}D$. The *horizontal dashed lines* delimit the frequencies where no transmission occurs at every angle, i.e., the band of omnidirectional reflection. (b) Experimental transmission spectra obtained for the three samples described in the text. The *gray areas* define the gaps calculated for each angle of incidence θ_i, and the *vertical dashed lines* describe the commune omnidirectional gap (After 58)

One can notice that the transmission is almost negligible in the regions where a gap (shadowed regions) is predicted by the band structure. The commune gap region indicated by vertical dashed lines has been found in very good agreement with the theoretical predictions.

3.6.2 Case of Solid–Fluid-Layered Media

Figure 3.19 gives the dispersion curves (gray areas) for an infinite SL made of Plexiglas and water layers. The gray areas represent the bulk bands. The dashed straight lines represent the transverse and longitudinal velocities of sound in Plexiglas, whereas the dashed dotted line gives the longitudinal velocity of sound in water. The thin solid and dotted curves represent the dispersion curves obtained from the reflection zeros (total transmission) for a finite SL composed of $N = 5$ Plexiglas layers inserted in water. The open circles curves show the positions of the transmission zeros (total reflection). One can notice a shrinking of the $N - 1$

Fig. 3.19 Dispersion curves
for a SL made of Plexiglas
and water layers. The curves
give $\Omega = \omega D/v_{t(\text{plexiglas})}$ as a
function of $k_{//}D$. The widths
of fluid and solid layers are
supposed equal:
$d_f = d_s = D/2$. The *gray
areas* represent the bulk
bands for an infinite SL. The
thin solid lines and *dotted
curves* show the positions of
the reflection zeros (total
transmission). Whereas the
open circles give the positions
of the transmission zeros
(total reflection). The *dashed
straight lines* represent the
transverse and longitudinal
velocities of sound in
Plexiglas. The *dashed-dotted
line* represents the
longitudinal velocity of sound
in water

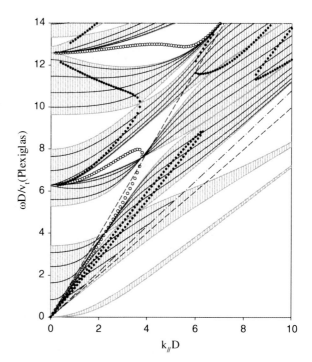

branches when they intercept the transmission zero branch around ($\Omega = 4.07$,
$k_{//}D = 2.3$) and ($\Omega = 7.64$, $k_{||}D = 3.8$). This phenomenon reproduces for other
values of the couple (Ω, $k_{||}D$) not shown here. This property of the shrinking of the
modes is a characteristic of solid-fluid SLs and is without analogue in their
counterpart solid–solid SLs (see Fig. 3.16). It is worth noting that the transmission
zeros occur only for incidence angles θ such that $0° < \theta < \theta_{\text{cr}}$, where θ_{cr} is a
critical angle depending on the velocities of sound in solid and fluid layers (see [95]
for more details). In the example considered here $\theta_{\text{cr}} = 39°$.

Figure 3.20 gives the variation of the transmission rates T (Fig. 3.20a–c, e–g and
i–k) as a function of the reduced frequency Ω for a finite SL composed of $N = 1$,
2 and 5 Plexiglas layers immersed in water. The left, middle and right panels
correspond to incident angles: $\theta = 0°$, 25° and 40° respectively. At the bottom of
these panels we plotted the corresponding dispersion curves (i.e., Ω versus the
Bloch wave vector k_3) (Fig. 3.20d, h and l). As predicted above, for $\theta = 0°$ (left
panel) and $\theta > \theta_{\text{cr}}$ (right panel), the transmission exhibits dips at some frequency
regions which transform into gaps as far as N increases. These gaps are due to the
periodicity of the system (Bragg gaps) and coincide with the band gap structure
of the infinite SL shown in Fig. 3.20d and l. For an incident angle $0° < \theta < \theta_{\text{cr}}$
(middle panel), one can notice the existence of a transmission zero around $\Omega = 7.64$
(Fig. 3.20e) which is due to the insertion of one Plexiglas layer ($N = 1$) in water.
This transmission zero transforms to a large gap when N increases. Besides this gap
there exists a dip around $\Omega = 5$ for $N = 2$ (Fig. 3.20f) which also transforms to a

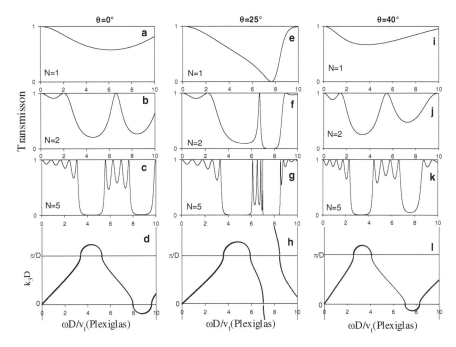

Fig. 3.20 Variation of the transmission coefficients as a function of the reduced frequency Ω for a finite SL composed of $N = 1$ [(**a**), (**e**) and (**i**)], $N = 2$ [(**b**), (**f**) and (**j**)] and $N = 5$ [(**c**), (**g**) and (**k**)] Plexiglas layers immersed in water. The left, middle, and right panels correspond to incident angles: $\theta = 0°$, $25°$, and $40°$ respectively, (**d**), (**h**) and (**l**) give the dispersion curves (i.e., Ω versus the Bloch wave vector k_3) inside the reduced Brillouin zone $0 < k_3 < \pi/D$. Outside this zone are represented the imaginary parts of k_3

gap when N increases; this gap is due to the periodicity of the structure. The transmission gaps map the band gap structure of the infinite SL (Fig. 3.20h), where one can notice that the imaginary part of the Bloch wave vector (responsible of the attenuation of the waves associated to defect modes) is finite in the Bragg gaps and tends to infinity inside the gaps due to the transmission zeros. These latter gaps can be used to localize strongly defect modes within the structure (see below).

From all the above results, one can conclude that for an incident angle $0° < \theta < \theta_{cr}$ (middle panel) there exists two types of gaps: Bragg gaps which are due to the periodicity of the structure and gaps which are induced by the transmission zeros. However, at normal incidence ($\theta = 0°$) (left panel) and for $\theta > \theta_{cr}$ (right panel) all the gaps are due to the periodicity of the system. The existence of these two types of gaps has been discussed also by Shuvalov and Gorkunova [122] in periodic systems of planar sliding-contact interfaces.

Now, we shall examine the condition for the existence and behavior of omnidirectional band gaps in finite solid–fluid-layered media. Let us first come back to the band gap structure given in Fig. 3.19 for a SL made of Plexiglas and water with the same thickness $d_s = d_f = D/2$. One can notice that the band gap structure of the infinite Plexiglas-water SL does not display any absolute gap, this means a gap existing for every value of the wave vector $k_\parallel$. Fig. 3.21a reproduces the results given in Fig. 3.19 for a finite Plexiglas-water SL made of $N = 8$ cells. The discrete modes are obtained from the maxima of the transmission rate that exceeds a threshold fixed to 10^{-3}. One can notice that any wave launched from water will display a partial gap for an incident angle $0° < \theta < 35°$ in the frequency region $4.015 < \Omega < 5.105$ indicated by horizontal lines. However, waves with incident angles $35° < \theta < 90°$ will be totally transmitted through the discrete modes of the SL [95]. These results remain valid for any incident liquid medium as, in general, the velocities of sound in most liquids are of the same order or less than water. In order to overcome this limitation or at least facilitate the existence of an omnidi-rectional gap, we proposed, like in the previous subsection on solid–solid SLs, two solutions. The first one consists to clad the SL on one side by a buffer layer of high acoustic velocities, which can act as a barrier for the propagation of phonons. The second solution consists to associate in tandem two SLs in such a way that their band structures do not overlap.

In the following, we shall give an example concerning the first solution. Figure 3.21b gives the discrete modes associated to the cladded-SL structure, i.e., the frequency domains in which the transmission rate exceeds a threshold of 10^{-3}. In this example the clad layer is made of Al with transverse and longitudinal velocities of sound (dashed and straight lines) higher than the SL bulk modes lying in the frequency region $4.015 < \Omega < 5.105$ (Fig. 3.21a, b). The thickness of the Al layer is $d_0 = 7D$ and the SL contains $N = 8$ cells of Plexiglas-water. By combining these two systems, the allowed modes of the SL and the guided modes induced by the Al clad layer above its velocities of sound do not overlap over the frequency range of the omnidirectional gap. This means that each system acts as a barrier for phonons of the other system. In such a way, one obtains an omnidirectional band gap indicated by the two horizontal lines in Fig. 3.21b in the frequency region $4.015 < \Omega < 5.105$. By comparing Fig. 3.21a, b, one can notice clearly that the presence of the clad layer has two opposite effects. It decreases the transmittance in some frequency domains (essentially below the sound line defined by the transverse velocity of sound in the clad), but also introduces new modes that can contribute themselves to transmission. The transmission by the latter modes is prevented by the SL when the corresponding branches fall inside the minigaps. In the allowed frequency regions belonging to both the SL and the clad layer, one can notice an interaction and an anticrossing of the modes associated to these two systems.

It is well known that the introduction of a defect layer (cavity) in a periodic structure can give rise to defect modes inside the band gaps [43–51, 81]. These modes appear as well defined peaks in the DOS; however, their contribution to the transmission rate depends strongly on the position of these defects inside the structure. Indeed, as it was shown before, a defect layer placed at the contact between the SL and the substrate (clad layer) induces guided modes in the band gap of the SL but without contributing to the transmission. However, the transmission through these modes can be significantly enhanced if the cavity layer is placed at the middle of the structure [46, 49–51, 67].

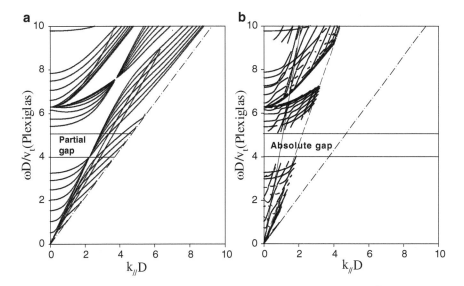

Fig. 3.21 (**a**) Dispersion curves of a finite SL composed of $N = 8$ Plexiglas layers immersed in water. The thicknesses of Plexiglas and water layers are equal. The discrete modes correspond to the frequencies obtained from the maxima of the transmission rate that exceeds a threshold of 10^{-3}. (**b**) The same as (**a**) but here the SL is cladded with an Al layer of thickness $d_0 = 7D$ on one side

In general, a periodic structure made of N cells ($N > 2$) is needed to create a transmission gap in which a defect mode is then introduced for filtering. We show that contrary to solid–solid SLs, it is possible to achieve large gaps as well as sharp resonances inside these gaps with a solid–fluid structure as small as a solid–fluid–solid sandwich triple layers (i.e., $N = 2$). This property is associated with the existence of zeros of transmission. Figure 3.22a gives the transmission rate as a function of the reduced frequency Ω for a finite Plexiglas-water SL composed of $N = 2$ (solid curves) and $N = 4$ (dotted curves) cells and for an incidence angle $\theta = 35°$. The fluid and solid layers have the same width $d_f = d_s = D/2$. One can notice that the transmission rate exhibits a large dip in the frequency region $4 < \Omega < 8$ around the transmission zero indicated by an open circle on the abscissa. This transmission gap maps the band gap of the infinite system indicated by solid circles on the abscissa. As it was discussed above, the transmission gap becomes well defined as far as N increases. Now, if a fluid cavity layer of thickness $d_0 = D$ is inserted in the middle of the structure, then a resonance with total transmission can be introduced in the gap (Fig. 3.22b). This resonance falls at almost the same frequency and its width decreases when N increases. Let us mention that the structure depicted in Fig. 3.22a, b with $N = 2$ consists on a sandwich system made of two Plexiglas layers separated by a water layer. Therefore, such a small size structure clearly show the possibility of obtaining a large gap and a sharp resonance inside the gap by just tailoring the width of these three layered media. This property is specific to solid–fluid structures and is without

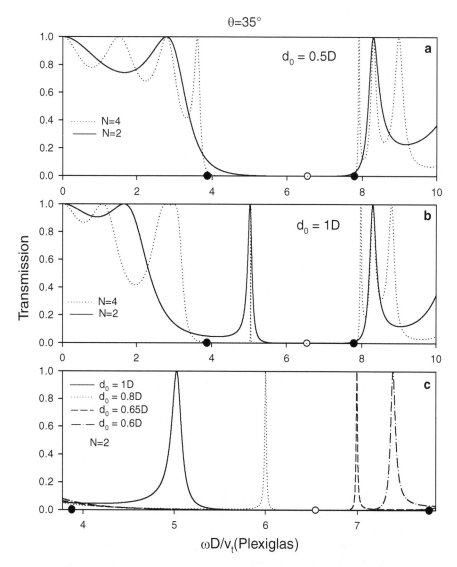

Fig. 3.22 (a) Transmission rate for a finite SL composed of $N = 2$ (*solid curves*) and $N = 4$ (*dotted curves*) Plexiglas layers immersed in water at an incidence angle $\theta = 35°$. The *solid* and *open circles* on the abscissa indicate the positions of the band gap edges and transmission zeros respectively. (b) Same as in (a) but in presence of a defect fluid layer of thickness $d_0 = D$ at the middle of the structure. (c) Same as in (b) for $N = 2$ and different values of the thickness d_0 of the cavity fluid layer as indicated in the *inset*

analogue for their counterparts solid–solid systems where at least a number $N > 2$ of layers is needed to achieve well-defined gaps and cavity modes. In what follows, we shall focus on the simple case of sandwich system (i.e., $N = 2$).

An important point to notice in Fig. 3.22b is the shape of the resonance lying in the vicinity of the transmission zero. Such a resonance is called Fano resonance [123]. The origin and the asymmetry Fano profile of this resonance was explained as a result of the interference between the discrete resonance and the smooth continuum background in which the former is embedded. The existence of such resonances in 2D and 3D phononic crystals, the so-called locally resonant band gap materials [124, 125], has been shown recently [126–128]. Some analytical models have been proposed to explain the origin and the behavior of these resonances [126–128]. In the case of 1D model proposed here, the Fano resonance in Fig. 3.22b is just an internal resonance induced by the discrete modes of the fluid layer when these modes fall at the vicinity of the transmission zeros induced by the surrounding solid layers. By decreasing the width of the fluid layer from $d_0 = 1D$ to $d_0 = 0.6D$ (Fig. 3.22c), one can notice that the position of the Fano resonance moves to higher frequencies, its width decreases and vanishes for a particular value of $d_0 = 0.71D$ before increasing again. At exactly $d_0 = 0.71D$, the transmission vanishes and the resonance collapses giving rise to the so-called ghost Fano resonance [129]. Around $d_0 = 0.71D$, the asymmetric Fano profile of the resonance becomes symmetric and changes the shape.

In Fig. 3.22c, the two solids surrounding the fluid layer have the same widths d_s, therefore the transmission zeros induced by the solid layers fall at the same frequency. Now, if the two solids have different widths (labeled, e.g., d_{s1} and d_{s2}), then one can obtain two transmission zeros and a resonance that can be squeezed between these two dips if d_{s1} and d_{s2} are chosen appropriately. In this case a symmetric Fano resonance can be obtained whose width can be tuned by adjusting the frequencies of the zeros of transmission. Such resonances have been found also for acoustic and magnetic circuits formed by a guide inserted between two dangling resonators [130, 131].

3.7   Conclusion

In this chapter, we have presented a theoretical analysis of the propagation and localization of phonons in one-dimensional crystals in both atomic (discrete) and elastic (continuum) approximations. In the case of continuous media, solid–solid and solid–fluid-layered materials have been considered. In general, we have limited ourselves to the case of isotropic materials for which shear-horizontal waves are decoupled from sagittal waves polarized in the plane defined by the normal to the surface and the wave vector parallel to the surface. This study has been performed within the framework of the Green's function method which enables us to deduce the dispersion curves, densities of states as well as the transmission and reflection coefficients. The Green's function approach used in this work is also of interest for studying the scattering of light by bulk and surface phonons. The advantage of the 1D-layered media treated here in comparison with their 2D and 3D counterparts systems, resides in obtaining, in general, closed-form expressions that enables us to discuss deeply different physical properties related to band gaps in such systems.

Despite the problem of acoustic waves in solid and fluid materials has been intensively studied since the beginning of the last century, this subject still attracts attention of researchers because of the high quality level of control and perfection reached in the growth techniques of microstructures and nanostructures, but also due to the sophisticated experimental techniques used to probe different modes of these systems in different frequency domains. In addition, these systems may present several applications in guiding, stopping and filtering waves.

First, we have treated pure longitudinal waves (normal incidence), where we have shown that the eigenmodes of a finite SL constituted of N cells with free-stress surfaces are composed of $N-1$ modes in each band and one mode by gap which is associated to one of the two surfaces surrounding the system. These latter modes are independent of N and coincide with the surface modes of two complementary SLs obtained from the cleavage of an infinite SL along a plane parallel to the interfaces. Then, we have shown the possibility of enhanced transmission between two media through surface modes. Some experimental results and the interest of the Green's function calculation in explaining the Raman spectra are also reviewed.

The application of multilayered media as acoustic mirrors and selective filters at oblique incidence has been shown. The transmission and reflection coefficients of wave propagation through these systems has shown several new properties as the existence of transmission zeros in solid–fluid SLs and therefore new gaps in addition to Bragg gaps, as well as the existence of Fano resonances.

Acknowledgments This work was realized within the framework of a scientific convention between the universities of Lille 1 and Oujda.

References

1. M.S. Kushwaha, P. Halevi, L. Dobrzynski, B. Djafari-Rouhani, Phys. Rev. Lett. **71**, 2022 (1993)
2. M. Sigalas, E.N. Economou, Solid State Commun. **86**, 141 (1993)
3. M.S. Kushwaha, Int. J. Mod. Phys. B **10**, 977 (1996)
4. L. Esaki, R. Tsu, IBM J. Res. Dev. **14**, 61 (1970)
5. L. Colvard, R. Merlin, M.V. Klein, A.C. Gossard, Phys. Rev. Lett. **45**, 298 (1980)
6. For a recent review, see: E.H.El Boudouti, B. Djafari-Rouhani, A. Akjouj and L. Dobrzynski, Surf. Sci. Rep. **64**, 471 (2009)
7. J. Sapriel, B. Djafari-Rouhani, Surf. Sci. Rep. **10**, 189 (1989)
8. B. Jusserand, M. Cardona, in *Light Scattering in Solids*, ed. by M. Cardona, G. Güntherodt (Springer, Berlin, 1989), p. 49
9. S. Tamura, *Proceedings of the 6th International Conference on Phonon Scattering in Condensed Matter*, Heidelberg, 1989 (World Scientific, Singapore, 1990), p.703
10. D.J. Lockwood, J.F. Young, *Light Scattering in Semiconductor Structures and Superlattices* (Plenum, New York, 1991)
11. S. Tamura, Phys. Rev. B **39**, 1261 (1989)
12. S. Mizuno, S. Tamura, Phys. Rev. B **45**, 13423 (1992)

13. M.J.S. Lowe, IEEE Trans. Ultrason. Ferroelectr. Freq. Control **42**, 525 (1995)
14. E.H. El Boudouti, B. Djafari-Rouhani, E.M. Khourdifi, L. Dobrzynski, Phys. Rev. B **48**, 10987 (1993)
15. E.H. El Boudouti, B. Djafari-Rouhani, A. Akjouj, L. Dobrzynski, Phys. Rev. B **54**, 14728 (1996)
16. F. Garcia-Moliner, V.R. Velasco, *Theory of Single and Multiple Interfaces. The Method of Surface Green Function Matching* (World Scientific, Singapore, 1992)
17. B. Jusserand, D. Paquet, F. Mollot, F. Alexandre, G. LeRoux, Phys. Rev. B **35**, 2808 (1987)
18. D.J. Lockwood, R.L.S. Devine, A. Rodriguez, J. Mendialdua, B. Djafari-Rouhani, L. Dobrzynski, Phys. Rev. B **47**, 13553 (1993)
19. A.S. Barker, J.L. Merz, A.C. Gossard, Phys. Rev. B **17**, 3181 (1978)
20. V. Narayanamurti, H.L. Stormer, M.A. Chin, A.C. Gossard, W. Wiegmann, Phys. Rev. Lett. **43**, 2012 (1979)
21. O. Koblinger, J. Mebert, E. Dittrich, S. Dottinger, W. Eisenmenger, P.V. Santos, L. Ley, Phys. Rev. B **35**, 9372 (1987)
22. D.J. Dielemen, A.F. Koenderink, M.G.A. van Veghel, A.F.M. Arts, H.W. de Wijn, Phys. Rev. B **64**, 174304 (2001)
23. S. Tamura, D.C. Hurley, J.P. Wolf, Phys. Rev. B **38**, 1427 (1988)
24. P.J. Shull, D.E. Chimenti, A. Safaeinili, J. Acoust. Soc. Am. **95**, 99 (1994)
25. A. Safaeinili, D.E. Chimenti, J. Acoust. Soc. Am. **98**, 2336 (1995)
26. H.T. Grahn, H.J. Maris, J. Tauc, B. Abeles, Phys. Rev. B **38**, 6066 (1988)
27. A. Yamamoto, T. Mishina, Y. Masumoto, M. Nakayama, Phys. Rev. Lett. **73**, 740 (1994)
28. A. Bartels, T. Dekorsy, H. Kurz, Appl. Phys. Lett. **72**, 2844 (1998)
29. A. Bartels, T. Dekorsy, H. Kurz, K. Köhler, Phys. Rev. Lett. **82**, 1044 (1999)
30. P. Sondhauss, J. Larsson, M. Harbst, G.A. Naylor, A. Plech, K. Scheidt, O. Synnergren, M. Wulff, J.S. Wark, Phys. Rev. Lett. **94**, 125509 (2005)
31. B. Djafari-Rouhani, L. Dobrzynski, O. Hardouin Duparc, R.E. Camley, A.A. Maradudin, Phys. Rev. B **28**, 1711 (1983)
32. T. Aono, S. Tamura, Phys. Rev. B **58**, 4838 (1998)
33. S. Mizuno, S. Tamura, Phys. Rev. B **53**, 4549 (1996)
34. M. Hammouchi, E.H. El Boudouti, A. Nougaoui, B. Djafari-Rouhani, M.L.H. Lahlaouti, A. Akjouj, L. Dobrzynski, Phys. Rev. B **59**, 1999 (1999)
35. H.J. Trodahl, P.V. Santos, G.V.M. Williams, A. Bittar, Phys. Rev. B **40**, R8577 (1989)
36. W. Chen, Y. Lu, H.J. Maris, G. Xiao, Phys. Rev. B **50**, 14506 (1994)
37. B. Perrin, B. Bonello, J.C. Jeannet, E. Romatet, Physica B **219–220**, 681 (1996)
38. B. Bonello, B. Perrin, E. Romatet, J.C. Jeannet, Ultrasonics **35**, 223 (1997)
39. N-W Pu, J. Bokor, Phys. Rev. Lett. **91**, 076101 (2003)
40. N-W Pu, Phys. Rev. B **72**, 115428 (2005)
41. E.M. Khourdifi, B. Djafari-Rouhani, Surf. Sci. **211/212**, 361 (1989)
42. D. Bria, E.H. El Boudouti, A. Nougaoui, B. Djafari-Rouhani, V.R. Velasco, Phys. Rev. B **60**, 2505 (1999)
43. E.M. Khourdifi, B. Djafari-Rouhani, J. Phys. Condens. Matter **1**, 7543 (1989)
44. D. Bria, E.H. El Boudouti, A. Nougaoui, B. Djafari-Rouhani, V.R. Velasco, Phys. Rev. B **61**, 15858 (2000)
45. K.-Q. Chen, X.-H. Wang, B.-Y. Gu, Phys. Rev. B **61**, 12075 (2000)
46. S. Mizuno, Phys. Rev. B **65**, 193302 (2002)
47. S. Tamura, H. Watanabe, T. Kawasaki, Phys. Rev. B **72**, 165306 (2005)
48. G.P. Schwartz, G.J. Gualtieri, W.A. Sunder, Appl. Phys. Lett. **58**, 971 (1991)
49. M. Trigo, A. Bruchhausen, A. Fainstein, B. Jusserand, V. Thierry-Mieg, Phys. Rev. Lett. **89**, 227402 (2002)
50. P. Lacharmoise, A. Fainstein, B. Jusserand, V. Thierry-Mieg, Appl. Phys. Lett. **84**, 3274 (2004)
51. N.D. Lanzillotti Kimura, A. Fainstein, B. Jusserand, Phys. Rev. B **71**, 041305(R) (2005)

52. P.X. Zhang, D.J. Lockwood, J.M. Baribeau, Can. J. Phys. **70**, 843 (1992)
53. P.X. Zhang, D.J. Lockwood, H.J. Labbe, J.M. Baribeau, Phys. Rev. B **46**, 9881 (1992)
54. M. Trigo, A. Fainstein, B. Jusserand, V. Thierry-Mieg, Phys. Rev. B **66**, 125311 (2002)
55. A. Bousfia, E.H. El Boudouti, B. Djafari-Rouhani, D. Bria, A. Nougaoui, V.R. Velasco, Surf. Sci. **482–485**, 1175 (2001)
56. D. Bria, B. Djafari-Rouhani, A. Bousfia, E.H. El Boudouti, A. Nougaoui, Europhys. Lett. **55**, 841 (2001)
57. D. Bria, B. Djafari-Rouhani, Phys. Rev. E **66**, 056609 (2002)
58. B. Manzanares-Martinez, J. Sanchez-Dehesa, A. Hakansson, F. Cervera, F. Ramos-Mendieta, Appl. Phys. Lett. **85**, 154 (2004)
59. G. Wang, D. Yu, J. Wen, Y. Liu, X. Wen, Phys. Lett. A **327**, 512 (2004)
60. L.C. Parsons, G.T. Andrews, Appl. Phys. Lett. **95**, 241909 (2009)
61. G.N. Aliev, B. Goller, D. Kovalev, P.A. Snow, Appl. Phys. Lett. **96**, 124101 (2010)
62. N. Gomopoulos, D. Maschke, C.Y. Koh, E.L. Thomas, W. Tremel, H.J. Butt, G. Fytas, Nano Lett. **10**, 980 (2010)
63. P.M. Walker, J.S. Sharp, A.V. Akimov, A.J. Kent, Appl. Phys. Lett. **97**, 073106 (2010)
64. T. Berstermann, C. Brggemann, M. Bombeck, A.V. Akimov, D.R. Yakovlev, C. Kruse, D. Hommel, M. Bayer, Phys. Rev. B **81**, 085316 (2010)
65. I.E. Psarobas, N. Papanikolaou, N. Stefanou, B. Djafari-Rouhani, B. Bonello, V. Laude, Phys. Rev. B **82**, 174303 (2010)
66. A.J. Kent, R.N. Kini, N.M. Stanton, M. Henini, B.A. Glavin, V.A. Kochelap, T.L. Linnik, Phys. Rev. Lett. **96**, 215504 (2006)
67. A. Bousfia, Ph.D. Thesis, University Mohamed I, Oujda, Morocco (2004)
68. H. Kato, Phys. Rev. B **59**, 11136 (1999)
69. S. Mizuno, Phys. Rev. B **63**, 035301 (2000)
70. D. Zhao, W. Wang, Z. Liu, J. Shi, W. Wen, Physica B **390**, 159 (2007)
71. L.M. Brekhovskikh, *Waves in Layered Media* (Academic, New York, 1981)
72. S.M. Rytov, Phys. Acoust. **2**, 68 (1956)
73. M. Schöenberg, Wave Motion **6**, 303 (1984)
74. M. Rousseau, J. Acoust. Soc. Am. **86**, 2369 (1989)
75. M.A. Biot, J. Acoust. Soc. Am. **28**, 168 (1962)
76. B. Gurevich, J. Acoust. Soc. Am. **106**, 57 (1999)
77. B. Gurevich, Goephysics **67**, 264 (2002)
78. R. Ciz, E.H. Saenger, B. Gurevich, J. Acoust. Soc. Am. **120**, 642 (2006)
79. T.J. Plona, K.W. Winkler, M. Schoenberg, J. Acoust. Soc. Am. **81**, 1227 (1987)
80. C. Gazanhes, J. Sageloli, Acustica **81**, 221 (1995)
81. R. James, S.M. Woodley, C.M. Dyer, F. Humphrey, J. Acoust. Soc. Am. **97**, 2041 (1995)
82. M. Shen, W. Cao, Appl. Phys. Lett. **75**, 3713 (1999)
83. H. Sanchis-Alepuz, Y.A. Kosevich, J. Sánchez-Dehesa, Phys. Rev. Lett. **98**, 134301 (2007)
84. G. Monsivais, R. Rodrguez-Ramos, R. Esquivel-Sirvent, L. Fernández-Alvarez, Phys. Rev. B **68**, 174109 (2003)
85. R. Rodriguez-Ramos, G. Monsivais, J.A. Otero, H. Calás, V. Guerra, C. Stern, J. Appl. Phys. **96**, 1178 (2004)
86. E.E. Mendez, F. Agullo-Rueda, J.M. Hong, Phys. Rev. Lett. **60**, 2426 (1988)
87. L. Dobrzynski, Surf. Sci. Rep. **6**, 119 (1986)
88. L. Dobrzynski, Surf. Sci. Rep. **11**, 139 (1990)
89. C. Kittel, *Introduction to Solid State Physics* (Wiley, New York, 1996)
90. L. Brillouin, *Wave Propagation in Periodic Structures* (Dover, New York, 1953)
91. S.-Y. Ren, *Electronic States in Crystals of Finite Size: Quatum Confinement of Bloch Waves* (Springer, New York, 2006). Chap.4
92. S.-Y. Ren, Phys. Rev. B **64**, 035322 (2001)
93. S.Y. Ren, Ann. Phys. (N.Y.) **301**, 22 (2002)
94. S.-Y. Ren, Y.-C. Chang, Phys. Rev. B **75**, 212301 (2007)

95. Y. El Hassouani, E.H. El Boudouti, B. Djafari-Rouhani, H. Aynaou, Phys. Rev. B **78**, 174306 (2008)
96. E.H. El Boudouti, Y. El Hassouani, B. Djafari-Rouhani, H. Aynaou, Phys. Rev. E **76**, 026607 (2007)
97. A.C. Hladky-Hennion, G. Allan, J. Appl. Phys. **98**, 054909 (2005)
98. B. Djafari-Rouhani, E.H. El Boudouti, A. Akjouj, L. Dobrzynski, J.O. Vasseur, A. Mir, N. Fettouhi, J. Zemmouri, Vacuum **63**, 177 (2001)
99. Y. El Hassouani, H. Aynaou, E.H. El Boudouti, B. Djafari-Rouhani, A. Akjouj, V.R. Velasco, Phys. Rev. B **74**, 035314 (2006)
100. M.L.H. Lahlaouti, A. Akjouj, B. Djafari-Rouhani, L. Dobrzynski, M. Hammouchi, E.H. El Boudouti, A. Nougaoui, B. Kharbouch, Phys. Rev. B **63**, 035312 (2001)
101. K.R. Subbaswamy, A.A. Maradudin, Phys. Rev. B **18**, 4181 (1978)
102. B. Djafari-Rouhani, E.M. Khourdifi, in *Light Scattering in Semiconductor Structures and Superlattices* (Ref. [11]) p 139.
103. J. He, B. Djafari-Rouhani, J. Sapriel, Phys. Rev. B **37**, 4086 (1988)
104. V. Lemos, O. Pilla, M. Montagna, C.F. de Souza, Superlattices Microstruct. **17**, 51 (1995)
105. J. Christensen, A.I. Fernandez-Dominguez, F. de Leon-Perez, L. Martin-Moreno, F.J. Garcia-Vidal, Nat. Phys. **3**, 851 (2007)
106. P. Yeh, *Optical Waves in Layered Media* (Wiley, New York, 1988)
107. T.E. Hartman, J. Appl. Phys. **33**, 3427 (1962)
108. C. Spielmann, R. Szipöcs, A. Stingl, F. Krausz, Phys. Rev. Lett. **73**, 2308 (1994)
109. P. Pereyra, Phys. Rev. Lett. **84**, 1772 (2000)
110. S. Esposito, Phys. Rev. E **64**, 026609 (2001)
111. G. Nimtz, A.A. Stahlhofen, Ann. Phys. (Berlin) **17**, 374 (2008)
112. J.P. Dowling, Science **282**, 1841 (1998)
113. Y. Fink, J.N. Winn, S. Fan, C. Chen, J. Michel, J.D. Joannopoulos, E.L. Thomas, Science **282**, 1679 (1998)
114. D.N. Chigrin, A.V. Lavrinenko, D.A. Yarotsky, S.V. Gaponenko, Appl. Phys. A Mater. Sci. Process. A **68**, 25 (1999)
115. D. Bria, B. Djafari-Rouhani, E.H. El Boudouti, A. Mir, A. Akjouj, A. Nougaoui, J. Appl. Phys. **91**, 2569 (2002)
116. X.-H. Wang, K.-Q. Chen, G. Ben-Yuan, J. Appl. Phys. **92**, 5113 (2002)
117. S. Mizuno, Phys. Rev. B **68**, 193305 (2003)
118. H. Kato, H.J. Maris, S.I. Tamura, Phys. Rev. B **53**, 7884 (1996)
119. B. Manzanares-Martínez, F. Ramos-Mendieta, Phys. Rev. B **76**, 134303 (2007)
120. Y. Tanaka, M. Narita, S.-I. Tamura, J. Phys. Condens. Matter **10**, 8787 (1998)
121. K. Imamura, Y. Tanaka, S. Tamura, Phys. Rev. B **65**, 174301 (2007)
122. A.L. Shuvalov, A.S. Gorkunova, J. Sound Vib. **243**, 679 (2001)
123. U. Fano, Phys. Rev. **124**, 1866 (1961)
124. Z. Liu, X. Zhang, Y. Mao, Y.Y. Zhu, Z. Yang, C.T. Chan, P. Sheng, Science **289**, 1734 (2000)
125. S. Yang, J.H. Page, Z. Liu, M.L. Cowan, C. Chan, P. Sheng, Phys. Rev. Lett. **88**, 104301 (2002)
126. C. Goffaux, J. Sánchez-Dehesa, A.L. Yeyati, P. Lambin, A. Khelif, J.O. Vasseur, B. Djafari-Rouhani, Phys. Rev. Lett. **88**, 225502 (2002)
127. Z. Liu, C.T. Chan, P. Sheng, Phys. Rev. B **71**, 014103 (2005)
128. Y.A. Kosevich, C. Goffaux, J. Sánchez-Dehesa, Phys. Rev. B **74**, 012301 (2006)
129. M.L. Ladrón de Guevara, F. Claro, P.A. Orellana, Phys. Rev. B **67**, 195335 (2003)
130. E.H. El Boudouti, T. Mrabti, H. Al-Wahsh, B. Djafari-Rouhani, A. Akjouj, L. Dobrzynski, J. Phys. Condens. Matter **20**, 255212 (2008)
131. H. Al-Wahsh, E.H. El Boudouti, B. Djafari-Rouhani, A. Akjouj, T. Mrabti, L. Dobrzynski, Phys. Rev. B **78**, 075401 (2008)

Chapter 4
2D–3D Phononic Crystals

A. Sukhovich, J.H. Page, J.O. Vasseur, J.F. Robillard, N. Swinteck, and Pierre A. Deymier

Abstract This chapter presents a comprehensive description of the properties of phononic crystals ranging from spectral properties (e.g., band gaps) to wave vector properties (refraction) and phase properties. These properties are characterized by experiments and numerical simulations.

4.1 Introduction

In this chapter, we focus on 2D and 3D phononic crystals, which, thanks to their spatial periodicity, allow the observation of new unusual phenomena as compared to the 1D crystals discussed in the previous chapter. In experimental studies, 2D crystals usually employ rods as scattering units, while 3D crystals are realized as arrangements of spheres. It is common in theoretical studies of phononic crystals to investigate crystals with scattering units that are simply air voids (e.g., empty cylinders) in a matrix. Although there are many different ways of realizing the

A. Sukhovich (✉)
Laboratoire Domaines Océaniques, UMR CNRS 6538, UFR Sciences et Techniques, Université de Bretagne Occidentale, Brest, France
e-mail: sukhovich@univ-brest.fr

J.H. Page
Department of Physics and Astronomy, University of Manitoba, Winnipeg, MB, Canada R3T 2N2
e-mail: jhpage@cc.umanitoba.ca

J.O. Vasseur • J.F. Robillard
Institut d'Electronique, de Micro-électronique et de Nanotechnologie, UMR CNRS 8520, Cité Scientifique 59652, Villeneuve d'Ascq Cedex, France
e-mail: Jerome.Vasseur@univ-lille1.fr; jean-francois.robillard@isen.fr

N. Swinteck • P.A. Deymier
Department of Materials Science and Engineering, University of Arizona, Tucson, AZ 85721, USA
e-mail: swinteck@email.arizona.edu; deymier@email.arizona.edu

P.A. Deymier (ed.), *Acoustic Metamaterials and Phononic Crystals*,
Springer Series in Solid-State Sciences 173, DOI 10.1007/978-3-642-31232-8_4,
© Springer-Verlag Berlin Heidelberg 2013

phononic crystal theoretically and experimentally (by varying material of the scattering units and the host matrix), one condition is always observed: the characteristic size of a scattering unit (rod or sphere) and a lattice constant should be on the order of the wavelength of the incident radiation to ensure that the particular crystal features arising from its regularity affect the wave propagating through the crystal. In other words, the frequency range of the crystal operation is set by the characteristic dimensions of the crystal (i.e., the size of its unit scatterer and its lattice constant). The exception from this rule, however, is resonant sonic materials, which exhibit a profound effect on the propagating radiation, whose wavelength can be as much as two orders of magnitude larger than the characteristic size of the structure, as was shown by Liu et al. [1, 2].

As described in Chap. 10, the regularity of the arrangement of scattering units of the phononic crystal gives rise to Bragg reflections of the acoustic or elastic waves that are multiply scattered inside the crystal. Their constructive or destructive interference creates ranges of frequencies at which waves are either allowed to propagate (pass bands) or blocked in one (stop bands) or any direction (complete band gaps). The width of the band gap obviously depends on the crystal structure and increases with the increase of density contrast between the material of the scattering unit and that of a host matrix. Switching from a liquid matrix to the solid one, e.g., from water to epoxy, which can support both longitudinal and transverse polarizations, results in even larger band gaps, as was shown by Page et al. [3].

As an example of a 2D phononic crystal, consider a crystal made of cylinders assembled in a *triangular* Bravais lattice, whose points are located at the vertices of the equilateral triangles. Figure 4.1 presents the diagram of the direct and reciprocal lattices with corresponding primitive vectors $\vec{a}_1, \vec{a}_2$ and $\vec{b}_1, \vec{b}_2$. Since $|\vec{a}_1| = |\vec{a}_2| = a$, where a is a lattice constant, it follows from the usual definition of reciprocal lattice vectors $\vec{a}_i \cdot \vec{b}_j = 2\pi\delta_{ij}$, where δ_{ij} is the Kronecker delta symbol, that $|\vec{b}_1| = |\vec{b}_2| = 4\pi/\sqrt{3}a$. By working out components of $\vec{b}_1$ and $\vec{b}_2$, one can be convinced that the reciprocal lattice of a triangular lattice is also a triangular lattice but rotated through $30°$ with respect to a direct lattice. Both direct and reciprocal lattices possess six-fold symmetry. The first Brillouin zone has a shape of a hexagon with two main symmetry directions, which are commonly referred to as ΓM and ΓK (Fig. 4.1).

As an example of a 3D crystal, let us consider a collection of spheres assembled in a face-centered cubic (FCC) structure, which is obtained from the simple-cubic lattice by adding one sphere to the center of every face of the cubic unit cell. Because of its high degree of symmetry, phononic crystals with this structure have been extensively investigated, both theoretically and experimentally. Figure 4.2 shows the direct lattice of the FCC structure along with the corresponding reciprocal lattice, which turns out to be a body-centered cubic (BCC) crystal structure (obtained from the simple-cubic structure by adding one atom in the center of its unit cell). Also displayed are the sets of primitive vectors $\vec{a}_1, \vec{a}_2, \vec{a}_3$ and $\vec{b}_1, \vec{b}_2, \vec{b}_3$ of both lattices. It can be easily seen from Fig. 4.2 that with this particular choice of the primitive vectors of the direct lattice we have $|\vec{a}_1| = |\vec{a}_2| = |\vec{a}_3| = a/\sqrt{2}$, and

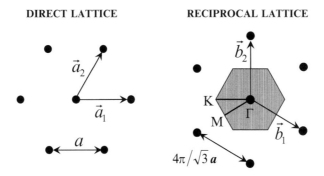

Fig. 4.1 The direct and reciprocal lattices of the 2D phononic crystals, which were investigated experimentally. The *shaded hexagon* indicates the first Brillouin zone. In the actual phononic crystal the rods were positioned at the points of the direct lattice (perpendicular to the plane of the figure)

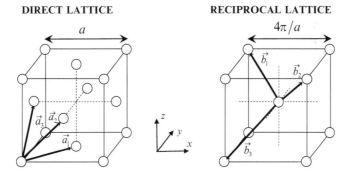

Fig. 4.2 The direct (FCC) and reciprocal (BCC) crystal lattices of the 3D phononic crystals

$\left|\vec{b}_1\right| = \left|\vec{b}_2\right| = \left|\vec{b}_3\right| = 2\sqrt{3}\pi/a$, where a is the length of the cube edge in the direct lattice.

The first Brillouin zone of the FCC lattice is a truncated octahedron and coincides with the Wigner-Seitz cell of the BCC lattice. It is presented in Fig. 4.3 along with its high symmetry directions. With respect to the coordinate system in Fig. 4.2, the coordinates of the high symmetry points (in units of $2\pi/a$) are: Γ [000], X [100], L [½ ; ½ ; ½], W [½ ; 1; 0], and K [¾; ¾; 0]. The investigation of the figure reveals that direction ΓL coincides with the direction also known as the [111] direction, i.e., a direction along the body diagonal of the conventional FCC unit cell, shown in Fig. 4.2.

A simple way of realizing a 3D crystal with the FCC Bravais lattice is by stacking the crystal layers along the [111] direction. The touching spheres are close packed in an ABCABC... sequence, which is shown in Fig. 4.4. The spheres

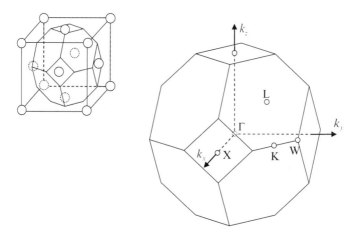

Fig. 4.3 The first Brillouin zone of the FCC lattice and its high symmetry points

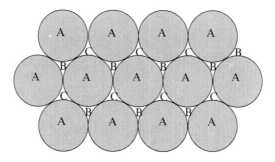

Fig. 4.4 Schematic diagram explaining the formation of a 3D crystal in a ABCABC... sequence

belonging to the first layer are denoted by the letter A. The next layer is formed by placing the spheres in the interstitials indicated by the letter B, and the third layer is formed by placing spheres in the interstitials of the second layer, which are denoted by the letter C. The sequence is then repeated again with the fourth layer beads to occupy interstitials in the third layer, which are positioned directly above beads denoted by the letter A. This packing results in the highest filling ratio of 74 %.

In this chapter, the dramatic effects of lattice periodicity on wave transport in 2D and 3D phononic crystals will be illustrated using these two representative crystal structures. Section 4.2 summarizes how such effects can be investigated experimentally, with emphasis on measurement techniques in the ultrasonic frequency range. Section 4.3 discusses the various mechanisms that can lead to the formation of band gaps, a topic that has been of central interest since the first calculations and experimental observations in phononic crystals. The rest of the chapter is concerned with phenomena that occur in the pass bands, starting with negative refraction in Sect. 4.4, the achievement of super-resolution lenses in Sect. 4.5 and band structure design and its impact on refraction in Sect. 4.6.

4.2 Experiments: Crystal Fabrication and Experimental Methods

4.2.1 Sample Preparation

4.2.1.1 2D Phononic Crystals

In this section we will consider the practical aspects of phononic crystal fabrication for the examples of 2D and 3D phononic crystals used by Sukhovich et al. [4, 5] and Yang et al. [6, 7] during their experiments on wave transport, negative refraction and focusing of ultrasound waves (see Sects. 4.3 and 4.4.). The 2D crystals were made of stainless steel rods assembled in a triangular crystal lattice and immersed in a liquid matrix. To ensure that the operational frequency of the crystals was in the MHz range, the characteristic dimensions of the crystals, lattice constant and rod diameter (1.27 mm and 1.02 mm correspondingly), were chosen to be comparable to the wavelength of ultrasound in water at this frequency range (Fig. 4.5).

For reasons that will be explained in more detail later, the crystals were made in two different shapes. A rectangular-shaped crystal had 6 layers stacked along the ΓM direction (Fig. 4.6a). A prism-shaped crystal was also made; it had 58 layers, whose length was diminishing progressively to produce sides forming angles of 30°, 60° and 90°. In this geometry, the shortest and longest sides are perpendicular to the ΓM directions (Fig. 4.6b), and the third intermediate-length side is perpendicular to the ΓK direction.

The filling fraction was 58.4 %. The particular details of crystal design depended on the type of liquid, which filled the space between the rods. For the crystals immersed in and filled with water, the rods were kept in place by two parallel polycarbonate plates in which the required number of holes was drilled; the crystal could then be easily assembled by sliding the rod's into the holes in these top and bottom templates (Fig. 4.7a, b). The rectangular crystal was 14 cm high while the prism-shaped crystal height was 9 cm.

Since key properties of the phononic crystals follow from their periodicity, the quality of the samples is critically dependent on the accuracy with which their geometry is set. For example, special care must be taken to use as straight rods as possible. At the same time, the holes defining the rods' positions should be precisely drilled, preferably using an automated programmable drilling machine.

Another rectangular-shaped crystal (with all parameters identical to those of the first crystal) was constructed to enable the liquid surrounding the rods (methanol) to be different to the medium outside the crystal (water), and consequently its design was more complicated. First of all, all plastic parts were made of an alcohol-resistant plastic (PVC). The crystal was encapsulated in a cell, whose face walls were made of a very thin (0.01-mm) plastic film tightly wrapped around the crystal (plastic film produced commercially and available as a food wrap worked very

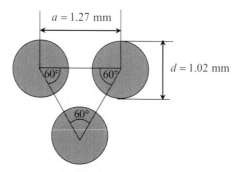

Fig. 4.5 Unit cell of a 2D phononic crystal

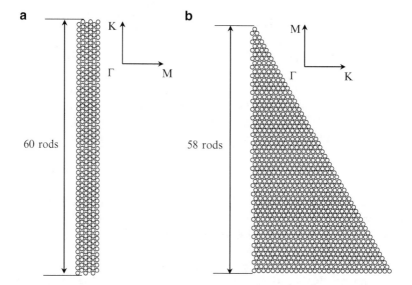

Fig. 4.6 Geometry of the 2D crystals. (**a**) Rectangular crystal. (**b**) Prism-shaped crystal

well). Finally, the edges of the cell were sealed from the surrounding water by two rubber O-rings. The design of the crystal is shown in Fig. 4.8.

The choice of the phononic crystal materials provided high density and velocity contrast, thus ensuring that most of the sound energy was scattered by the rods and concentrated in the host matrix. Table 4.1 provides values of the densities and sound velocities for the constituent materials of the 2D crystals.

4.2.1.2 3D Phononic Crystals

3D phononic crystals, used in the experiments by Yang et al. [6, 7] and by Sukhovich et al. [8, 9], were made out of very monodisperse tungsten carbide

Fig. 4.7 2D crystals filled with and immersed in water: (**a**) rectangular crystal, (**b**) prism-shaped crystal

beads, 0.800 ± 0.0006 mm in diameter, immersed in reverse osmosis water. The beads were manually assembled in the FCC structure, with layers stacked along the cube body diagonal (the [111] direction) in an ABCABC... sequence. To ensure the absence of air bubbles trapped between the beads, the whole process of assembling crystals was conducted in water. To support the beads in the required structure, acrylic templates were used. The template consisted of a thick substrate with plastic walls attached to it (Fig. 4.9).

One can show that in order to keep beads in the FCC crystal lattice two kinds of walls should be used with sides inclined at angles $\alpha = 54.74°$ and $\beta = 70.33°$ above the horizontal, and with inner side lengths L_A and L_B. The values of L_A and L_B depend on the number of beads n along each side of the first crystal layer and the bead diameter d. These lengths are given by the following expressions:

$$L_A = \left(n - 1 + \tan\frac{\alpha}{2}\right)d$$

$$L_B = \left(n - 1 + \tan\frac{\beta}{2}\cot 75°\right)d \qquad (4.1)$$

With 49 beads on each side of the bottom layer, (4.1) gives $L_A = 38.814$ mm and $L_B = 38.552$ mm.

In the experiments on the resonant tunneling of ultrasound pulses, the samples consisted of *two* 3D phononic crystals with the same number of layers and separated by an aluminum spacer of constant thickness. For brevity, these samples will be referred to as *double* 3D crystals. After the lower crystal was assembled, the spacer was placed on the top without disturbing beads of the crystal. The upper crystal was then assembled on the surface of the spacer. Spacer edges were machined at angles

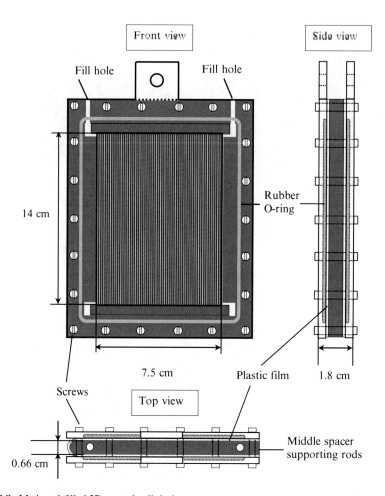

Fig. 4.8 Methanol-filled 2D crystal cell design

Table 4.1 Comparison of the physical properties of the constituent materials used for 2D phononic crystals [49]

Material	Density (g/cm^3)	Longitudinal velocity (mm/μs)	Shear velocity (mm/μs)
Stainless steel	7.89	5.80	3.10
Water	1.00	1.49	–
Methanol	0.79	1.10	–

matching the angles of the walls of the template. Also, the thickness of the spacer was calculated such that it replaced precisely an integer number of layers of the single crystal. This ensured that the beads resting on the spacer filled the entire available surface without leaving any gaps, enabling high-quality crystals to be

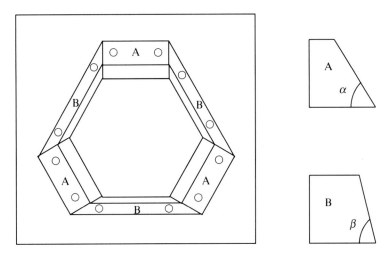

Fig. 4.9 Template for 3D phononic crystal (top view) with side views of walls A and B. Note that $\tan\alpha = \sqrt{2}$ and $\tan\beta = 2\sqrt{2}$

constructed. In most of the experiments, the thickness of the spacer was chosen to be 7.05 ± 0.01 mm.

The base of the template was made fairly thick (84.45 mm) to allow temporal separation between the ultrasonic pulses that was directly transmitted through the crystal, and all of its subsequent multiple reflections inside the substrate. The density and velocity mismatch in the case of 3D crystals was even larger than for 2D crystals, as tungsten carbide has density of 13.8 g/cm^3, longitudinal velocity of 6.6 mm/μs and shear velocity of 3.2 mm/μs. The actual sample (single 3D crystal) is shown in Fig. 4.10, while the close-up of its surface is presented in Fig. 4.11.

4.2.2   Experimental Methods

In the sonic and ultrasonic frequency ranges, the properties of phononic crystals are best studied experimentally by directing an incident acoustic or elastic wave towards the sample and measuring the characteristics of the outgoing wave, which was modified while propagating through the crystal. In practice, pulses are preferred to continuous monochromatic waves since pulses are much more convenient to work with. Due to their finite bandwidth, in a single experiment they allow information to be obtained over a wide frequency range. The use of pulses also facilitates the elimination of stray sound from the environment surrounding the crystal. In what follows, we describe two types of experiments, each used to investigate different aspects of phononic crystals.

Fig. 4.10 3D single phononic crystal assembled in the supporting template

Fig. 4.11 Close-up view of the surface of the crystal, which is shown in Fig. 4.10

4.2.2.1 Transmission Experiments

In transmission experiments one measures the coherent ballistic pulse emerging from the output side of the sample after a short pulse (often with a Gaussian envelope) was normally incident on the input side. Usually, crystals with two flat surfaces are used and crystal properties are investigated along the directions for which the direction of the output pulse is not expected to change with respect to that of the input pulse. In this case the far-field waveforms are spatially uniform in a plane parallel to the crystal faces, and thus the outgoing pulse can be accurately detected using a planar transducer, whose active element's characteristic dimensions are many times larger than the wavelength of the measured pulse. (The diffraction orders that appear at high frequencies are effectively eliminated by measuring the transmitted field over the finite transverse width that is set by the diameter of the detecting transducer.) Such a transducer averages any field fluctuations (for example due to imperfections inside the sample) and provides information on the average transport properties of the crystal. Another benefit of such averaging is an increase of the signal-to-noise ratio. Note also that to ensure the best possible approximation of the incident pulse by a plane wave, the sample should be placed in the far-field of the generating transducer. In the ultrasonic frequency range, the most convenient reference material in which the transducers and crystal can be located is water.

The analysis of the recorded pulse is done by comparing it with a reference pulse, obtained by recording a pulse propagating directly between generating and receiving transducers (with the sample removed from the experimental set-up). To allow the transmission properties to be determined from a direct comparison between the reference and measured pulses, the reference pulse should be shifted by the time $\Delta t = L/v_{\mathrm{wat}}$, where L is the crystal thickness and v_{wat} is the speed of sound in the medium between source and receiver. Since the attenuation in water is negligibly small, the time-shifted reference pulse accurately represents the pulse that is incident on the input face of the sample.

Figures 4.12a and 4.12b shows a typical example of incident and transmitted pulses for a 3D phononic crystal of tungsten carbide beads in water. The effects on the transmitted pulse of multiple scattering inside the crystal are clearly seen by the considerable dispersion of the pulse shape. Since the full transmitted wave function is measured, complete information on both *amplitude* and *phase* can be determined using Fourier analysis. The amplitude transmission coefficient as a function of frequency is given by the ratio of the magnitudes of the Fourier transforms of the transmitted and input pulses:

$$T(f) = \frac{A_{\mathrm{trans}}(f)}{A_{\mathrm{ref}}(f)} \qquad\qquad (4.2)$$

Figure 4.12c shows the Fourier transform magnitudes corresponding to the pulses in Figs. 4.12b and 4.12b, demonstrating not only the large effect that phononic crystals can have on the amplitude of transmitted waves but also the wide range of frequencies that can be probed in a single pulsed measurement.

In addition to the transmission coefficient, ballistic pulse measurements also provide information on the transmitted phase, from which the wave vector can be obtained. This phase information is also directly related to the phase velocity v_{phase}

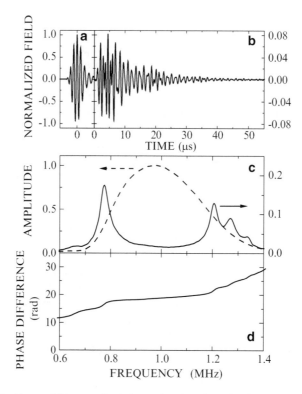

Fig. 4.12 (a) Incident and (b) transmitted ultrasonic pulses through a 6-layer 3D phononic crystal of tungsten carbide beads in water. The crystal structure is FCC, and the direction of propagation is along the [111] direction. (c) The amplitude of the incident (*dashed line*; left axis) and transmitted pulses (*solid line*; right axis), obtained from the fast Fourier transforms of the waves in (a) and (b). Their ratio yields the frequency dependent transmission coefficient [(4.2)]. (d) The phase difference between the transmitted and reference pulses, from which frequency dependence of the wave vector can be determined [(4.3)]. The large decrease in transmitted amplitude near 1 MHz and the nearly constant phase difference of $n\pi$, where $n = 6$ is the number of layers in the crystal, are characteristics of a Bragg gap

of the component of the Bloch state with wave vector in the extended zone scheme. These parameters are measured by analyzing the cumulative phase difference $\Delta\varphi$ between transmitted and input pulses (obtained from Fourier transforms of both signals—see Fig. 4.12d). This phase difference can be expressed as follows:

$$\Delta\varphi = kL = \frac{2\pi L}{v_{phase}}f \tag{4.3}$$

where L is the crystal thickness. The ambiguity of 2π in the phase can be eliminated by making measurements down to sufficiently low frequencies, since the phase difference must approach zero as the frequency goes to zero. From (4.3) it is possible to obtain directly the wave vector as function of frequency in the extended zone scheme; the corresponding wave vector in the reduced zone scheme is obtained by subtracting the appropriate reciprocal lattice vector. Thus, (4.3) allows the dependence of the angular frequency ω on the wave vector k to be determined, yielding the dispersion curve and hence the band structure.

Finally, the experiments on the transmission of ballistic pulses allow the group velocity, which is the velocity of Bloch waves in the crystal, to be measured. By its definition, the group velocity is the velocity with which a wave packet travels as a whole. Since the transmitted pulse may get distorted from its original Gaussian shape as it passes through the crystal, especially if the pulse bandwidth is wide (as in Fig. 4.12), the group velocity may lose its meaning in this case [10]. However, it is still possible to recover two essentially Gaussian pulses by digitally filtering the input and output pulses with a narrow Gaussian bandwidth centered at the frequency of interest. The group velocity at that frequency is then found by the ratio of the sample thickness L to the time delay Δt_g between two filtered pulses:

$$v_g = L \big/ \Delta t_g. \tag{4.4}$$

This direct method of measuring the group velocity is illustrated by Fig. 4.13, which shows input and transmitted pulses filtered at the central frequency of 0.95 MHz with a bandwidth of 0.05 MHz, for a 12-layer 3D crystal of tungsten carbide beads in water. The delay time is also indicated. By repeating this procedure for different frequencies, the frequency dependence of the group velocity can be found.

4.2.2.2 Field Mapping Experiments

In certain cases, the outgoing field is not expected to be spatially uniform and the direction of the outgoing pulse might not be perpendicular to the crystal's output face (as in focusing and negative refraction experiments). To investigate the field distribution a transducer whose size is larger than the wavelength cannot be used as it smears out the spatial variations of the field by detecting the average pressure across the transducer face. To resolve subwavelength details and map the field accurately one needs an ultrasound detector with physical dimensions less than a wavelength. For example, Yang et al. [7] and Sukhovich et al. [5] used a small hydrophone with an active element diameter of 0.4 mm to investigate spatial properties of the output sound field. This detector was appropriate since in their experiments the wavelength in water ranged from 0.5 to 3 mm. In practice, the ultrasound field was measured at every point of a rectangular grid by mounting the hydrophone on a 3D motorized translation stage. In case of the experiments by Sukhovich et al. [5], the plane of the grid was perpendicular to the rods and intersected them approximately in their mid-points (to avoid edge effects). Fig. 4.14 illustrates the experimental geometry used to map the outgoing field in negative refraction experiments with the prism-shaped crystal.

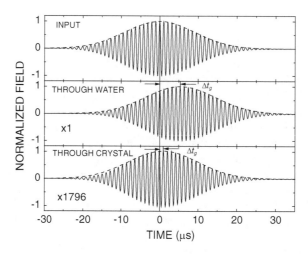

Fig. 4.13 The group velocity is measured directly from the time delay Δt_g between the peaks of the filtered input and transmitted pulses for a sample of given thickness L. The *bottom* and *middle panels* compare pulses with a bandwidth of 0.05 MHz transmitted through a 12-layer 3D phononic crystal of tungsten carbide beads in water and through the same thickness of water. Relative to the input pulse (*top panel*), the time delay for the transmitted pulse through the crystal is much shorter than through water, showing directly that large values of the group velocity are measured in this phononic crystal at this frequency, which falls inside the first bandgap. The pulses are normalized to unity for ease of comparison, with the normalization factors being indicated in the figures. The *dashed lines* indicate that the Gaussian envelope of the input pulse is well preserved for both transmitted pulses

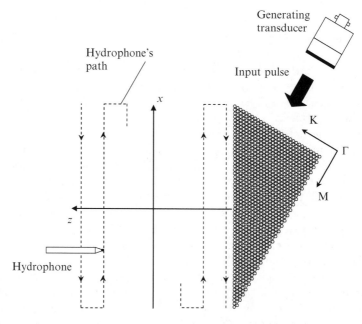

Fig. 4.14 Field mapping to investigate negative refraction for a prism-shaped 2D phononic crystal

The recorded waveforms (acquired at each point of the grid) can be analyzed in either time or frequency domains. In the time domain, the value of the field at each grid point is read at some particular time and then used to create an image plot, which is essentially a snapshot of the field at this particular moment of time. By creating several image plots for different times, one can also investigate the time evolution of the transmitted sound. The video in the supplementary information to [5] shows an example of such time-evolving field maps. In the frequency domain, one first calculates the Fourier transforms (FTs) of the acquired waveforms. The magnitude of each FT is read at a single frequency and these values are used to make the image plot. The image plot in this case represented an amplitude map (proportional to the square root of intensity), which would be obtained from the field plot if continuous monochromatic wave were used as an input signal instead of a pulse. Examples of field and amplitude distributions measured in the negative refraction and focusing experiments by Sukhovich et al. [5, 11] are shown in Sects. 4.4 and 4.5.

4.3   Band Gaps and Tunneling

Lattice periodicity in phononic crystals leads to large dispersive effects in wave transport, which are shown by band structure plots that depict the relationship between frequency and wave vector along certain high symmetry directions. Representative examples of the band structures of 2D and 3D phononic crystals are illustrated in Figs. 4.15 and 4.16 for the structures described in Sect. 4.1. In both these examples, the continuous medium surrounding the inclusions is water, with the scattering inclusions being 1.02-mm-diameter steel rods for the 2D case and 0.800-mm-diameter tungsten carbide spheres for the 3D case. The solid curves in these figures show the band structures calculated using Multiple Scattering Theory (MST), which is ideally suited for determining the band structures of mixed crystals consisting of solid scatterers embedded in a fluid matrix (see Chap. 10). The symbols represent experimental data, determined from measurements of the transmitted cumulative phase $\Delta\phi$, as described in the previous section. To compare with the theoretical band structure plots, the measured wave vectors ($k = \omega/v_p = \Delta\phi/L$) are folded back into the first Brillouin zone by subtracting a reciprocal lattice vector ($\mathbf{k}_{reduced} = \mathbf{k}_{extended} - \mathbf{G}$). Excellent agreement between theory and experiment is seen, showing that experiments on relatively thin samples (6 layers for the 2D case, and 12 layers for the 3D case) are sufficient to reveal the dispersion relations of waves in the pass bands of an infinite periodic medium.

For both phononic crystals, there is a large velocity and density difference between the scattering inclusions and the continuous embedding medium, facilitating the formation of band gaps due to Bragg scattering. It is well known that Bragg gaps are caused by destructive interference of waves scattered from planes of periodically arranged scatterers. The lowest frequencies at which such band gaps may occur satisfy the condition that the separation between adjacent

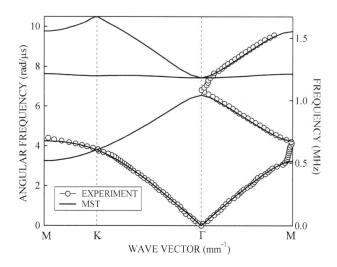

Fig. 4.15 Band structure of a 2D phononic crystal of 1.02-mm-diameter steel rods arranged in a triangular lattice and surrounded by water. The lattice constant $a = 1.27$ and the steel volume fraction is 0.584. *Solid curves* are predictions of the MST and *open circles* are experimental data. There are no data points for the second band along ΓK as this is a "deaf" band to which an incident plane wave cannot couple

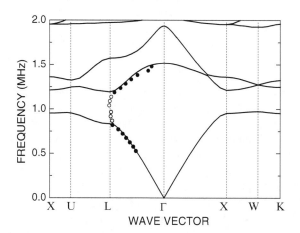

Fig. 4.16 Band structure of a 3D phononic crystal made from 0.800-mm-diameter tungsten carbide spheres arranged in the FCC lattice and surrounded by water at a volume fraction of 0.74. *Solid curves* are predictions of the MST and *circles* are experimental data

crystal planes is approximately half the wavelength in the embedding medium. In the 2D crystal, the lowest "gap" is only a stop band along the ΓM direction, with the lowest complete gap occurring between the 2nd and 3rd pass bands. For the 3D crystal, the lowest band gap near 1 MHz is wide and complete, with the complete gap width being nearly 20 % and the width along the [111] direction extending to approximately 40 %. These results show that phononic crystals with relatively simple structures can exhibit wide gaps, which are easier to achieve for phononic crystals than their optical counterparts because of the ability to manipulate large scattering contrast *via* velocity and density differences. Indeed, there is an extensive literature on how to create large band gaps for phononic crystals with a wide variety of structures, with the important role of density contrast now being well established (see the special edition on phononic crystals in Zeitschrift fur Kristallographie for many examples and references [12]).

The existence of band gaps in phononic crystals of finite thickness is shown clearly through measurements of the transmission coefficient. Results for the 2D and 3D crystals are plotted in Figs. 4.17 and 4.18, where the symbols represent experimental data and the solid curves are theoretical predictions using the layer MST [5, 9]. At low frequencies below the first band gap, the transmission exhibits small oscillations due to an interference effect resulting from reflections at the crystal boundaries; there are $n-1$ oscillations, where n is the number of layers, and the peaks in these oscillations correspond to the low frequency normal modes of the crystal. At band gap frequencies, the amplitude transmission coefficient shows very pronounced dips which became deeper in magnitude as the number of layers in the crystal increases. The sample-thickness dependence of the transmission coefficient in the middle of the gap (at 0.95 MHz) is plotted for the 3D crystals in Fig. 4.19. This figure shows that the transmitted amplitude A decreases exponentially with thickness in the gap, $A(L) = A_0 \exp[-\kappa L]$, consistent with evanescent decay of the amplitude, with κ being the imaginary part of the wave vector. The value of κ is 0.93 mm^{-1} in the middle of the gap, quantifying how quickly the transmission drops as the thickness increases. Thus, wave transport crosses over from propagation with virtually no losses outside the gap to evanescent transmission inside the gap. This evanescent character of the transmission at gap frequencies suggests that ultrasound is transmitted through crystals of finite thickness by tunneling, whose dynamics can be investigated by measuring the group velocity v_g and predicting its behavior using the MST [6]. Figure 4.20 shows that the group velocity increases linearly with sample thickness in the absence of dissipation (solid line), an unusual result that is the classic signature of tunneling in quantum mechanics [13], implying that the group time ($t_g = L/v_g$) is independent of thickness in sufficiently thick samples. This behavior is clearly seen in Fig. 4.20 by the theoretical predictions without absorption for thicknesses greater than 5 layers of beads. The dashed line in this figure implies a constant value of the tunneling time through the phononic crystal given by $t_g = 0.54$ μs, as expected for tunneling when $\kappa L \gg 1$. The experimentally measured group velocities are less than this theoretical prediction but are still remarkably fast, being greater than the speed of sound in water (1.5 mm/μs) for all crystal thicknesses, and greater than the velocities of elastic waves in tungsten carbide (6.66 and 3.23 mm/μs for longitudinal and shear waves, respectively) for the largest thicknesses. These experimental results for v_g are smaller than the dashed line in Fig. 4.20 because of absorption, which can be taken into account in the MST by allowing the moduli of the constituent materials

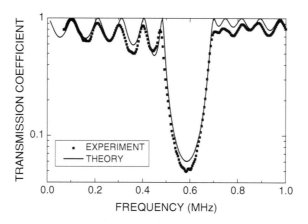

Fig. 4.17 Amplitude transmission coefficient as a function of frequency for a 6-layer 2D phononic crystal along the ΓM direction. *Squares* and *lines* represent experimental data and MST predictions, respectively

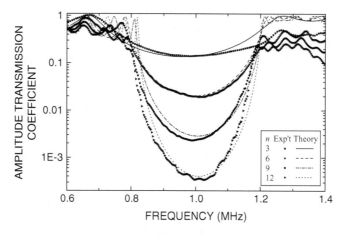

Fig. 4.18 Amplitude transmission coefficient as a function of frequency for 3-, 6-, 9- and 12-layer 3D phononic crystals of tungsten carbide beads in water along the ΓL direction. *Symbols* and *lines* represent experimental data and MST predictions, respectively

to become complex. The predictions of the theory with absorption are shown by the dashed curve and give a satisfactory description of the experimental results, indicating how dissipation, which has no counterpart in the quantum tunneling case, significantly affects the measured tunneling time.

The effect of dissipation on tunneling was interpreted using the two-modes model (TMM), which allows the role of absorption to be understood in simple physical terms [6]. Absorption in the band gap of a phononic crystal cuts off the long multiple scattering paths, making the destructive interference that gives rise to the band gap incomplete. As a result, a small-amplitude propagating mode exists in

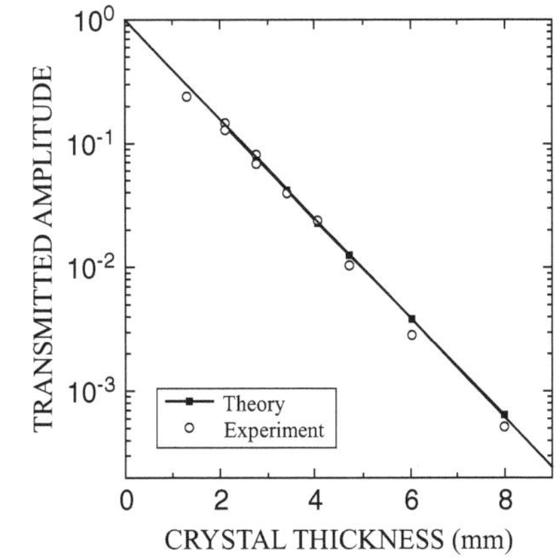

Fig. 4.19 Amplitude transmission coefficient as a function of thickness in the middle of the Bragg gap at 0.95 MHz for FCC phononic crystals of tungsten carbide beads in water

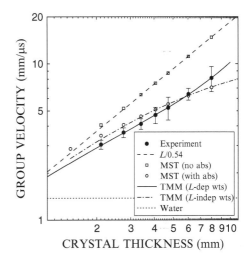

Fig. 4.20 Group velocity as a function of thickness in the middle of the Bragg gap at 0.95 MHz for FCC phononic crystals of tungsten carbide beads in water

parallel with the dominant tunneling mode, so that the group velocity can be calculated from the weighted average of the tunneling time t_{tun} and the propagation time $t_{\text{prop}} = L/v_{\text{prop}}$. Thus, $\bar{v}_g = L/(w_{\text{tun}}t_{\text{tun}} + w_{\text{prop}}L/v_{\text{prop}})$, where w_{tun} and w_{prop} are the weighting factors, which depend on the coupling coefficients and attenuation factors of each mode [6, 14]. The best fit to the data, shown by the solid curve in Fig. 4.20, was obtained with a coupling coefficient to the tunneling mode of 0.95, confirming the dominance of the tunneling mechanism, and with a contribution from the propagating component that diminished gradually with thickness, consistent with decreased dissipation in the thicker crystals—a physically reasonable result [14]. It is also interesting to note that with thickness-independent weight factors, the predictions of the TMM and the MST with absorption are very similar. These results show that the TMM successfully account for the effects of absorption on the tunneling of ultrasonic waves in phononic crystals, thereby providing a simple physical picture of the underlying physics.

The demonstration of the tunneling of ultrasound through the band gap of a phononic crystal raises an interesting question: Can resonant tunneling, analogous to the resonant tunneling of a particle through a double barrier in quantum mechanics, be observed in phononic crystals? This effect is intriguing since on resonance the transmission probability of a quantum particle through a double barrier is predicted to be unity, even though the transmission probability through a single barrier is exponentially small. This question has been addressed through experiments and theory on the transmission of ultrasound through pairs of phononic crystals separated by a uniform medium, which formed a cavity between them [8]. Evidence for resonant tunneling was revealed by large peaks in the transmission coefficient on resonance, which occurs at frequencies in a band gap when the cavity thickness approaches a multiple of half the ultrasonic wavelength. However, the transmission was less than unity on resonance because of the effects of dissipation in the phononic crystals, an effect that has a simple interpretation in the two modes model as a consequence of leakage due to the small propagating component in the band gap. Thus, the subtle effects of absorption on resonant tunneling in acoustic systems could also be studied. In addition, the use of pulsed experiments enabled the dynamics of resonant tunneling to be investigated. Very slow ("subsonic") sound was observed on resonance, while at neighboring frequencies, very fast ("supersonic") speeds were found. In contrast to the quantum case, ultrasonic experiments on resonant tunneling in double phononic crystals enable the full wave function to be measured, allowing both phase and amplitude information, in addition to static and dynamic aspects, to be investigated.

While the most commonly studied type of band gap in phononic crystals arises from Bragg scattering, band gaps may also be caused by mechanisms, such as hybridization and weak elastic coupling effects, which do not rely on lattice periodicity. Hybridization gaps are caused by the coupling between scattering resonances of the individual inclusions and the propagating modes of the embedding medium [15]. Their origin may be viewed as a level repulsion effect. Band gaps due to this hybridization mechanism were first observed, and have also been studied more recently, in random dispersions of plastic spheres in a liquid matrix [16–20]. Such gaps are of particular importance in the context of acoustic and elastic metamaterials, where the coupling of strong low frequency resonances with the surrounding medium may lead to negative values of dynamic mass density and modulus [21]. In phononic crystals, it is the possibility of designing structures in which both hybridization and Bragg effects occur in the same frequency range that is especially interesting [22]. For example, the combination of Bragg and hybridization effects has been invoked to explain the remarkably wide bandgaps that have been found both experimentally and theoretically in three dimensional (3D) crystals of dense solid spheres (e.g., steel, tungsten carbide) in a polymeric matrix (e.g., epoxy, polyester) [14, 23]. Other examples of band gaps that are enhanced by the combined effects of resonances and Bragg scattering have been

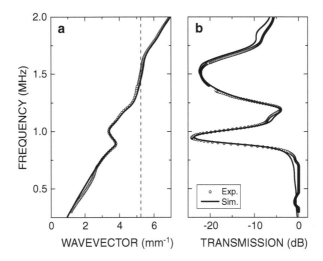

Fig. 4.21 Dispersion relation (**a**) and transmission coefficient (**b**) for a 6-layer 2D crystal of nylon rods in water at a nylon volume fraction of 0.40. *Symbols* and *solid curves* represent experimental data and finite element simulations respectively. The lower band gap near 1 MHz is an example of a pure hybridization gap, characterized by a sharp dip in transmission and a range of frequencies in the dispersion curve for which the group velocity is negative. The broader second gap centered near 1.5 MHz has the character of a Bragg gap, with a large positive group velocity, and occurs at the edge of the first Brillouin zone, indicated by the *vertical dashed* line

demonstrated in two-dimensional crystals of glass rods in epoxy and three dimensional arrays of bubbles in a PDMS matrix [24, 25].

We illustrate the characteristic features of hybridization gaps by showing results of experiments and finite element simulations on a two-dimensional hexagonal phononic crystal of nylon rods in water [26]. Figure 4.21 shows the dispersion relation and transmission coefficient in the vicinity of the lowest scattering resonance of nylon rods for a crystal with a nylon volume fraction of 40 %. The resonance occurs near 1 MHz for the 0.46-mm-diameter rods used in this crystal. Near this frequency, the dispersion relation exhibits a negative slope, corresponding to a range of frequencies with negative group velocity. Direct measurements of the negative group velocity were performed from transmission experiments using narrow-bandwidth pulses in the time domain, where the peak of the transmitted pulse was observed to exit the crystal before the peak of the input pulse entered the crystal. The negative time shift arises from pulse reshaping due to anomalous dispersion and does not violate causality. This property of negative group velocity is characteristic of resonance-related band gaps, and can be used to distinguish them from Bragg gaps, for which the group velocity is large and positive, as shown above. At higher frequencies, a second gap is observed for this crystal near 1.5 MHz; this gap is dominated by Bragg effects, with large positive group velocities inside the gap.

A third mechanism leading to the formation of band gaps occurs in three-dimensional single-component phononic crystals with the opal structure: spherical particles that are bonded together by sintering to form a solid crystal without a second embedding medium. Band gaps in such phononic crystals have been observed at both hypersonic and ultrasonic frequencies [26, 27]. They have also been seen in disordered structures of randomly positioned sintered spherical particles [28, 29]. The origin of the band gaps is associated with resonances of the spheres, but the underlying mechanism is quite different to the formation of hybridization gaps. Indeed the physics is more analogous to the tight-binding model of electronic band structures, with the resonant frequencies of the spheres corresponding to the electronic energy levels of the atoms. The coupling between the individual resonances of the spherical particles, due to the necks that form between the particles during sintering, leads to the formation of bands of coupled resonances with high transmission (pass bands). However, if the mechanical coupling between the spheres is sufficiently weak, these pass bands have limited bandwidth, and band gaps form in between them. These band gaps can be quite wide and are omnidirectional.

Up to now, the theory and experiments we have described in this chapter have been related to absolute band gap properties of phononic crystals. These results on sound attenuation and tunneling have proved phononic crystals meaningful in the perspective of building-up artificial materials with frequency dependent properties. However, the periodic structure of phononic crystals similarly impacts propagation of elastic waves in the frequency range of the passing bands. More specifically, the zone folding effects imply the existence of negative group velocity bands. Such bands offer the opportunity of negative refraction. In the next sections, theoretical and practical aspects of negative refraction are discussed.

4.4   Negative Refraction in 2D Phononic Crystals

The periodicity of the phononic crystals makes them markedly different from the homogeneous materials since wave propagation now depends on the direction inside the crystal. It was shown in the previous section that the periodicity is the fundamental cause for the existence of the stop bands and band gaps. In this section, we will consider some other remarkable properties of phononic crystals not found in regular materials: negative refraction and sound focusing. It will be shown that both phenomena are essentially *band structure* effects.

It is well known that reflection and refraction of waves of any nature (acoustic, elastic or electromagnetic) occurring at the interface between two different media are governed by *Snell's* law. According to Snell's law, the component of the wavevector, which is tangential to the interface, must be *conserved* as the wave propagates from one medium to another. Let us consider, for example, the simple case of a plane wave obliquely incident from a liquid with Lamé coefficients λ_1 and $\mu_1 = 0$ on an isotropic solid characterized by Lamé coefficients λ_2 and μ_2 (Fig. 4.22). As a result of the wave interaction with the boundary, part of the energy of the incident wave is reflected back into the liquid in the form of a reflected

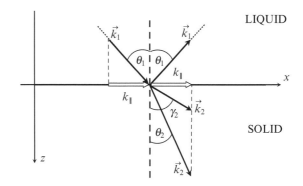

Fig. 4.22 Reflection and refraction of a plane wave incident obliquely on the liquid/solid interface from the liquid. Note the conservation of the wavevector component $k_{||}$

wave, which propagates with the phase velocity $c_1 = \sqrt{\lambda_1/\rho}$. The rest of the incident wave is transmitted into the solid and generates two outgoing waves, longitudinal and transverse, which propagate with phase velocities $c_2 = \sqrt{(\lambda_2 + 2\mu_2)/\rho}$ and $b_2 = \sqrt{\mu_2/\rho}$ respectively. Snell's law requires that parallel (to the interface) components of the wavevectors of the incident wave, $k_1 = \omega/c_1$, and of both refracted waves, $k_2 = \omega/c_2$ and $k_{2t} = \omega/b_2$ be equal (note that k_1 lies in the x–z plane and so do k_2 and k_{2t}). Mathematically, this means that the following conditions must be satisfied:

$$k_1 \sin \theta_1 = k_2 \sin \theta_2 = k_{2t} \sin \gamma_2 \tag{4.5}$$

where angles θ_1, θ_2 and γ_2 are indicated in Fig. 4.22. By introducing the notion of the index of refraction n and n', where $n = k_2/k_1$ and $n' = k_{2t}/k_1$, Snell's law is frequently written in the following form:

$$\begin{aligned} \sin \theta_1 &= n \sin \theta_2 \\ \sin \theta_1 &= n' \sin \gamma_2 \end{aligned} \tag{4.6}$$

With the help of Snell's law (4.5), one can easily calculate the refraction angles θ_2 and γ_2 when the parameters of the two media and the angle of incidence θ_1 are known (it is clear from Snell's law that the angle of reflection must be equal to the angle of incidence). Physically, Snell's law implies that refraction and reflection occur in the same way at any point of the interface between two media (i.e., independent of the x coordinate in Fig. 4.22).

The refraction of the wave from one medium to another can be conveniently visualized with the help of the *equifrequency* surfaces (or contours in case of 2D systems). Equifrequency surfaces are formed in k-space by all points whose wavevectors correspond to plane waves of the same frequency ω. Physically, they display the magnitude of the wavevector $\vec{k}$ of a plane wave propagating in the given medium as a function of the *direction* of propagation. For any isotropic medium the equifrequency surfaces are perfect spheres (circles in 2D), since the

Fig. 4.23 Equifrequency
surface of an isotropic
medium

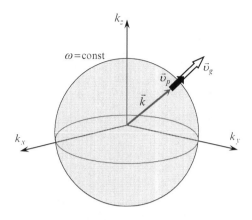

magnitude of the wavevector is independent of the direction of propagation, as
illustrated in Fig. 4.23.

Another extremely important property of equifrequency surfaces is that at its
every point the direction of the group velocity $\vec{v}_g$ (or equivalently the direction of the
energy transport) in the medium at a given frequency coincides with the direction of
the normal to the equifrequency surface (pointing towards the increase of ω). In
other words, $\vec{v}_g$ is given by the gradient of ω as a function of the wavevector $\vec{k}$:

$$\vec{v}_g = \vec{\nabla}_{\vec{k}}\,\omega(\vec{k}) \tag{4.7}$$

On the other hand, the direction of the phase velocity $\vec{v}_p$ (or the direction of the
propagation of constant phase) is set by the direction of the wavevector $\vec{k}$. As shown
in Fig. 4.23, in an isotropic medium both phase and group velocities point in the
same direction. This is however not the case in an *anisotropic* medium (e.g., GaAs
or CdS), in which magnitude of the wavevector is direction dependent and thus
equifrequency surfaces will not be perfect spheres anymore.

Having introduced the notion of the equifrequency surfaces/contours, let us use
them to illustrate the refraction of a plane wave in Fig. 4.24. This is accomplished
by drawing the equifrequency contours (since all wavevectors lie in the x–z plane)
for each medium on the scale that would correctly represent the relative magnitudes
of the wavevectors of the incident and refracted waves. By projecting the parallel
component of the incident wavevector $\vec{k}_i$ (which must be conserved according to
Snell's law) on the contours of the solid, one is able to find the direction of
propagation (i.e., refraction angles) of both waves in the solid (Fig. 4.24). As was
explained in the preceding paragraph, group velocities $\vec{v}_g$ and wavevectors $\vec{k}$ are
parallel to each other (because of the spherical shape of the equifrequency
contours) and also point in the *same* direction, since ω increases as the magnitude
of the wavevector increases, meaning that $\vec{\nabla}_{\vec{k}}\,\omega(\vec{k})$ points along the *outward* normal
to the equifrequency contour. The significance of the last observation will become
apparent when the refraction in 2D phononic crystals will be discussed.

Fig. 4.24 Refraction of a plane wave in Fig. 4.1 is illustrated with the help of the equifrequency contours (the same diagram holds for the transverse wave, which is omitted for simplicity)

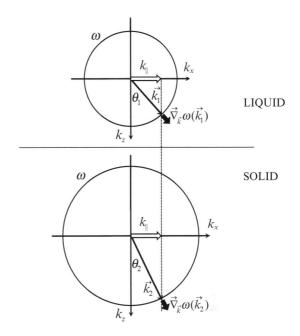

    The periodicity of the phononic crystal makes it an anisotropic medium, in which the magnitude of the wavevector depends on the direction inside the crystal and equifrequency contours are, in general, not circular. However, the frequency ranges still might exist where the equifrequency are almost perfect circles as is the case of a 2D crystal made of solid cylinders assembled in a triangular crystal lattice in a liquid matrix. For example, for a crystal made of stainless steel rods immersed in water the MST predicts the existence of circular equifrequency contours in the 2nd band for the frequencies that are far enough from the Brillouin zone edges (ranging from 0.75 MHz to 1.04 MHz, which is the top frequency of the 2nd band). The equifrequency contours for the several frequencies are presented in Fig. 4.25 [5].

    Note that in this frequency range the wavevector $\vec{k}_{cr}$ and the group velocity $\vec{v}_g$ (which defines the direction of the energy transport inside the crystal) are *antiparallel* to each other. This is the consequence of the fact that ω increases with the *decreasing* magnitude of the wavevector, meaning that $\vec{\nabla}_{\vec{k}}\omega(\vec{k})$ points along the *inward* normal to the equifrequency contour, as explained in Fig. 4.26. It is also obvious that, because of the circular shape of the equifrequency contours in the 2nd band, $\vec{k}_{cr}$ and $\vec{v}_g$ are antiparallel *irrespective* of the direction inside the crystal.

    Let us investigate the consequence of this fact by considering the refraction into such a phononic crystal of a plane wave incident on the liquid/crystal interface from the liquid and having frequency lying in the 2nd band of the crystal (Fig. 4.27). The parallel component of the wavevector in both media must be conserved just as it was in the case displayed in Fig. 4.24. What is different however is that the wave vector inside the crystal and the direction of the wave propagation inside the crystal

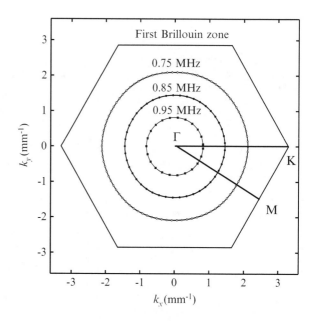

Fig. 4.25 Equifrequency contours predicted by the MST for the several frequencies in the 2nd band of the 2D phononic crystal made of stainless steel rods in water

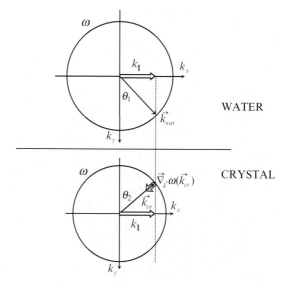

Fig. 4.26 Refraction of a plane wave at the water/crystal interface. The choice of the upward direction of the wavevector $\vec{k}_{cr}$ provides a wave propagating inside the crystal

are now *opposite* to each other. As a result, both incident and refracted waves (rays) stay on the *same* side of the normal to the water/crystal interface (compare with Fig. 4.24 in which incident wave crosses the plane though the normal as it refracts into the lower medium).

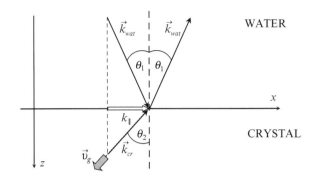

Fig. 4.27 Negative refraction of a plane wave incident obliquely on the water/crystal interface. Note the conservation of the wavevector component $k_\parallel$

Since the refracted wave happens to be on the negative side of the normal, this unusual refraction can also be described by assigning an effective *negative* index of refraction to the crystal. In this case we say that the incident wave is *negatively* refracted into the crystal and use the term "*negative* refraction" to indicate this phenomenon. Before we proceed further with discussion of sound wave refraction in phononic crystals, it is worth noting that the negative refraction considered above is fundamentally different from negative refraction in double negative materials, as originally envisaged for electromagnetic waves by Veselago [30] in materials with negative values of both electric permittivity ε and magnetic permeability μ. Although both phenomena look similar, it is a *band structure* effect in case of phononic crystals whereas in case of doubly negative materials it is brought about by the negative values of the *local* parameters of the medium (ε and μ for the electromagnetic wave case). It is also important to recognize that the negative direction of refraction is always given by the direction of the group velocity in phononic crystals.

Let us now consider the question of the *experimental* observation of the negative refraction in phononic crystals. First, it should be mentioned, that the same effect must occur when the direction of the wave in Fig. 4.27 is reversed, i.e., when the wave is incident on the crystal/water interface from the crystal. One might contemplate an experiment in which a plane wave would be incident obliquely on a *flat* phononic crystal with *parallel* sides. According to the previous discussion, it should be refracted negatively twice before it finally appears on the output side of the crystal, as shown in Fig. 4.28.

This type of experiment, however, is not able to provide conclusive evidence of the negative refraction, as the direction of the propagation of the output wave will be the same whether it refracts negatively inside the phononic crystal or positively in a slab of a regular isotropic material (Fig. 4.28). In case of an input beam of finite width, one can look for evidence of either negative or positive refraction inside the slab by measuring the position of the output beam with respect to the input beam and comparing it to the predicted value. In practice, this shift in position of the finite width beam may be difficult to resolve. Another type of experiment, which is able to

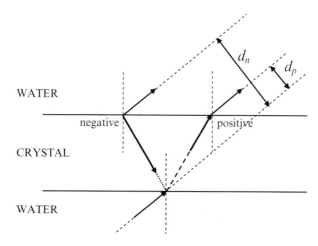

Fig. 4.28 Propagation of the sound wave through a flat crystal with parallel surfaces. Both negatively and positively refracted waves leave the crystal's surface in the same direction. Also indicated are distances d_p and d_n by which positively and negatively refracted beams are displaced with respect to the input beam.

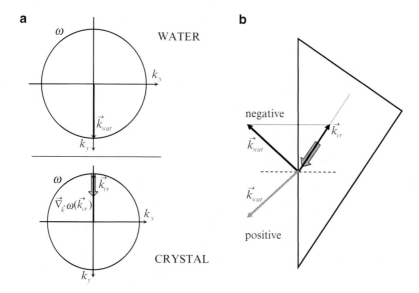

Fig. 4.29 Negative refraction experiment with the prism-shaped phononic crystal. (a) Equifrequency contours in water and in the crystal. In (b), the directions of positive and negative refraction at the output face of the prism crystal are shown. The *thick arrow* indicates the direction of wave propagation inside the crystal

provide *direct* verification of whether positive or negative refraction takes place, employs a *prism-shaped* phononic crystal (Fig. 4.29).

For the prism-shaped crystal, the input plane wave is incident normally on the shortest side of the crystal and propagates into the crystal *without* any change in its original direction, just as it would do in the case of a prism made out of a regular material (see Fig. 4.29a). Recall that the ensuing wave inside the crystal will have its wavevector $\vec{k}$ opposite to the direction of its propagation. This wave, however, will be incident *obliquely* on the output side of the crystal and must undergo negative refraction upon crossing the crystal/water interface (Fig. 4.29b), whereas in the case of a prism of a regular material the output wave will be positively refracted. Therefore, by recording on which side of the normal the outgoing wave appears as it leaves the crystal, one is able to directly observe negative refraction of the sound waves. From the predictions of the MST, one would expect the outgoing wave to emerge on the negative side of the normal. This prediction was tested in the experimentally by Sukhovich et al. [5]. The 2D phononic crystal was made in a shape of a right-angle prism which is shown in Fig. 4.30. along with the high symmetry directions of the triangular crystal lattice.

In the experiment, the input signal was normally incident on the shortest side of the crystal, and the wavefield was scanned at the output side of the crystal (Fig. 4.29b). Figure 4.31 presents the snapshot of the wavefield on which the negatively refracted outgoing wave is clearly observed.

The angle at which the negatively refracted wave emerges with respect to normal, $-21° \pm 1°$, was found to be in good agreement with the one predicted by the MST and Snell's law ($-20.4°$).

4.5   Flat Lenses and Super Resolution

In 2000, Pendry [31] has proposed to use "Double-negative" metamaterials, which means composite systems exhibiting both negative permittivity and dielectric constant, as a building material for potentially perfect lenses that beat the Rayleigh diffraction limit. This is possible thanks to the contribution of two phenomena. First intrinsic properties of negative index metamaterials provide self-focusing capabilities to a simple slab of these materials. The second effect requires the evanescent part of the spectra of a source to couple with the lens and being resonantly "amplified" in order to reach the image without losses. From this time, experimental and theoretical demonstrations of acoustic metamaterials and phononic crystals have been reported.

Early results by Yang et al. [7] in 2004 have shown the applicability of phononic structures for sound focusing. They have realized phononic crystals made of 0.8 mm-diameter tungsten carbide beads surrounded by water. The face centered cubic structure of the closed packed beads exhibits a complete band gap in the 0.98 to 1.2 MHz range. From the analysis of the equifrequency surfaces summarized in Fig. 4.32a, b, the authors have shown that significant negative refraction effects are expected due to the highly anisotropic properties of the dispersion relations.

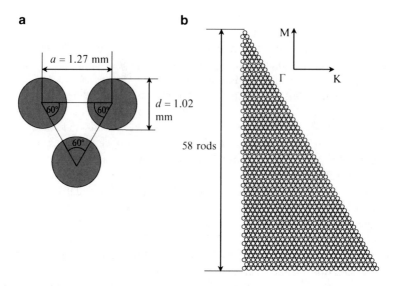

Fig. 4.30 Geometry of the 2D prism-shaped crystal. (**a**) Unit cell. (**b**) View from above. High symmetry directions, indicated as ΓM and ΓK, correspond to those shown in Fig. 4.1

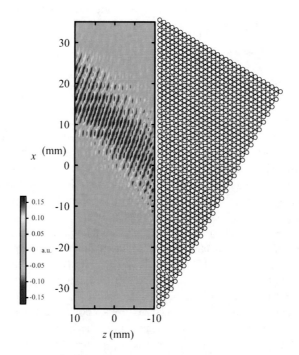

Fig. 4.31 Outgoing pulses in the negative refraction experiment (after Sukhovich et al. [5])

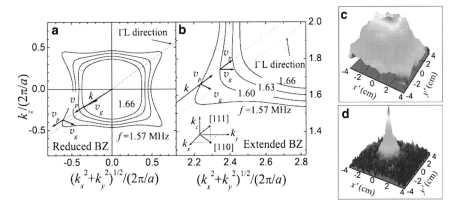

Fig. 4.32 Focusing of sound in a 3D phononic crystal after Yang et al. [7]. (**a**) Cross section of the equifrequency surfaces at frequencies near 1.60 MHz in the reduced (**a**) end extended (**b**) Brillouin zones. The cross section plane contains the [001], [110] and [111] directions. (**c**) Experimental field patterns measured a 1.57 MHz without the phononic crystal in place. (**d**) same as (**c**) with the phononic crystal in place

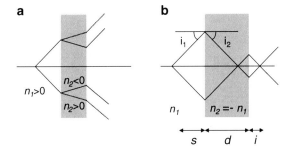

Fig. 4.33 Illustration of the refraction properties of a negative index material slab. (**a**) In the usual case of a positive material a source gives only divergent beams. If the slab is made of a negative index metamaterial then the beams are convergent in the extent of the slab. (**b**) If the slab is thick enough (or the index has sufficient magnitude), the incoming rays focus twice in the thickness of the slab and on the output side. Here the index is supposed to be opposite to the index of the embedding media. Two images are produced, inside the slab and on the output side (If the slab is too thin (Fig. 4.33a) then a single virtual image exists)

Experiments have been carried out to study the transmission of sound across a stack made of the phononic crystal mounted onto a thick substrate. As will be discussed in Fig. 4.33, negative refraction through a phononic crystal slab is expected to produce a focus inside the crystal and on the output medium. This later focus was observed by Yang et al. at the right distance on the substrate surface. They used a pinducer that produce ultrasonic pulses and a hydrophone mounted on a 3D translation stage. The recorded data was then treated by Fourier transform in order to recover the components at the frequency of interest. The field patterns in Fig. 4.32c, d show the focusing effect in the presence of the phononic crystal.

In 2009, evidences of an acoustic super-lensing effect have been provided by Sukhovich et al. [11]. Here we describe the principles of acoustic super-resolution and go into details about these recent results.

4.5.1 Sound Focusing by a Slab of Negative Index Material

Among the numerous consequences of negative refraction, the most promising in terms of applications is the ability for a slab of negative index material to produce an image from any point source. Indeed, in the extent of an equivalent homogeneous negative index material, the Snell's law simply applies using the negative index.

$$n_1 \sin i_1 = n_2 \sin i_2 \qquad (4.8)$$

Here, n_1 and n_2 are the indexes and i_1 and i_2 the incident and refracted angles. The negative value of i_2 accounts for both refracted and incident beams being on the same side with respect to the normal plane. Let us consider a sound source that emits waves in a usual positive medium in front of a slab of another material. As depicted on Fig. 4.33a, geometric ray tracing predicts that, if both materials are positive, every beam from the source will cross the two interfaces between the two materials and diverge as well on the output side of the slab. By contrast, if the slab is made of a negative index material then, any diverging beam will converge in the thickness of the slab. In the latter case, provided that the slab is sufficiently thick, the beam will focus twice (Fig. 4.32b).

This way a simple parallel slab of negative material performs by itself the focusing of an image as a lens would do. It is worthy to note that the principle of such a lens does not rely on the effect of shaping the material but rather on the intrinsic properties of negative index materials. The properties of these lenses are completely different from their usual counterparts. First, a simple geometrical analysis shows that the link between the respective positions of the image and source points is:

$$i = d \frac{\tan i_2}{\tan i_1} - s = d \frac{-n_1/n_2 \cos i_1}{\sqrt{1 - (n_1/n_2)^2 \sin^2 i_1}} - s, \qquad (4.9)$$

where d is the slab thickness, s the distance from the point source to the input side and i the distance from the output side to the image. The consequence of this relation is that rays with different angles of incidence focus at different distances from the output side. This is a drawback since producing an image from a point source requires that all the angular components of the incident signal are focused to a same point, which is called stigmatism. Here this requirement is fulfilled only if:

$$n_2 = -n_1,$$

(4.10)

which is the condition for All Angles Negative Refraction (AANR). This first condition is a strong yet possible condition for imaging with a negative metamaterial slab. In that case (4.9) reduces to:

$$i = d - s.$$

(4.11)

4.5.2   Origin of the Rayleigh Resolution Limit: Toward Super Resolution

This condition being satisfied, one can hope to build a lens whose resolution at a wavelength λ is at best $\Delta = \lambda/2$. This limitation, known as Rayleigh resolution limit, holds even in the case of no-loss materials and with a lens of infinite aperture. As pointed out by Pendry [31], its origin lies in the loss of the near field, evanescent, components from the source. If we consider the field emitted by a point source one must consider components with real wave-vectors (propagating waves) and pure imaginary wave-vectors due to the finite extension of the source. The former components are evanescent waves whose decay occurs over the distance of a few wavelengths. In the following we describe by means of a Green's function formalism [32] how the loss of these components leads to the Rayleigh resolution limit.

Let assume an infinite slab of thickness d made of a homogeneous double negative material immersed in a positive medium. Despite Green's functions are well suited to describe the response of any medium (possibly inhomogeneous) to a point source stimulus, for the sake of simplicity, both media are treated as homogeneous fluids. This assumption will be discussed further on a practical case. However, this description is still suitable to show how to enhance the resolution thanks to the integration of evanescent components. The notations and geometry used in the following parts are depicted on Fig. 4.34, where ρ_1, ρ_2, c_1 and c_2 are the densities and the sound waves velocities (phase velocities) of media 1 and 2 respectively.

The Green's function $G(\vec{x}, \vec{x}\prime)$ describes the field generated at $\vec{x}$ by a Dirac source located at $\vec{x}\prime$. Due to the axial symmetry of the problem and the aim to introduce the concept of wave-vectors we shall write this function as a two-dimensional spatial Fourier transform in the plane parallel to the fluid/slab interface:

$$G(\vec{x}, \vec{x}') = \int \frac{d^2 \vec{k}_{//}}{(2\pi)^2} e^{i\vec{k}_{//}\left(\vec{x}_{//} - \vec{x}'_{//}\right)} g(\vec{k}_{//}, x_3, x'_3),$$

(4.12)

Fig. 4.34 Notations used in the Green's function analysis of the Rayleigh resolution limit and super-resolution phenomena

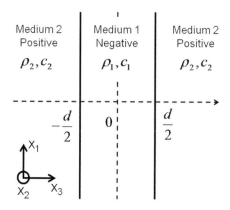

where $\vec{k}_{//}$ and $\vec{x}_{//}$ are the components of the wavevector and position vector parallel to the (x_1, x_2) plane. This is the function of a composite medium composed of the flat lens (medium 1) of thickness d (with faces centered on $-d/2$ and $d/2$) immersed between two semi-infinite media 2. Following the notions developed in Chap. 3 about composition of Green's functions, this Fourier Transform can be expressed by:

$$g(\vec{k}_{//}, x_3, x_3') = \frac{2\rho_1 c_1^2 \alpha_1 e^{-\alpha_2(x_3 - x_3' - d)}}{\left(\rho_1 c_1^2 \alpha_1 + \rho_2 c_2^2 \alpha_2\right)^2 e^{\alpha_1 d} - \left(\rho_1 c_1^2 \alpha_1 - \rho_2 c_2^2 \alpha_2\right)^2 e^{-\alpha_1 d}} \qquad (4.13)$$

$$\text{for} \quad x_3' < -d/2 \quad \text{and} \quad x_3 > d/2$$

Here, $\alpha_i = -ik_{3,(i)}$, is the component of the wave-vector perpendicular to the interface between medium 1 and medium 2.[1] This wave-vector is the key parameter since its value will account for the propagating or evanescent nature of the waves and its sign depends on the positive or negative index of the material. The component $k_{3,(i)}$ of the wave-vector is fully determined at a given frequency and $k_{//}$ by the dispersion relation of a homogeneous fluid:

$$\left(\frac{\omega}{c_i}\right)^2 = k_{//}^2 + k_{3,(i)}^2 \qquad (4.14)$$

In addition, the conservation of the parallel component of the wave-vector implies that $k_{//}$ is the same in both media. One can see that (4.14) admits real solutions for $k_{3,(i)}$ (i.e., propagating waves) only if $\omega \leq k_{//}c_i$. But we have to consider the opposite case when $\omega > k_{//}c_i$ and $k_{3,(i)}$ is pure imaginary (i.e., evanescent waves). Finally, in the case of a double negative material, the wave-vector is

---

[1] One can note that the zeros of the denominator in (4.13) correspond to all propagating and bound modes of the system.

Table 4.2 Normal to the slab component of the wave vector is defined depending on the evanescent or propagating nature of the wave and of the sign of the medium index

	Medium 1 Negative index	Medium 2 Positive index
Propagating $\omega \geq k_{//}c_i$	$k_{3,(1)} = -\sqrt{\frac{\omega^2}{c_1^2} - k_{//}^2}$	$k_{3,(2)} = \sqrt{\frac{\omega^2}{c_2^2} - k_{//}^2}$
Evanescent $\omega < k_{//}c_i$	$k_{3,(1)} = i\sqrt{k_{//}^2 - \frac{\omega^2}{c_1^2}}$	$k_{3,(2)} = i\sqrt{k_{//}^2 - \frac{\omega^2}{c_2^2}}$

anti-parallel to direction of propagation which is accounted for by the minus sign for the real $k_{3,(1)}$. This choice of a negative sign in the case of propagating waves ensures causality as pointed out by Veselago [32]. These considerations about the wave-vectors are summarized in Table 4.2.

As shown above, in order to achieve sound focusing by means of a negative index slab, the All Angles Negative Refraction condition has to be satisfied. Since the index is defined by $n_i = 1/c_i$, it implies that $c_1 = c_2 = c$. We will further simplify the model with some loss of generality by assuming that $\rho_1 = -\rho_2 = -\rho \leq 0$. The negative sign of the density is due to the fact that medium 1 is a double negative material which means that both bulk modulus and density are negative. Therefore, the Fourier transform of the Green's function from (4.13) reduces to:

$$g(k_{//}, x_3, x'_3) = \frac{e^{-\alpha(x_3 - x'_3 - 2d)}}{2\rho c^2 \alpha} \quad \text{for} \quad x'_3 < -d/2 \quad \text{and} \quad x'_3 > d/2 \quad (4.15)$$

This function has to be summed over the parallel components range $k_{//}$ of the source. This range will determine the resolution of the image. Indeed, if we assume that both propagating and evanescent modes contribute to the formation of the image (i.e., the integral is carried out for $k_{//}$ from zero to infinity[2]) then:

$$G(\vec{x}, \vec{x}') = \int_0^\infty \frac{d^2 \vec{k}_{//}}{(2\pi)^2} e^{i\vec{k}_{//}(\vec{x}_{//} - \vec{x}'_{//})} g(k_{//}, x_3, x'_3) = \frac{e^{i(\omega/c)|\vec{x} - \vec{x}_i|}}{4\pi\rho c^2 |\vec{x} - \vec{x}_i|}, \quad \text{where}$$

$$\vec{x}_i = \left(0, 0, \frac{d}{2} + d - s\right). \quad (4.16)$$

This expression is that of a spherical wave originating at the point $\vec{x}_i$. The spatial extent of this image is zero and therefore represents the perfectly reconstructed image of the point source. Comparing this results to the notations of Fig. 4.33b, we retrieve the relationship $i = d - s$. On the opposite, if we consider the usual far field situation, evanescent waves do not contribute to the image reconstruction and at a given frequency ω, the upper limit for $k_{//}$ is ω/c. Then, the Green's function

---

[2] The formulae: $\sqrt{\frac{\pi}{2}} \frac{e^{-a|\xi|}}{a} = \frac{1}{\sqrt{2\pi}} \int_{-\infty}^\infty \frac{e^{i\xi x}}{a^2 + x^2} dx$ is used to calculate the Green's function.

describes an image similar to (4.9) convoluted by a Gaussian profile whose half-width is:

$$\Delta = 2\pi c/\omega = \lambda/2 \qquad (4.17)$$

This latter case accounts for the Rayleigh resolution limit. Beating this resolution limit requires to achieve reconstruction of the image with at least a part of the evanescent spectrum from the source. Furthermore, we see that the actual resolution of an image is defined by the upper bound of the integral in (4.16). If any mechanism enables the integration of components with wave vectors up to $k_{//m} > \omega/c$, then the resolution is:

$$\Delta = 2\pi/k_m < \lambda/2, \qquad (4.18)$$

which demonstrates that the system achieves super resolution.

4.5.3 Design of a Phononic Crystal Super Resolution Lens

Sub wavelength resolution imaging has been a topic of considerable interest over the past decade. As seen above, this effect requires negative refraction and the ability of a system to transmit the entire spatial Fourier spectrum from a source, including evanescent components. Here, we discuss the possibility to implement such an acoustic super-lens and go into details about the recent experimental and theoretical demonstration by Sukhovich et al. [11] using a structure consisting of a triangular lattice of steel cylinders in methanol, all surrounded by water (Fig. 4.35a).

First, negative refraction can arise from one of two mechanisms. Double negative metamaterials consist of systems including locally resonant structures which exhibit a negative effective mass and negative bulk modulus [33, 34]. Other suitable systems are phononic crystals, consisting of a periodic array of inclusions in a physically dissimilar matrix [5, 6, 9, 11, 35, 36]. Negative refraction in phononic crystals relies on Bragg scattering that induces bands with a negative group velocity. It should be noticed that, since both metamaterials and phononic crystals have complex dispersion curve, the approximation of a homogeneous media is unlikely to be satisfied over the whole frequency range. However, it is possible to design these systems such that in a narrow frequency band, they can be considered as double negative materials with an effective negative index. In order to achieve AANR, one has to design the phononic crystal such that at a given frequency, the equifrequency contour is similar to an isotropic media, i.e., is a circle. In addition, at this frequency, in order to satisfy condition $c_1 = c_2 = c$, this circle must have the same diameter as the equifrequency contour of the media that surrounds the phononic crystals lens. This requirement explains the choice of methanol as the fluid medium surrounding the steel rods in the phononic crystal so that, at a frequency in the second band, the size of the circular equifrequency contours of the crystal could be tuned to match the equifrequency contours of water outside the crystal (Fig. 4.36). Thus, one of the important conditions for good focusing could be achieved with this combination of materials. Indeed, any liquid with a sound velocity that is small enough relative to water would have sufficed, with methanol being a convenient choice not only because it is a low-loss fluid with a low velocity (approximately two thirds the velocity in water) but also because it is readily available. In this case, in the vicinity of the frequency of 544 kHz (the operation frequency), the methanol-steel lens behaves as a negative index medium whose index is opposite to the index of water, thus achieving the AANR condition.

The second requirement to obtain sub wavelength imaging is to keep the contribution of the source evanescent modes. Following Sukhovich et al. [11], sub wavelength imaging of acoustic waves has also been shown to be possible using a square lattice of inclusions on which a surface modulation is introduced [38], a steel slab with a periodic array of slits [39], and an acoustic hyperlens made from brass fins [40]. In these demonstrations, the mechanism by which this phenomenon occurs has been attributed to amplification of evanescent modes through bound surface or slab modes of the system. In these systems, bound acoustic modes whose frequency falls is the vicinity of the operation frequency exist. In that case, provided that the lens is located in the close field of the source, some energy radiated by the evanescent modes will couple in a resonant manner to these bounded modes. The whole phononic crystal slab is excited and reemits the evanescent components necessary to the perfect image reconstruction. It is worthy to note that the amplification mechanism does not violate the conservation of energy since evanescent waves does not carry energy as pointed out by Pendry [31]. In this case, couplings with bounded modes play the role of the amplification mechanism. These modes can be studied by means of a Finite Difference Time Domain (FDTD) (see Chap. 10) simulation as shown on Fig. 4.37. If we look at the dispersion graph in the direction parallel to the water/lens interface (i.e., in the ΓK direction of the phononic crystal first Brillouin zone), we see a number of branches that corresponds to waves whose displacement is confined in the phononic crystal or at the surface of the slab. More specifically, at the operation frequency of 544 kHz, some nearly horizontal branches extend outside the water dispersion cone. These modes are likely to couple with wave vectors outside the cone at this frequency in accordance with the scheme described by Luo et al. [41].

4.5.4 Experimental and Theoretical Demonstration

Experiments have been carried out by Sukhovich [11] on a 2D phononic crystal made of 1.02-mm-diameter stainless steel rods arranged in a triangular lattice with lattice parameter of $a = 1.27$ mm. The surface of the crystal was covered by a very thin (0.01 mm) plastic film and the crystal was filled with methanol. A rectangular lens was constructed from 6 layers of rods, with 60 rods per layer, stacked in the

ΓM direction of the Brillouin zone, i.e., with the base of the triangular cell parallel to the surface. The experiments were conducted in a water tank. The ultrasound source was a narrow subwavelength piezoelectric strip, oriented with its long axis parallel to the steel rods; it was therefore an excellent approximation to a 2D point source. The spatiotemporal distribution of the acoustic field on the output side of the lens was detected with a miniature 0.40-mm-diameter hydrophone mounted on a motorized stage, which allowed the field to be scanned in a rectangular grid pattern. This setup ensures that the widths of the source and detector are smaller than the wavelength in water ($\lambda = 2.81$ mm) at the frequency of operation (530 kHz). The pressure field, shown on Fig. 4.38, exhibits a focal spot on the axis of the lens at a distance of approximately 3 mm from the output side. The resolution of this image is defined as the half-width of the pressure peak corresponding to the image. This value is determined by locating the maximum amplitude and fitting a vertical cut of the pressure field through this point by a sinus cardinal function ($\mathrm{sinc}(2\pi x/\varDelta)$). The half width $\varDelta/2$ is taken to be the distance from the central peak to the first minimum. The resolution at 530 kHz was found to be 0.37λ, where $\lambda = 2.81$ mm. This value is significantly less than the value of 0.5λ that corresponds to the Rayleigh diffraction limit, demonstrating that the phononic crystal flat lens achieves super-resolution.

These experimental results are supported by FDTD simulations. The FDTD method is based on a discrete formulation of the equations of propagation of elastic waves in the time and space domains on a square grid. The method is described in further details in Chap. 10. Here the whole methanol/steel phononic crystal is meshed as well as a part of the surrounding water. The limits of the simulation cell are treated under the Mur absorbing boundary condition that prevents reflections. The simulated phononic crystal slab has only 31 rods per layer in order that calculations remain compatible with computational resources. However tests have shown low influence of the reduced length. The acoustic source is simulated by a line source (0.55 mm wide) of mesh points emitting a sinusoidal displacement at frequency $v = 530$ kHz in accordance with the best experimental result. Their displacement has components parallel and perpendicular to the surface of the lens. The contour map on Fig. 4.38b shows the field of the time-averaged absolute value of the pressure. It can be seen in that an image exists on the right side of the crystal accompanied by lobes of high pressure that decay rapidly with distance from the surface of the crystal. The similarity between the experimental scheme and the FDTD mesh enables direct comparison of both experimental and simulated pressure fields. The FDTD results confirm the observation of super resolution with an image resolution of 0.35λ in excellent agreement with experiments. Both experimental and FDTD field patterns of Figs. 4.35b and 4.38 exhibit intense excitation inside the lens which is consistent with the role that bound modes are expected to play in the resonant transmission of the acoustic spectra. Theses modes, near the operating frequency, are bulk modes of the finite slab, not surface modes that decay rapidly inside the slab.

As seen above, the Rayleigh resolution limit originates from the upper limit of the Fourier spectrum transmitted to the image point which is at best ω/c in the far field regime of an imaging device. Here, since re-emitted evanescent waves can

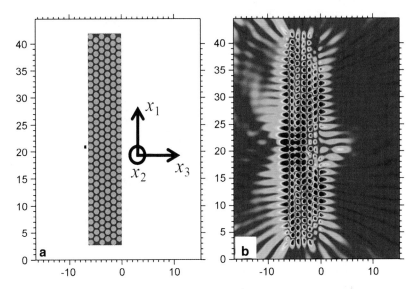

Fig. 4.35 Scheme of the system studied by Sukhovich et al. [11]. The radius of the steel inclusions is $r = 0.51$ mm with a lattice parameter of $a = 1.27$ mm. (**a**) FDTD grid used for the numerical study. The black line in front of the input side represents the source. (**b**) Averaged pressure field obtained through FTDT simulation. Note the image on the output side whose resolution (0.35λ) is below the Rayleigh limit

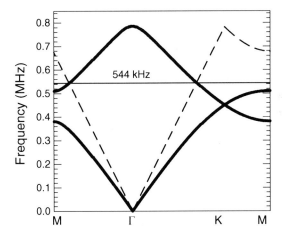

Fig. 4.36 Band structure of the methanol-steel phononic crystal after [37]. The *solid lines* represent the dispersion curves. The dispersion relation of the surrounding medium (water) is drawn as *dashed lines*. The second band exhibits a negative group velocity and intersects the water cone on a circular equifrequency at 544 kHz

contribute to the image, the resolution beats this criterion. By this mechanism one can virtually build an image up to an arbitrary resolution provided all evanescent modes are amplified and a sufficient time is available to reach the steady state regime for all evanescent modes. However, despite the absence of losses in the

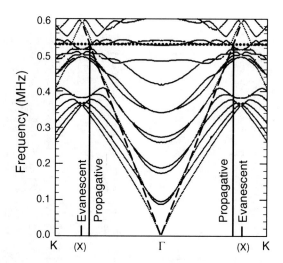

Fig. 4.37 Band structure diagram of the whole phononic crystal slab and water system in the direction parallel to the lens surface (FDTD calculation after [37]). Each curve corresponds to an acoustic mode propagating either in the phononic crystal slab or at the water/slab interface. The operation frequency (544 kHz) is indicated as a *horizontal dotted line*. The *straight dashed lines* are the dispersion relation of water. The x-axis range has been extended to the fist Brillouin zone ΓK of the triangular lattice

simulation scheme, the simulated resolution value is only 0.35λ. This fact indicates that the transmission of evanescent waves does not occur over the full Fourier spectra but rather up to a limiting cut-off value k_m. The previous analysis of super resolution in term of Green's function assumed the constituent material of the lens to be a homogeneous negative index material and did not discuss the possible origins of limitations to the transmitted Fourier spectra of the source. In the practical case when a phononic crystal, which is an inhomogeneous periodic material, is used as the lens, only modes with wave vector k parallel to the lens surface that is compatible with the periodicity of the phononic crystal in that same direction can couple to the sound source. In other words, all evanescent modes cannot contribute to the reconstruction of the image. The upper bound of the integration is determined by the largest wave vector k_m parallel to the lens surface that is compatible with the periodicity of the phononic crystal in that same direction and that can be excited by the sound source. In Fig. 4.37, the dispersion curves of the slab immersed in water are shown in the direction parallel to the lens surface. The dashed diagonal lines are the dispersion curves of acoustic waves in water and the dotted horizontal line represents the operating frequency. At this frequency, the wave vector components of the incident wave with $k_{//} < \omega/c$ can propagate in the crystal; they will form an image according to classical geometric acoustics. Components with $k_{//} > \omega/c$ will couple to the bound modes of the slab provided that these bound modes dispersion curves are in the vicinity of the operating frequency. In this way, the existence of many modes of the slab with nearly flat dispersion curves in the vicinity of the operating frequency is beneficial for achieving super resolution, as mentioned in [37]. One might imagine that

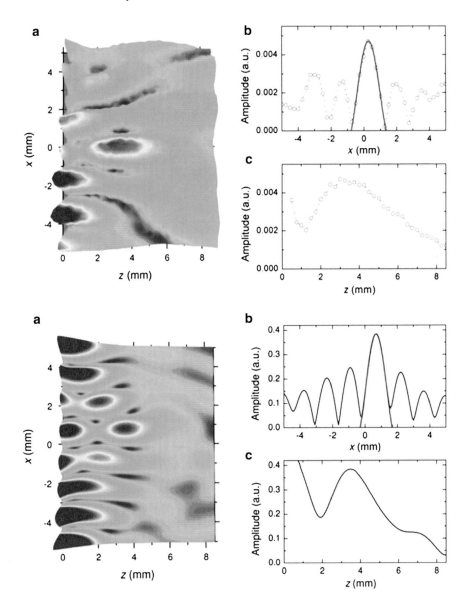

Fig. 4.38 Comparison of experimental (*top*) and FDTD simulation (*bottom*) results after Sukhovich et al. [11] showing the averaged pressure filed and pressure profiles along the lens axis and the output side

evanescent waves with transverse wave vector of any magnitude above ω/c could couple with bound modes. However, the modes that propagate through the thickness of the lens must resemble those of the infinite periodic phononic crystal. The symmetry of the waves inside the lens must therefore comply with the triangular symmetry of the phononic crystal. More precisely, the modes of the crystal are periodic in k-space with a period equal to the width of the first 2D triangular

Brillouin zone. This is the reason why the x-axis of Fig. 4.37 has been extended up to the K point of the hexagonal lattice reciprocal space. If an incident wave has a wave vector above the first Brillouin zone boundary, then it will couple to a mode having a wave vector that can be written as $\vec{k}_{//}' = \vec{k}_{//} + \vec{G}$ where $\vec{G}$ is a reciprocal lattice vector and $\vec{k}$ lies in the first Brillouin zone. In our case, since the first Brillouin zone of a triangular lattice extends from $-4\pi/3a$ to $4\pi/3a$ in the ΓK direction (parallel to the lens surface), the information carried by incident evanescent waves with transverse wave vector components,

$$k_{//}<k_{\mathrm{m}} = 4\pi/3a, \tag{4.19}$$

will contribute to the formation of the image. According to (4.11), with this definition, one finds that the best possible image resolution is:

$$\frac{\Delta}{2} = \frac{3a}{4}, \tag{4.20}$$

Applying this estimate to our phononic crystal with $a = 1.27$ mm, and a wavelength in water at 530 kHz of 2.81 mm, the minimum feature size that would be resolvable with this system is 0.34λ. This estimate matches results very well for the best resolution found for this system (0.34λ) presented in Sect. 4.2, and with experiment (0.37λ).

4.5.5 Effects of Physical and Operational Parameters on Super Resolution

In this section, we explore the effects of several factors on the image resolution of the phononic crystal flat lens. These factors include operational parameters such as the source frequency and the position of the source and geometrical factors such as the width and thickness of the lens. By exploring modifications to the system, we aim to shed light on the parameters that have the greatest impact on the imaging capabilities of the phononic crystal lens and understand their effects as they deviate from the best operating conditions.

4.5.5.1 Operating Frequency

Up to now, the operating frequency of the source was chosen to be 530 kHz, as in [6], this value was chosen as a compromise between proximity with bounded modes required for evanescent waves coupling and the AANR frequency in order to achieve the best experimental resolution. We now focus on the effects of the operating frequency in the 510 to 560 kHz range by means of numerical simulations and experimental measurements. Figure 4.39a shows the image resolution and

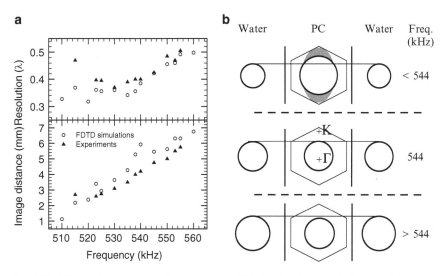

Fig. 4.39 Effects of the operating frequency after [37]. (**a**) Resolution and distance of the image as a function of the operating frequency. Results from experiments (*triangles*) are compared to FDTD simulation (*circles*). (**b**) Schematic representation of the transmission through the phononic crystal lens based on the equifrequency contours shapes. The equifrequency contour of the phononic crystal lens is represented as a *circle* inside the first Brillouin zone of the hexagonal infinite crystal. The gray areas illustrate the existence of bound modes with frequency very close to the operating frequency

distance of the focus from the exit surface of the lens as a function of the operating frequency. Experiments and calculations are in reasonable agreement from 523 to 560 kHz. Experiments exhibit an optimum resolution (0.37λ) at 530 kHz as discussed above. As expected, experimental values are higher than the computed values since practical imperfections in the lens fabrication and measurement noise lower the resolution of the focus. However, the difference does not exceed 0.05λ which is excellent. For increasing frequencies, the image lateral width increases up to the Rayleigh value (0.5λ) while the focus forms farther from the lens output side. These trends are confirmed in both experiments and FDTD results. However, no clear minimum of the resolution is observed in the simulations.

The observation of an optimum resolution has been interpreted in terms of a trade off between the AANR condition and the excitation of bound modes of the phononic crystal [37]. Figure 4.39b depicts the EFC in water and in the phononic crystal for different frequencies as circles of different diameters. The occurrence of super-resolution is discussed with respect to the operating frequency of 544 kHz which is the frequency of AANR expected from simulations.

First, if the source frequency is tuned *lower than 544 kHz*, super resolution is achieved with a resolution below 0.39λ. Since the operating frequency is lower than 544 kHz, the equifrequency contour of water is a smaller circle than the EFC inside the crystal. All components of the incident wave vectors corresponding to propagating modes can be negatively refracted by the crystal, i.e., the AANR condition is satisfied. However, the mismatch of the equifrequency contours diameters leads to a negative effective index of refraction with magnitude greater than one, causing the different components from the source to focus at different places. On another hand, operating frequencies well below 544 kHz are close to the flat bands of bound modes in the phononic crystal slab, allowing for efficient excitation by the evanescent waves from the source (Fig. 4.37). These modes are depicted as a gray region on the EFC of the slab in Fig. 4.39b. Thus, the gain from the amplification of evanescent modes is retained and super-resolution is achieved.

At the frequency of 544 kHz the EFC of water and the phononic crystal have the same diameter resulting in an effective index of -1. This condition implies a perfect focusing of all propagating components of the source into a single focal point. However Fig. 4.37 shows that the flat bands of bound modes of the lens are now well below the operating frequency, which means that coupling with these modes and amplification of the evanescent waves during transmission is now inefficient. The experimental optimum of the lateral resolution at 530 kHz occurs between the bound mode frequencies (510 kHz) and the perfect matching of the equifrequency contours (544 kHz).

In the case of *frequencies above 544 kHz*, the EFC of water has now a greater diameter than the EFC of the phononic crystal and the AANR condition is not matched. A part of the propagating components experience total reflection at the water/lens interface and the resolution worsens up to 0.5λ at 555 kHz.

These results confirm the importance of the design of the phononic crystal super-lens with respect to two conditions. First, one has to meet the AANR condition, which requires that the phononic crystal be a negative refraction medium with a circular EFC matching the EFC of the outside medium (water). Second, bound modes must exist in the phononic crystal whose frequencies are close to the operating frequency so that amplification of evanescent components may occur. The optimum frequency is found as the best compromise between those two parameters.

Finally, the effect of the operating frequency on the image distance can be understood according to acoustic ray tracing. Here, since the magnitude of the effective acoustic index of the phononic crystal decreases as the frequency increases, the image appears farther from the lens exit surface for higher frequencies [see (4.9)]. This trend, confirmed by experiments as well as simulations (see Fig. 4.39a) shows the high sensitivity of the image location to changes in frequency. Here, tuning the frequency from 523 to 555 kHz shifts the image from 2.6 to 5.75 mm. A change over 6 % in the frequency is able to tune the focal spot distance over 220 %.

4.5.5.2 Distance from the Source to the Lens

Here, we consider the effects of the position of the source with respect to the phononic crystal surface. Super resolution requires coupling of the evanescent waves from the source to bound modes in the phononic crystal in order to achieve

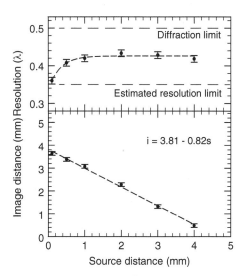

Fig. 4.40 Plot of the resolution and image distance as a function of the source distance

amplification and re-emission. This process is thus only possible if the phononic crystal lies in the near field of the point source at a distance where evanescent components are not too much attenuated. In terms of sizes, this means that distance from source to lens and the period of the phononic crystal (lattice constant) are comparable in magnitude. A question arises whether or not the detailed heterogeneous structure of the phononic crystal can be ignored and replaced by a continuous model of a negative index material. This question was addressed from a numerical point of view by varying the distance between the point source and the surface at the optimum frequency of 530 kHz. The measured effects are the position of the source with respect to the exit face and the lateral resolution of the focus as shown on Fig. 4.40. Indeed, if the phononic crystal can be modeled by a homogeneous negative index material slab, geometrical ray tracing implies that the distance from lens to focus is described by (4.11). The dashed horizontal lines represent the Rayleigh diffraction limit (0.5λ) and the estimated maximum resolution limit (0.34λ) calculated in Sect. 4.5. It results that as the distance between the source and the face of the lens is increased, excitation of the bound modes is less and less effective and the resolution decreases. For this range of image distances, the resolution remains smaller than the Rayleigh diffraction limit. The fact that this limit is not reached on the plot is related to the close distances which range from 0.036λ to 1.4λ. The lens is always in the near field of the source for the studied range. One expects that for larger distances the resolution will reach the Rayleigh diffraction limit, accompanied by loss of super-resolution. It should be noted that the source cannot be placed farther than one lens thickness from the lens itself in order to get a real image. Thus, to observe the complete loss of super-resolution would require to significantly increase the lens thickness as well as the source distance which is demanding for a computational point of view.

The Green's function model described in Sect. 4.5.2 shows that if all evanescent and propagating modes are contributing the image is perfectly reconstructed as a point source at a distance $d-s$ from the exit face of the lens. This position is in accordance with geometric rays tracing in two media with opposite refraction indices. It could be shown that if the two media had some low acoustic index mismatch or if the lens media had uniaxial anisotropy in normal incidence axis direction [32], the relation would still be linear. This linear behavior is indeed observed thanks to simulation data on Fig. 4.40 where the focus location fits a linear relation with a slope of -0.82 with respect to source location. However, the intercept of this curve is not exactly the thickness of the lens ($d = 6.52$ mm), as expected from (4.11). We have seen that the operating frequency could change dramatically the focus location since it defines the effective index of the phononic crystal. Here the results are presented at the frequency of 530 kHz which is not the exact value of the AANR condition when an index of $n = -1$ is achieved. The value for 530 kHz is rather $n = -1.07$. Acoustic ray tracing predicts that if n is the effective index of the phononic crystal relative to water, then the focus position for a source placed very close to the lens is $d/|n|$. This would predict an intercept at 6.09 mm, still far from the observed value. Thus, the frequency effect over index alone is insufficient to explain completely the discrepancy. This discrepancy is therefore most probably due to the fact that the assumption of a homogeneous negative medium is poorly valid in the case of a phononic crystal because of the similar length scales between the lattice parameter, lens thickness, wavelength and the source distance. At least, it is less valid than in the case of metamaterial slabs [42] where the resonant inclusions have sizes well below the wavelength.

4.5.5.3 Geometry of the Phononic Crystal Lens

The geometry of the lens itself has been studied in terms of its effects on resolution and the location of the image. The respective effects of the thickness and width of the phononic crystal lens are discussed successively. The width of the lens has been studied from the experimental and computational point of views. The picture of a semi infinite slab (in the x_1 and x_2) directions used for the Green's function model is quite different in the context of simulations and experiments where the width of the lens is measured along x_1 by the number of rod inclusions in each layer parallel to the surface. The question raised by the limited width of the lens is similar to what is called aperture in the context of optics. The spatially limited transmission due to the finite extent of a lens is responsible for a loss of resolution due to the convolution of any image by an Airy function. Thus, a sufficient width has to be chosen so that this limitation is low enough in order to demonstrate the super resolution effect. Sukhovich et al. [5] have used lenses of 15, 31 and 61 rods per layer in crystals made of 6 layers, all other parameters being constant. The behavior of the lenses with 31 and 61 rods per layer are similar and suitable to exhibit super resolution. The position of the image and resolution as a function of the position of the source

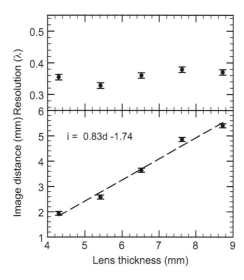

Fig. 4.41 Effects of the number of layers. Lateral resolution and distance of the image as a function of the number of layers

(Fig. 4.41) were almost identical. By contrast, the results for the narrower 15 rods per layer are significantly different. This effect was attributed to the small aspect ratio (2.5) of this lens inducing significant distortions. For lenses wider than 31 rods, the aspect ratio is greater than 5 and does not affect the results.

For what concerns the thickness of the lens, it can be varied by changing the number of layers of inclusions. Robillard et al. [37] simulated thicknesses of 4, 5, 6, 7 and 8 layers for the case with a width of 31 rods per layer. The distance from the source to the surface was maintained at 0.1 mm and the corresponding results are shown in Fig. 4.41. It follows that, within the range of measurement error, the resolution does not change with width as expected. This fact is also confirmed by the authors by the existence of similar bound modes in the vicinity of the operating frequency whatever the lens thickness. The frequencies of the bound modes that are responsible for super-resolution do not vary significantly as the thickness changes. Last point, always according to the ray tracing and Green's models, the distance of the image is expected to be linearly dependent on the lens thickness. This fact is observed as well but the fitted value of this slope is not one, as expected in the case of a homogeneous negative medium, but 0.83. As discussed earlier in this paragraph, the lens made of an effective homogeneous medium may not be a valid hypothesis in these conditions. Again, the discrepancy between the slope of 0.83 compared to one indicates the thickness mismatch between effective homogeneous slabs and phononic crystal slabs [42].

4.5.5.4 Location of the Source in the Direction Parallel to the Lens

The position of the source in a direction parallel to the slab input face plays a role that is linked to the amplification mechanism of evanescent components from the source. Necessary couplings with bound modes of the phononic crystal slab and near field proximity implies that this mechanism is sensitive the heterogeneous structure of the phononic crystal. Especially, efficient coupling requires that displacement fields of the bound modes and evanescent waves overlap in space. Since the lens excitation exhibits high pressure lobes in front of each steel cylinder when super resolution is achieved, it is assumed that the bound modes involved have similar displacement patterns. Thus, by shifting the source in a direction parallel to the slab the efficiency of the couplings is expected to change and result in modification of the super resolution effect. This process was simulated by a source facing the gap midway between two cylinders of the phononic crystal. In this case, the resolution falls to 0.54λ as can be seen by the wider focus on Fig. 4.42a, b.

Experiments confirm these results are in accordance with experimental results; moving the source parallel to the surface from the position opposite a cylinder (best resolution) by only a quarter of its diameter caused the image resolution to degrade from 0.37λ to 0.47λ.

Thus, looking at Fig. 4.42a gives an understanding of the bound modes displacement. The pressure exhibits lobes of maximum amplitude between cylinders and consequently the displacement amplitude would show maxima in front of each cylinder and nodes between them. Placing the source at any of the nodes of the displacement field prevents evanescent waves from coupling efficiently with the bound modes.

4.5.5.5 Disorder

The properties of Phononic Crystals rely on the coherent summation of the Bragg scattered components of acoustic waves on the successive planes of the crystal. Because of this coherent character, any deviation from perfect order inside the crystal structure is expected to introduce diffusion effects that are detrimental to imaging properties. Especially, the super-resolution effect that is described in this section should be sensitive to such defects. This hypothesis has been verified from both the experimental and numerical point of view [11]. Figure. 4.42c, d show FDTD results that assume some random deviation in the rods position from the perfect triangular lattice configuration. This positional disorder in the numerical model has a standard deviation of 5 %, which corresponds to an upper limit for the experimental crystal. The experimental measurements were found to be very sensitive to disorder in the position of the steel rods. These results confirm in that disorder in the phononic crystal is detrimental to the quality of the image and for some random realizations can even eliminate the focusing property of the lens.

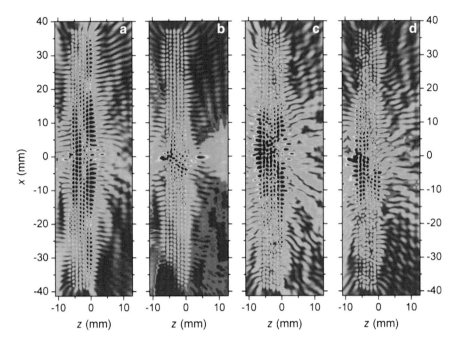

Fig. 4.42 Influence of the location of the source in the direction parallel to the lens and of the disorder after [11]. The contour maps of the normalized average absolute value of pressure calculated via FDTD at a frequency of 530 kHz for the phononic crystal lens imaging are plotted. (a) The line source is located at 0.1 mm from the left lens surface and centered with respect to a surface cylinder at $x = 0$. (b) Same simulation as (a) but with the source shifted down by $a/2$ in the direction parallel to the surface of the lens. (c) and (d) show two lenses with positional disorder of the steel rods showing imperfect focusing (c) and loss of focusing (d)

4.6   Band Structure Design and Impact on Refraction

As shown before, 2D and 3D phononic crystals have been extensively studied and implemented for their frequency dependent (ω-space) effects on sound or elastic wave propagation. Especially, absolute band gaps have led to a variety of guiding, confinement and filtering designs. The astonishing demonstration of sound tunneling is also related to the presence of band gaps. On the other hand, negative bands and the subsequent negative refraction that occurs at the interface of some phononic crystals and the surrounding media is a property related to the shape of the EquiFrequency Contour (EFC) of the dispersion curves in the wave-vectors plane (k-space). For the purpose of achieving super resolution imaging with a phononic crystal lens, one has to design a phononic crystal with circular EFCs. These two effects, band gaps and all-angle negative refraction, have received much attention from the community since the first reports on sonic crystals. However, as expected from the behavior of elastic waves in genuine crystals, a wider variety of properties should result from the periodic arrangement of phononic crystals constituents. The propagation of waves is always fully understandable by means of the dispersion relations, i.e. the ω and k-spaces. Since dispersion curves are determined by geometrical (sizes, symmetry) and material (stiffness, density) parameters, phononic crystals can be designed in order to exhibit advanced spectral (ω) and directional (k) properties based on the analysis of the dispersion relations. In this section we show how the design, especially the symmetry, of a phononic crystal, can lead to strongly anisotropic effects such as positive, negative and even zero angle refraction at a single frequency. Other effects such as collimation, beam splitting and phase controlling are also predicted. Eventually, we discuss the opportunity to control the respective phases between different acoustic beams (φ-space) and its possible implementation on acoustic logic gates.

4.6.1 Square Equifrequency Contours in a PVC/Air Phononic Crystal

In 2009, Bucay et al. [43] have described theoretically and computationally the properties of a phononic crystal made of polyvinylchloride (PVC) cylinders arranged as a square lattice embedded in a host air matrix. We will develop this section of Chap. 4 from the properties of this representative system. This PVC/air system exhibits an absolute band gap in the 4–10 kHz range followed by a band exhibiting negative refraction. The band structure for the infinite periodic phononic crystal is generated by the Plane Wave Expansion (PWE) method and plotted in Fig. 4.43b. In the 13.5 kHz equifrequency plane, the second negative band defines a contour of nearly square shape centered on the M point of the first Brillouin zone. This shape appears clearly in Fig. 4.43c which shows a contour map of the dispersion surface taken between frequency values 13.0 and 16.0 kHz extended to several Brillouin zones. Though the properties of such an arrangement can be reproduced in other systems of suitable symmetry and material parameters, we describe here the parameters used in that particular demonstration. The spacing between the cylinders (lattice parameter) is $a = 27$ mm and the radius of the inclusions is $r = 12.9$ mm. The PVC/Air system parameters are: $\rho_{PVC} = 1364$ kg/m^3, $c_{t,PVC} = 1000$ m/s, $c_{l,PVC} = 2230$ m/s, $\rho_{Air} = 1.3$ kg/m^3, $c_{t,Air} = 0$ m/s, and $c_{l,Air} = 340$ m/s (ρ is density, c_t is transverse speed of sound, and c_l is longitudinal speed of sound). The PVC cylinders are considered as infinitely rigid and of infinite height. This assumption of rigidity simplifies the band structure calculation and is justified by a large contrast in density and speed of sound between the solid inclusions and the matrix medium. Again, the results gathered from this analysis are applicable to other solid/air phononic crystals of the same filling fraction because, in reference to other solids, air has extremely small characteristic acoustic impedance.

Bucay et al. [43] have focused on the consequences on acoustic propagation in the passing bands with such square shaped EFCs. Here we summarize these effects and their possible applications in acoustic imaging and information processing. The next paragraphs use the schematic of Fig. 4.43a on which a PVC/air phononic crystal slab is surrounded by air. This schematic corresponds to the FDTD simulation space. One

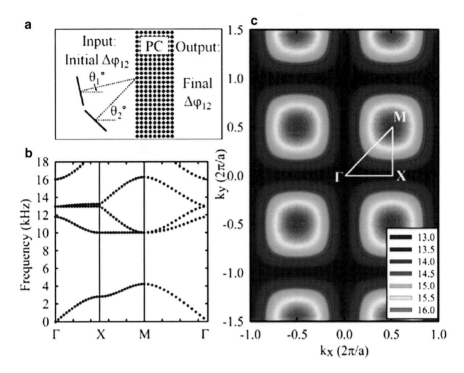

Fig. 4.43 Schematic and band structure of the PVC/air system after Bucay et al. [183]. (**a**) Schematic illustration of the FDTD simulation cell. The acoustic sources can assume any incident angle to the phononic crystal face and be set with any relative phase difference. (**b**) Band structure generated by PWE method along the edges of the first Brillouin zone (pictured in (**c**)). (**c**) EFCs (extended zone scheme of irreducible Brillouin zone) in range of 13.0–16.0 kHz

or several beams impinge on the input side. Each source on the input side of the simulation space is modeled by a slanted line of grid points consistent with the desired incidence angle of the source. The nodes along this line are displaced in a direction orthogonal to the source line as a harmonic function of time. These sources can assume any incident angle to the phononic crystal face and can be ascribed any relative phase difference, thus allowing for complete analysis of the phononic crystal wave vector space (k-space) and phase-space (φ-space). The output side is reserved for the detection of exiting acoustic signals.

4.6.2 Positive, Zero, and Negative Angle Refraction, Self-Collimation

First, looking at the EFC contour at a given frequency of 13.5 kHz, it appears that the square symmetry of the phononic lattice has a strong impact on the band structure (Fig. 4.44). Indeed, while at very low frequencies the dispersion relations are linear (low frequency parts of the acoustic branches), the higher order branches

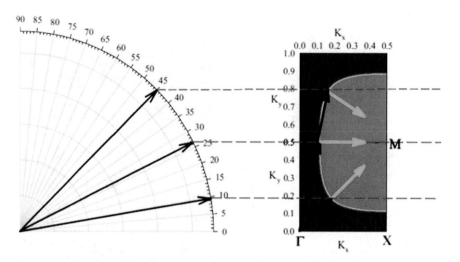

Fig. 4.44 Determination of the refraction angles of several incident beams from the EFC of the PVC/air system

considered at the frequency of 13.5 kHz have direction dependent properties. These k-dependent properties appear themselves in the almost square shape of the equifrequency contour. The equivalent media formed by the PVC/air has to be considered as anisotropic. This particular EFC is plotted in Fig. 4.44 along with the EFC in air at the same frequency. The EFC of the PVC/air system has been extended over another Brillouin zone in the K_y direction on this plot in order to exhibit one complete face of the square which is centered on the M point. Since the surrounding medium is linear and isotropic in the operating frequency range its EFC is simply circular. Let us now discuss the different cases of the beam refraction induced by the unusual shape of the EFCs.

In order to clearly describe these cases, we remind the reader how the wave vector and group velocity of a refracted beam is determined from the angle of an incident beam.

The conservation of frequency and parallel to surface ($k_{//}$) component of wave vector is required. These rules are written in Eqs.4.21 and 4.22 where the subscripts i and r stand for *incident* and *refracted*.

$$\omega_i = \omega_r, \tag{4.21}$$

$$\overrightarrow{k_{//i}} = \overrightarrow{k_{//r}} + \overrightarrow{G}, \tag{4.22}$$

The presence of a vector $\overrightarrow{G}$ of the reciprocal lattice will be discussed later, in the non-periodic media it is a zero vector. In other words, the normal component of the wave vector $k_\perp$ is determined such that the wave vector $k_r = k_{\perp r} + k_{//r}$ in the second medium matches a dispersion curve at the frequency ω_i. If such a matching

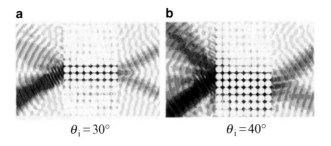

$\theta_i = 30°$ $\theta_i = 40°$

Fig. 4.45 Zero refraction and negative refraction occur at the same frequency, in the same PVC/air system depending only on the incidence angle. The incident beams are oriented upward at angles of (**a**) 30° and (**b**) 40° respectively

point exists, a refracted beam exists, otherwise the incident beam undergoes total reflection. The couple (k_r, ω_r) defines a point of the Brillouin zone at which the group velocity can be determined by (4.7). It must be noted that, contrary to the case of an isotropic media, the wave vector and group velocity might not be collinear in the general case. This can be seen in Fig. 4.44 where the wave vectors are depicted by black arrows and the group velocity vectors by blue arrows.

From these rules and Fig. 4.44 it follows that any beam that impinges on the phononic crystal with an incidence angles lower that 5° cannot couple to any propagation mode of the phononic crystal and thus will be completely reflected. This can be seen as a directional band gap. Between 5° and 55° waves are refracted and propagate in the phononic crystal and several cases are distinguished. Below 28°, refracted waves have a group velocity vector (blue arrows) with a positive parallel (K_y) component. They undergo classical positive refraction. At the singular angle of 28°, the contour is flat in the K_y direction such that the group velocity will be perfectly oriented toward the x axis. Such behavior corresponds to a zero angle of refraction and is quite unusual. An illustration of this phenomenon is shown in Fig. 4.45a with a FDTD result of the averaged pressure field. An incident beam at 30° is oriented toward the surface of a PVC/air crystal slab. Since the incidence angle is very close to the predicted zero refraction angle (28°) it is refracted and the beam follows a path close to the x axis.

In Fig. 4.45b, a beam with higher incidence angle is negatively refracted, in accordance with the previous discussion. The ability of this system to achieve positive, negative and zero angle refraction at a single frequency has been successfully tested experimentally and theoretically by Bucay et al. [43]. One should note that the vicinity of the 28° incidence angle coincides with small degrees of refraction. One could define an incidence range that gives rise to refracted angles reasonably close to zero. As an example, for incidence angle between 20° and 30° the angle of refraction is within in the −2° to 2° range. Thus, from this point of view this system is able to combine a wide angle input wave into a nearly collimated beam. This ability called *self-collimation* is pretty unusual and could have significant uses in the field of acoustic imaging. The discussed system can also enable the propagation in the same volume of the phononic crystal of two non-collinear incident beams. This spatial overlapping of two waves carrying non-identical signals offers interferences conditions that might be useful for information processing as we shall see later.

4.6.3 Beam Splitting

Another striking property of such a system is the presence of two output beams as seen on Fig. 4.45. The incident beam impinges from the bottom part of the simulation cell. The upper beam on the input side is a partial reflection. On the exit side, the beam splits into balanced parts. This phenomenon, confirmed experimentally [43], is striking since Snell's law of refraction does not account for such behavior. Optical analogues of such an effect are birefringent crystals which discriminate light into several beams with respect to its polarization or beam splitters that share incident energy into two output beams. Again, this analogy does not account for the radically different origins of this effect in optics and acoustic phononic crystals. Indeed, while optic beam splitters take advantage of balanced transmission and reflection coefficients by means of suitable surface coatings, the phononic crystal beam splitter produces two identical refracted beams, that both have propagated through the phononic crystal following the same path. In the latter case, the splitting effect relies only on the properties of wave coupling between periodic (phononic crystal) and homogeneous (air) media. Potential applications of this spontaneous beam splitting effects are discussed in the following sections. Here we describe its origins.

The schemes in Fig. 4.46 show the equifrequency planes in a system composed of a phononic crystal slab similar to the PVC/Air system immersed in a fluid medium (air). The plot extends over two Brillouin zones. The operating frequency is 13.5 kHz, which corresponds to a square EFC of the phononic crystal. Note that the circular EFC in air is larger than the first Brillouin zone of the phononic crystal. Let us now apply coupling rules for an incident wave to propagate inside the phononic crystal. In (4.22) we have introduced an additional vector $\vec{G}$ that belongs to the reciprocal lattice. Indeed, in crystalline structures as in any periodic structure the momentum conservation can be satisfied modulo a certain vector $\vec{G}$. This conservation rule for sonic waves is analogous to the one governing phonon diffusion in solids [44]. The processes which involve a zero G vector are called natural processes. They ensure complete conservation of the crystal momentum, while non-zero G vector processes (Umklapp) ensure momentum conservation due to the contribution of the crystal total momentum. The latter involve a wave vector outside of the first Brillouin zone. From this rule follows that for a given incidence angle, the incident beam can couple to several modes inside the phononic crystal. The wave vectors of these modes lie in distinct Brillouin zones. Since the extent of the EFC in air is twice as large as the Brillouin zone, two of $k_{//}$ are possible for the propagation into the phononic crystal. On the output side these two different modes

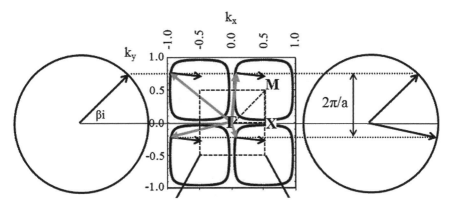

Fig. 4.46 Determination of the refraction angles of several incident beams from the EFC of the PVC/air system. The central scheme depicts an extended zone EFC contour of the phononic crystal. *Gray arrows* are wavevectors while *black arrows* are the group velocity vector

couple back to the surrounding media according to the same rules, which account for the presence of two beams.

The remarkable property of the multiple modes inside the phononic crystal is that they have similar group velocity vectors (black arrows) but different wavevectors (gray arrows). It results that they will only split on the output side but share the exact same path inside the crystal.

Additionally, Fig. 4.46 shows that, on the opposite side of the zero incidence line, another beam might couple with exactly the same set of wave vectors inside the crystal. Then, two beams can produce exactly the same effects and are called complementary. Complementary waves will have incidence angles $\theta_0 + \Delta\theta$ and $-\theta_0 + \Delta\theta$ with θ_0 being the zero-refraction angle.

4.6.4   Phase Control

Except for the case of complementary incident waves, any couple of incident beams will be refracted at different refraction angles and thus accumulate a certain phase difference while propagating through the crystal. One should remark that, here again, refracted waves in the phononic crystal have somewhat uncommon properties since their group velocity is nearly parallel to the normal to the crystal/air interface (Fig. 4.46) while their k-vector, has a wide range of possible orientations due to the incidence angle. Group velocity and wave vector being non-collinear simply means that energy and phase propagates in different directions. In the vicinity of the zero angle refraction, a wide span of Bloch waves exists with group velocities that coincide with small degrees of refraction, allowing refraction to occur between propagating waves within the nearly same volume of crystal. This is shown through the high slope around $0°$ in Fig. 4.47a which represents the angle of incidence of the input beam as a function of the angle of refraction in the bulk of the phononic crystal slab.

A fine analysis of the square EFC shows that, while the group velocity of different refracted beams have nearly the same zero angle of refraction, their wave vectors quite different. Since group velocity describes the propagation of the energy while the wave vector k is related to the propagation of phase, this fact shows that beams propagating in close directions in the phononic crystal might accumulate significantly different phase shifts.

To investigate this effect, Swinteck and Bringuier [45–47] have calculated the phase shift accumulated per unit length of a phononic crystal slab as a function of the incidence angle. Two impinging waves with wave vectors $\overrightarrow{k_1}$ (angle θ_1) and $\overrightarrow{k_2}$ (angle θ_2) excite several Bloch modes throughout the k-space of the phononic crystal. As seen in the *beam-splitting* effect, because the extent of the first Brillouin zone is smaller than the circular EFC in the surrounding media, each incident wave will couple to two Bloch modes that correspond to complementary waves. These two Bloch modes are noted $\overrightarrow{k_{1A}}$ and $\overrightarrow{k_{1B}}$ in Fig. 4.48a and are necessary to describe the wave physics in this phononic crystal in terms of phase. Each of these wave vector pairs has a unique refraction angle noted as α_1 and α_2. The following calculations will focus on the phase shift accumulated between Bloch modes $\overrightarrow{k_{1A}}$ and $\overrightarrow{k_{2A}}$ only (noted $\varphi_{1A,2A}$), though similar discussion would lead to compatible results for the second pair of modes. These two Bloch wave vectors are expressed as:

$$\overrightarrow{k_{1A}} = \frac{2\pi}{a}\left\{k_{1x}\vec{i} + k_{1y}\vec{j}\right\} \qquad (4.23)$$

$$\overrightarrow{k_{2A}} = \frac{2\pi}{a}\left\{k_{2x}\vec{i} + k_{2y}\vec{j}\right\} \qquad (4.24)$$

where k_{1x} and k_{1y} are the components of the wave vector $\overrightarrow{k_{1A}}$ and k_{2x} and k_{2y} are the components of the wave vector $\overrightarrow{k_{2A}}$ (in units of $2\pi/a$). $\vec{i}$ and $\vec{j}$ are unit vectors along axes x and y respectively.

Each incident beam $\vec{k}$ is refracted by an angle α and travels in the phononic crystal along a path that is simply:

$$\vec{r} = L\vec{i} + L\tan(\alpha)\vec{j} \qquad (4.25)$$

where L is the slab thickness. The phase accumulated at the exit face of the slab with respect to the input point is:

$$\varphi = \vec{k}\cdot\vec{r} = \frac{2\pi L}{a}\left\{k_x + \tan(\alpha)k_y\right\} \qquad (4.26)$$

It follows that the phase difference between Bloch modes with wave vectors $\overrightarrow{k_{1A}}$ and $\overrightarrow{k_{2A}}$ can be expressed as:

$$\varphi_{1A,2A} = \overrightarrow{k_{1A}} \cdot \overrightarrow{r_1} - \overrightarrow{k_{2A}} \cdot \overrightarrow{r_2} = \frac{2\pi L}{a}\left\{ k_{1x} + \tan(\alpha_1)k_{1y} - k_{2x} - \tan(\alpha_2)k_{2y} \right\} \quad (4.27)$$

Let us formulate a few remarks about this result. First, to evaluate this phase shift it is useful to plot it as a function of the incidence angle θ_1 of one input beam the other beam being a constant reference beam. The angle (28.1°) for which zero while angle refraction occurs is a preferred choice. Second, as expected, the result depends linearly on the thickness of the slab. Third, computing this phase shift can be done by extracting the components, (k_{1x}, k_{1y}) and (k_{2x}, k_{2y}), used in (4.27) from the EFC data in Fig. 4.45. Finally, the calculated phase shift per unit length is plotted in Fig. 4.47b along with FDTD results that agree very well with the above analysis. Looking closely at (4.27), one understands that the phase shift has two origins. First, the travel paths inside the phononic crystal for the both waves are different ($\overrightarrow{r_1} \neq \overrightarrow{r_2}$). The second effect comes from the difference in phase velocities ($\left|\overrightarrow{k_{1A}}\right| \neq \left|\overrightarrow{k_{2A}}\right|$). Waves of different phase velocities traveling different paths certainly will develop a phase shift. From Fig. 4.47b one can deduce the phase difference between a pair of beams which is of crucial importance since it determines how exiting beams interfere. It is worth noting that the steel/methanol system described in Sect. 4.5 exhibits, at the considered operating frequencies, circular EFCs centered on the Γ point. In such a configuration phase and group velocity are collinear and anti-parallel. Such a system wouldn't produce substantial phase shifts between two Bloch modes that are nearly collinear.

Figure 4.48b shows that outgoing beams intersect each other on the output side in two points. These points are places where the relative phase between two beams can be found by measure of the interference state. The choice of the two incidence angles higher and lower than 28.1 (the zero angle of refraction) is important. Indeed it ensures that one beam is refracted positively and the other one negatively, while forming the intersection points on the exit side. In the end, the incidence angles of the two beams determine wave vectors $\overrightarrow{k_1}$ and $\overrightarrow{k_2}$ and the angles of refraction α_1 and α_2 which give the phase shift. Therefore incidence angle selection is proposed as a leverage to modulate the relative phase between propagating acoustic beams.

4.6.5 Implementation of Acoustic Logic Gates

More recently, it has been proposed to use these interference effects to implement an acoustic equivalent of the so-called Boolean logic gates [47] on the basis of phase control by means of a phononic crystal slab. Here we discuss the example of the NAND gate which is identified as universal since the implementation of any other Boolean logic gate is feasible by associating several NAND gates [48]. The NAND gate is a two inputs function which truth table is described in Fig. 4.49a. The setup of Bringuier et al. [47] relies on a phononic crystal slab and two permanent sources S1

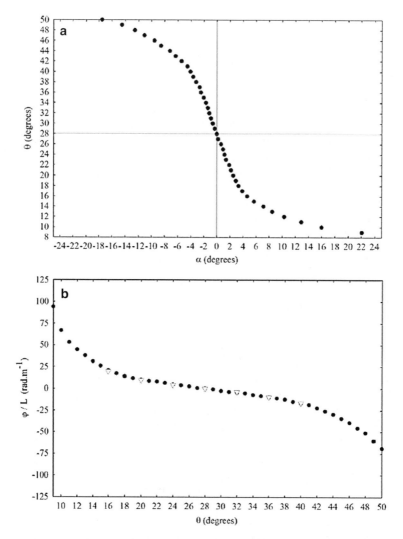

Fig. 4.47 (**a**) Angle of the incident beam as a function of its refraction angle. The graph can be read as follows: one obtains a 0° refracted beam inside the phononic crystal when the incidence angle is 28°. (**b**) Phase shift per unit length of phononic crystal as a function of the incidence angle. The phase shift is evaluated with respect to a zero refracted beam (*Circles*: analytical solution. *Triangles*: results from FDTD calculation)

and S2 impinging at the same point of the input face. The angles of incidence are such that these beams are not complementary waves, i.e., their paths do not perfectly overlap in the phononic crystal slab. The following demonstration is based on FDTD simulations on the PVC/air system described above. In this scheme it is straightforward to keep a given phase relation between the two permanent sources S1 and S2. In this particular case, they they impinge in-phase on the input side of the phononic

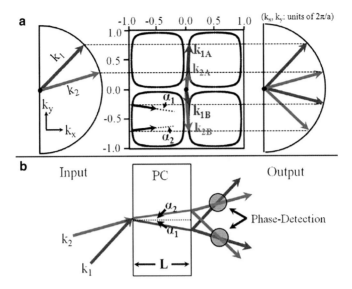

Fig. 4.48 (**a**) (*k*-space) Bloch modes excited in the PVC/air system by two waves having different angles of incidence. (**b**) (real space) paths of the corresponding waves in the phononic crystal slab. After Swinteck et al. [46]

crystal. Because their incidence angles are $10°$ and $38°$, the two sources will refract negatively and positively in accordance with the Fig. 4.48b. The phase shift on the output side is calculated thanks to (4.27) and is evaluated to be 2π radians. This results in constructive interference on the output side between the centers of the exiting beams. At this particular point where the interferences are constructive, a detector D is positioned. This "detector" simply indicates that the averaged pressure is recorded over a given cut which makes an angle $24°$ (i.e., in between $10°$ and $38°$). The corresponding pressure profile is presented on the left side of Fig. 4.49b. The position of the constructive interference point is indicated by a vertical dashed line which, indeed, corresponds to a maximum of the pressure. This state describes the zero inputs state of the NAND gates. In this regime the continuous high level of pressure is interpreted as a 1 output from the gate.

The authors model the inputs of the NAND by two additional beams I1 and I2 which are the corresponding complementary waves ($19°$ and $50°$) to the sources, S1and S2, respectively. As compared to the permanent sources, I1 and I2 are set such that their phases are π radians on the input side. It results from this condition that whenever I1 is turned on, it perfectly overlaps the path of S1 in the phononic crystal (because these are complementary waves) and since their phase difference is π, they interfere destructively. It results that only S2 contributes to the averaged pressure at the detector point as shown on Fig. 4.49c. The same analysis holds if I1 is off and I2 is on. The last case corresponds to having both inputs emitting waves simultaneously. In this case S1 and I1 as well as S2 and I2 interfere destructively

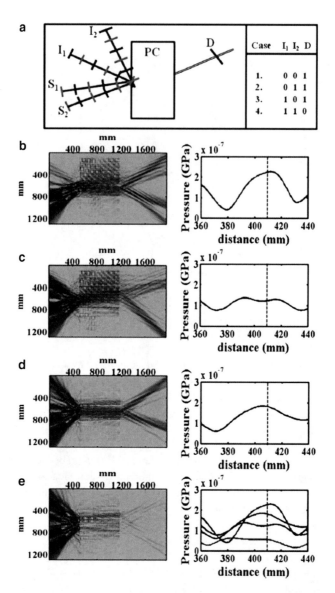

Fig. 4.49 Implementation of the NAND gate with phononic crystals. The system consists of a phononic crystal with the same square EFC characteristics as in the PVC/Air system and two permanent sources S_1 and S_2 are incident at different angles

and this case exhibits the minimal pressure at the detector point among all other cases.

The situation when "I1 is emitting" (or "I1 is not emitting") means that the first input of the NAND gate is at state 1 (or state 0). By establishing a threshold value just above the minimal pressure, the output is defined to be in state 1 if the pressure is above the threshold and in state 0 if the pressure is below the threshold. Finally, the only configuration that produces a 0 output state is the state with I_1 and I_2 both emitting waves. This complies with the truth table of the NAND gate. This study demonstrates another possible application of the full dispersion properties (frequency, wave vector and phase) of phononic crystals in the field of information processing.

4.7 Conclusion

In this chapter, we have focused on 2D and 3D phononic crystals and their unusual properties. After having introduced the necessary concepts of Bravais lattices and their corresponding Brillouin zones, we have summarized how phononic crystals properties can be investigated experimentally especially in the ultrasonic frequency range. The discussion then focused on spectral aspects of phononic crystals. The existence of band gaps is the first property of phononic crystals investigated theoretically and experimentally. Because of the evanescent character of waves whose frequency falls into the band gaps, tunneling of sound has been demonstrated. However, band gaps despite the wealth of applications they bring (sound isolation, wave guiding, resonators, filtering...) are not the only striking phenomena in phononic crystals. Other phenomena observed in the passing bands have been studied in details such as negative refraction. Negative refraction occurs when the wave vector and the group velocity are anti-parallel in a material. The similarities between negative refraction and the negative index metamaterials have been discussed. This chapter also provides a wealth of details about experimental conditions of negative refraction. Later sections have focused on the conditions required to use negative refraction in combination with close field coupling to a phononic crystal slab in order to achieve super-resolution, i.e., imaging a source point with a better than half-wavelength resolution. Finally, we have briefly described recent developments about the impact of the phononic crystal symmetry on refraction properties. A model system exhibiting anisotropic propagation properties has been described by its refraction properties as a function of their incidence angles. This type of system has been demonstrated in the context of self-collimation, beam-splitting, phase controlling and a possible implementation of logic gates.

Throughout the chapter it has been shown that, despite the variety of possible implementations of phononic crystals, their properties can always be described in the frame of Bragg reflections of the acoustic or elastic waves that interfere constructively or destructively. The consequences of periodicity manifest themselves in the dispersion relations that *fully* describe the spectral, directional and phase properties of propagation in phononic structures. From this point of view the analogy between phononic crystals and natural crystalline material is complete. It follows that, the complete spectrum of opportunities offered by periodic artificial structures is extremely large and still not fully explored.

References

1. Z. Liu, X. Zhang, Y. Mao, Y.Y. Zhu, Z. Yang, C.T. Chan, P. Sheng, Locally resonant sonic materials. Science **289**, 1734 (2000)
2. P. Sheng, X. Zhang, Z. Liu, C.T. Chan, Locally resonant sonic materials. Physica B **338**, 201 (2003)
3. J.H. Page, S. Yang, M.L. Cowan, Z. Liu, C.T. Chan, P. Sheng, 3D phononic crystals, in *Wave Scattering in Complex Media: From Theory to Applications*, (Kluwer Academic Publishers: NATO Science Series, Amsterdam, 2003) pp. 283–307
4. A. Sukhovich, J.H. Page, B. van Tiggelen, Z. Liu. Resonant tunneling of ultrasound in three-dimensional phononic crystals. *Physics in Canada* **60**(4), 245 (2004)
5. A. Sukhovich, L. Jing, J.H. Page, Negative refraction and focusing of ultrasound in two-dimensional phononic crystals. Phys. Rev. B **77**, 014301 (2008)
6. S. Yang, J.H. Page, Z. Liu, M.L. Cowan, C.T. Chan, P. Sheng, Ultrasound tunnelling through 3D phononic crystals. Phys. Rev. Lett. **88**, 104301 (2002)
7. S. Yang, J.H. Page, Z. Liu, M.L. Cowan, C.T. Chan, P. Sheng, Focusing of sound in a 3D phononic crystal. Phys. Rev. Lett. **93**, 024301 (2004)
8. F. Van Der Biest, A. Sukhovich, A. Tourin, J.H. Page, B.A. van Tiggelen, Z. Liu, M. Fink, Resonant tunneling of acoustic waves through a double barrier consisting of two phononic crystals. Europhys. Lett. **71**(1), 63–69 (2005)
9. J.H. Page, A. Sukhovich, S. Yang, M.L. Cowan, F. Van Der Biest, A. Tourin, M. Fink, Z. Liu, C.T. Chan, P. Sheng, Phononic crystals. Phys. Stat. Sol. (b) **241**(15), 3454 (2004)
10. J.H. Page, P. Sheng, H.P. Schriemer, I. Jones, X. Jing, D.A. Weitz, Group velocity in strongly scattering media. Science **271**, 634 (1996)
11. A. Sukhovich, B. Merheb, K. Muralidharan, J.O. Vasseur, Y. Pennec, P.A. Deymier, J.H. Page, Experimental and theoretical evidence for subwavelength imaging in phononic crystals. Phys. Rev. Lett. **102**, 154301 (2009)
12. Zeitschrift fur Kristallographie, vol, **220**, issues 9–10, ed. by I.E. Psarobas, pp. 757–911 (2005)
13. T.E. Hartman, Tunneling of a wave packet. J. Appl. Phys. **33**, 3427 (1962)
14. J.H. Page, S. Yang, Z. Liu, M.L. Cowan, C.T. Chan, P. Sheng, Tunneling and dispersion in 3D phononic crystals. Z. Kristallogr. **220**, 859–870 (2005)
15. M. Sigalas, M.S. Kushwaha, E.N. Economou, M. Kafesaki, I.E. Psarobas, W. Steurer, Classical vibrational modes in phononic lattices: theory and experiment. Z. Kristallogr. **220**, 765–809 (2005)
16. J. Liu, L. Ye, D.A. Weitz, P. Sheng, Phys. Rev. Lett. **65**, 2602 (1990)
17. X.D. Jing, P. Sheng, M.Y. Zhou, Phys. Rev. Lett. **66**, 1240 (1991)
18. R.S. Penciu, H. Kriegs, G. Petekidis, G. Fytas, E.N. Economou, J. Chem. Phys. **118**, 5224 (2003)
19. G. Tommaseo, G. Petekidis, W. Steffen, G. Fytas, A.B.. Schofield, N. Stefanou, J. Chem. Phys. **126**, 014707 (2007)
20. M.L. Cowan, J.H. Page, Ping Sheng, Phys. Rev. B **84**, 094305 (2011)
21. J. Li, C.T. Chan, Double-negative acoustic metamaterial. Phys. Rev. E **70**, 055602(R) (2004)
22. T. Still, W. Cheng, M. Retsch, R. Sainidou, J. Wang, U. Jonas, N. Stefanou, G. Fytas, Phys. Rev. Lett. **100**, 194301 (2008)
23. R. Sainidou, N. Stefanou, A. Modinos, Formation of absolute frequency gaps in three-dimensional solid phononic crystals. Phys. Rev. B **66**, 212301 (2002)
24. H. Zhao, Y. Liu, G. Wang, J. Wen, D. Yu, X. Han, X. Wen, Phys. Rev. B **72**, 012301 (2005)
25. V. Leroy, A. Bretagne, M. Fink, H. Willaime, P. Tabeling, A. Tourin, Appl. Phys. Lett. **95**, 171904 (2009)
26. C. Croënne, E.J.S. Lee, Hefei Hu, J.H. Page, Special Issue on Phononics, ed. by Ihab El Kady and Mahmoud I Hussein, in AIP Advances, **1**(4), (2011)
27. A.V. Akimov et al., Phys. Rev. Lett. **101**, 33902 (2008)
28. J.A. Turner, M.E. Chambers, R.L. Weaver, Acustica **84**, 628–631 (1998)

29. H. Hu, A. Strybulevych, J.H. Page, S.E. Skipetrov, B.A. van Tiggelen, Nature Phys. **4**, 945–948 (2008)
30. V.G. Veselago, The electrodynamics of substances with simultaneously negative values of ε and μ. Usp. Fiz. Nauk **92**, 517 (1964)
31. J.B. Pendry, Negative Refraction Makes a Perfect Lens. Phys. Rev. Lett. **85**, 3966 (2000)
32. P.A. Deymier, B. Merheb, J.O. Vasseur, A. Sukhovich, J.H. Page, Focusing of acoustic waves by flat lenses made from negatively refracting two-dimensional phononic crystals. Revista Mexicana De Fisica **54**, 74 (2008)
33. J. Li, K.H. Fung, Z.Y. Liu, P. Sheng, C.T. Chan, in *Physics of Negative Refraction and Negative Index Materials*, Springer Series in Materials Science, vol. 98, Chap. 8 (Springer, Berlin, 2007)
34. P. Sheng, J. Mei, Z. Liu, W. Wen, Dynamic mass density and acoustic metamaterials. Physica B **394**, 256 (2007)
35. X. Zhang, Z. Liu, Negative refraction of acoustic waves in two-dimensional phononic crystals. Appl. Phys. Lett. **85**, 341 (2004)
36. K. Imamura, S. Tamura, Negative refraction of phonons and acoustic lensing effect of a crystalline slab. Physical Review B **70**, 174308 (2004)
37. J.-F. Robillard, J. Bucay, P.A. Deymier, A. Shelke, K. Muralidharan, B. Merheb, J.O. Vasseur, A. Sukhovich, J.H. Page, Resolution limit of a phononic crystal superlens. Physical Review B **83**, 224301 (2011)
38. Z. He, X. Li, J. Mei, Z. Liu, Improving imaging resolution of a phononic crystal lens by employing acoustic surface waves. Journal of Applied Physics **106**, 026105 (2009)
39. F. Liu, F. Cai, S. Peng, R. Hao, M. Ke, Z. Liu, Parallel acoustic near-field microscope: A steel slab with a periodic array of slits. Physical Review E **80**, 026603 (2009)
40. P. Sheng, Metamaterials: Acoustic lenses to shout about. Nature Materials **8**, 928 (2009)
41. C. Luo, S.G. Johnson, J.D. Joannopoulos, J.B. Pendry, Subwavelength imaging in photonic crystals. Phys. Rev. B **68**, 045115 (2003)
42. V. Fokin, M. Ambati, C. Sun, X. Zhang, Method for retrieving effective properties of locally resonant acoustic metamaterials. Physical Review B **76**, 144302 (2007)
43. J. Bucay, E. Roussel, J.O. Vasseur, P.A. Deymier, A.-C. Hladky-Hennion, Y. Pennec, K. Muralidharan, B. Djafari-Rouhani, B. Dubus, Positive, negative, zero refraction, and beam splitting in a solid/air phononic crystal: theoretical and experimental study. Phys. Rev. B **79**, 214305 (2009)
44. G.P. Srivastava, *The Physics of Phonons*. (A. Hilger, 1990)
45. N. Swinteck, J.-F. Robillard, S. Bringuier, J. Bucay, K. Muralidharan, J.O. Vasseur, K. Runge, P.A. Deymier, Phase-controlling phononic crystal. Appl. Phys. Lett. **98**, 103508 (2011)
46. N. Swinteck, S. Bringuier, J.-F. Robillard, J.O. Vasseur, A.C. Hladky-Hennion, K. Runge, P.A. Deymier, Phase-control in two-dimensional phononic crystals. J. Appl. Phys. **110**, 074507 (2011)
47. S. Bringuier, N. Swinteck, J.O. Vasseur, J.-F. Robillard, K. Runge, K. Muralidharan, P.A. Deymier, Phase-controlling phononic crystals: Realization of acoustic Boolean logic gates. The Journal of the Acoustical Society of America **130**, 1919 (2011)
48. A.P. Godse, D.A. Godse, *Digital Logic Circuits* (Technical Publications Pune, Pune, India, 2010)
49. A.R. Selfridge, Approximate Material Properties in Isotropic Materials, *Transactions on Sonics and Ultrasonics* SU-32, 381 (1985)

Chapter 5
Dynamic Mass Density and Acoustic Metamaterials

Jun Mei, Guancong Ma, Min Yang, Jason Yang, and Ping Sheng

Abstract Elastic and electromagnetic waves are two types of classical waves that, though very different, nevertheless display many analogous features. In particular, for the acoustic waves, there can be a correspondence between the two material parameters of the acoustic wave equation, the mass density and bulk modulus, with the dielectric constant and magnetic permeability of the Maxwell equations. We show that the classical mass density, a quantity that is often regarded as positive definite in value, can display complex finite-frequency characteristics for a composite that comprises local resonators, thereby leading to acoustic metamaterials in exact analogy with the electromagnetic metamaterials. In particular, we demonstrate that through the anti-resonance mechanism, a locally resonant sonic material is capable of totally reflecting low-frequency sound at a frequency where the effective dynamic mass density can approach positive and negative infinities. The condition that leads to the anti-resonance thereby offers a physical explanation of the metamaterial characteristics for both the membrane resonator and the 3D locally resonant sonic materials. Besides the metamaterials arising from the dynamic mass density behavior at finite frequencies, we also present a review of other relevant types of acoustic metamaterials. At the zero-frequency limit, i.e., in the absence of resonances, the dynamic mass density for the fluid–solid composites is shown to still differ significantly from the usual volume-averaged expression. We offer both a physical explanation and a rigorous mathematical derivation of the dynamic mass density in this case.

J. Mei • G. Ma • M. Yang • J. Yang • P. Sheng (✉)
Department of Physics, The Hong Kong University of Science and Technology, Clear Water Bay, Kowloon, Hong Kong, China
e-mail: sheng@ust.hk

P.A. Deymier (ed.), *Acoustic Metamaterials and Phononic Crystals*,
Springer Series in Solid-State Sciences 173, DOI 10.1007/978-3-642-31232-8_5,
© Springer-Verlag Berlin Heidelberg 2013

5.1   Introduction

The novel characteristics of metamaterials represent an emergent phenomenon in which the basic mechanism of resonances, when considered in aggregate, can give rise to material properties that are outside the realm provided by Nature. In the case of acoustic metamaterials, the novel characteristics directly arise from the finite-frequency behavior of the two relevant material parameters—the mass density and bulk modulus. The focus of this chapter is on the dynamic mass density and its related metamaterial characteristics. For completeness, a brief review of other types of acoustic metamaterials is also presented.

It is well known that in the quantum mechanical band theory of solids, the effective mass of an electron can change sign depending on its energy within an energy band. However, as this is attributed to the electron's wave character, the classical mass density is usually regarded as a positive-definite quantity since the quantum mechanical effects are absent. In particular, for a two-component composite, the effective mass density is usually given by the volume-averaged value:

$$\rho_{\mathrm{eff}} = fD_1 + (1 - f)D_2, \tag{5.1}$$

where $D_{1(2)}$ denotes the mass density of the 1st (2nd) component, and f is the volume fraction of component 1. We denote the static mass density (5.1) ρ_{eff}.

An implicit assumption underlying the validity of the static mass density expression is that in the presence of wave motion, the two components of the composite *move in unison*. However, this assumption is not always true. For a composite comprising many identical local resonators embedded in a matrix material, if the local resonators' masses move out of phase with the matrix displacement (as when the wave frequency ω exceeds the resonance frequency of the resonators), then we have a case in which the matrix and the resonators' masses display *relative motion*. If, in addition, we assume that the local resonators occupy a significant volume fraction, then it is clear that within a particular frequency range, the overall effective mass density can appear to be negative [1–6]. This fact can be simply illustrated in a one-dimensional (1D) model [7, 8], where n cylindrical cavities of length d are embedded in a bar of rigid material. Within each cavity, a sphere of mass m is attached to the cavity wall by two identical springs with elastic constant K. An external force F acts on the rigid bar, which has a static mass M_0, as shown in Fig. 5.1.

For the first resonator, the displacements of the sphere and the right wall are denoted by u and U, respectively (Fig. 5.1). By assuming that $-f_1$ and $-f_2$ are the forces on the sphere exerted by the left and right springs, respectively, with f_2 along the same direction as F, and f_1 the opposite, then Hook's law tells us that $-f_1 + f_2 = -2K(U - u)$. From Newton's second law, we have $f_1 - f_2 = (-i\omega)^2 mu$. From these two relations, we obtain $u = \frac{2K}{2K - m\omega^2} U$. Applying Newton's second law to the rigid bar, we have $F + n(f_2 - f_1) = (-i\omega)^2 M_0 U$. Hence $F = (-i\omega)^2 [M_0 U + nmu] = (-i\omega)^2 (D_{\mathrm{eff}} V) U$. Here the effective *dynamic mass density* D_{eff} is defined as $F/(-\omega^2 U)$:

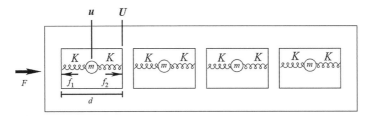

Fig. 5.1 A one-dimensional acoustic metamaterial composed of a series of local resonators embedded in a rigid bar. Here the directions of f_1 and f_2 are shown as that on the left and right walls of the cavity, respectively. Adapted from [7]

$$D_{\text{eff}}V = M_0 + nm\frac{u}{U} = M_0 + \frac{nm}{1 - (\omega^2/\omega_0^2)}, \qquad (5.2)$$

where $\omega_0^2 = 2K/m$ and V denotes the total volume of the system. Thus *negative* dynamic mass is possible at finite frequencies (when ω^2 is in the range of ω_0^2), and this phenomenon enables the realization of acoustic metamaterials. Equation (5.2) is also informative in showing that the dynamic mass density is generally defined as the averaged force density f divided by the averaged acceleration a, i.e.,

$$D_{\text{eff}} = \langle f \rangle / \langle a \rangle, \qquad (5.3)$$

where $\langle \rangle$ denotes averaging over interfaces with the external region of the observer. Obviously, this is precisely how (5.2) is obtained. The above simple example serves to illustrate the point that the dynamic mass density, in the presence of relative motion between the components, can differ from the volume-averaged static mass density. In more realistic models in which the matrix is an elastic medium, it will be shown below that the dynamic mass density's resonance-like behavior is directly associated with the *anti-resonance(s)* of the system.

In the limit of $\omega \to 0$ so that resonances can be excluded, the volume-averaged mass density holds true for most composites. However, the fluid–solid composites constitute an important exception. A well-known example is the fourth sound of liquid helium 4 in a porous medium [9], which arises from the *relative* motion between the liquid helium 4 and the solid frame—even at the low-frequency limit. More generally, it is well known that for a fluid–solid composite, there is a viscous boundary layer thickness $\ell_{\text{vis}} = \sqrt{\eta/\rho_f\omega}$ at the fluid–solid interface, where η denotes the fluid viscosity and ρ_f the fluid density. It is clear from the definition of ℓ_{vis} that the $\eta \to 0$ limit cannot be interchanged with the $\omega \to 0$ limit since in the former case $\ell_{\text{vis}} \to 0$ whereas in the latter case we have $\ell_{\text{vis}} \to \infty$. Thus the Biot slow wave, predicted as a second longitudinal wave in a fluid–solid composite [10] and eventually experimentally verified [11], may be viewed as a "fourth sound" for the viscous fluid, valid when the pore size ℓ of the porous medium is larger than ℓ_{vis} [12]. Thus the dynamic mass density of a fluid–solid composite is what governs the wave propagation when the dimensionless ratio $\sqrt{\eta/\rho_f\omega}/\ell \ll 1$.

In what follows, we describe in Sect. 5.2 the initial realization of acoustic metamaterials based on the concept of local resonators and their special characteristics. In particular, it is shown that such metamaterials can break the mass density law, which governs air-borne sound attenuation through a solid wall. This is followed by the presentation of the membrane-type metamaterials in Sect. 5.3 that may be regarded as the two-dimensional (2D) version of resonant sonic materials. The unifying characteristic of the anti-resonance and negative dynamic mass density is emphasized in both Sects. 5.2 and 5.3. In Sect. 5.4, we give a brief review of other types of acoustic metamaterials that have since been realized. Section 5.5 is devoted to the dynamic mass density in the low-frequency limit (for the fluid–solid composites), prefaced by a short review of the multiple-scattering theory (MST). We conclude in Sect. 5.6 with a brief summary and some remarks on the prospects and challenges.

5.2 Locally Resonant Sonic Materials: A Metamaterial Based on the Dynamic Mass Density Effects

In Fig. 5.2a we show a cross-sectional photo image of the basic unit for the locally resonant sonic material [1]. It comprises a metallic sphere 5 mm in radius coated by a layer of silicone rubber. Figure 5.2b is a picture showing a cube assembled from these basic units with epoxy, in a simple cubic structure with a lattice constant of 1.55 cm. It is clear that the metallic sphere of the basic unit acts as a heavy mass, with silicone rubber as the weak spring. Hence there must be a low-frequency resonance. Moreover, the resonance is local in character, to be distinguished from the structural resonances that are common to any mechanical object. Figure 5.2c, d show the transmission characteristics and band structure of the crystal shown in Fig. 5.2b, respectively. It is noted that there is a deep transmission dip at 380 Hz, followed by a transmission maximum at 610 Hz. This pattern is repeated at 1,340 Hz and 1,580 Hz. Here the solid line is the theory prediction calculated from the MST, and the solid circles are the measured data. They show good agreement. In Fig. 5.2d, the calculated band structure is shown. The flat band edges, at 380 Hz and 1,340 Hz, are characteristic of local (anti-)resonances that are very weakly coupled to each other.

It is seen that the structure shown in Fig. 5.2b has a complete bandgap between 380 Hz and 610 Hz. In contrast to phononic crystals where the relevant wavelength corresponding to the primary bandgap frequency must be comparable to the lattice constant, here the wavelength (in epoxy) at 380 Hz is ~300 times the lattice constant. That is, the locally resonant sonic materials can open phononic gaps at frequencies that are much lower than that derived from considerations of their structural length scales. In fact, since the effect is due to local resonances, and these resonances depend only on the rubber's elastic constants and metal sphere's mass, the bandgap frequency should be totally decoupled from structural considerations.

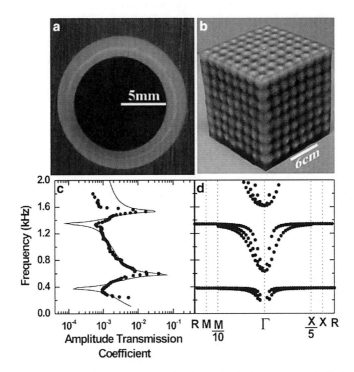

Fig. 5.2 (a) Cross section of a coated sphere that forms the basic structure unit (**b**) for an 8 × 8 ×8 sonic crystal. (**c**) Calculated (*solid line*) and measured (*circles*) amplitude transmission coefficients along the [100] direction are plotted as a function of frequency. The calculation is for a four-layer slab of simple cubic arrangement of coated spheres, periodic parallel to the slab. The observed transmission characteristics correspond well with the calculated band structure (**d**), from 200 to 2,000 Hz, of a simple cubic structure of coated spheres. Figure adapted from [1]

The fact that the locally resonant sonic materials can have bandgaps may be simply explained by using analogy with the tight binding approach for the electronic structure calculations, in which the starting point is the discrete electronic energy levels in individual atoms. Our local resonances also have a discrete spectrum. When the atoms interact with each other (through the hopping matrix element in the tight binding formulation), the discrete energy levels broaden into energy bands. If the interaction is weak, the bands may not completely overlap and what remain are exactly the bandgaps. Moreover, the band edges are usually flat just as what we see in Fig. 5.2d. From this analogy, it is plausible that since periodicity plays only an implicit role in the tight binding approach, it may not be a necessary requirement for the creation of bandgaps. Hence it was shown by Weaire [13] that in tetrahedrally bonded system (such as the amorphous silicon), the existence of bandgaps indeed does not require long-range periodic order. This is another aspect that differs from phononic crystals, in which the bandgap is the result of Bragg scattering.

Below we present the novel functionality of the locally resonant sonic material together with its relevant physics. It will be seen that the dynamic mass density behavior of the system naturally emerges as the dominant cause of its special characteristics.

5.2.1 Metamaterial Functionality

In Fig. 5.2c, it is seen that at 380 Hz, the locally resonant sonic material can have a sharp minimum in transmission. In order to appreciate the significance of this phenomenon, it is necessary to first review the law of acoustic attenuation by a solid wall, usually denoted the *mass density law*.

Consider a sound wave in air with angular frequency ω impinging normally on a solid wall of thickness d, mass density ρ_2 and bulk modulus κ_2. Sound transmission amplitude is given by

$$T = \frac{4v \exp(ik_2d)}{(1+v)^2 - (1-v)^2 \exp(2ik_2d)}, \tag{5.4}$$

where $k_2 = \omega/\sqrt{\kappa_2/\rho_2}$ is the wavevector in solid and $v = \sqrt{\kappa_2\rho_2/\kappa_1\rho_1}$ is the solid–air impedance ratio, with κ_1 and ρ_1 denoting the bulk modulus and mass density of air, respectively. For solid walls that are less than a meter in thickness, which is usually the case, we have $k_2d \ll 1$ and $v \gg 1$ for frequencies less than 1 kHz. In that limit, an accurate approximation to (5.4) is given by

$$T \cong i\frac{2\sqrt{\rho_1\kappa_1}}{\omega\rho_2d}. \tag{5.5}$$

It is seen that the bulk modulus of the wall does not appear in (5.5). That is, to a high degree of accuracy, the sound attenuation through a solid wall is independent of whether the wall is rigid or soft. Only the wall's mass per unit area (ρ_2d) matters. That is why (5.5) is called the mass density law. But perhaps the most important aspect of (5.5) is that T is inversely proportional to the sound frequency. Hence low-frequency sound is inherently difficult to attenuate. This is the reason why low-frequency noise is such a pernicious source of urban environmental pollution.

In Fig. 5.3, we plot the measured amplitude transmission coefficient (solid circles with the connecting solid line) for a 2.1-cm slab of composite material containing 48 vol% of *randomly dispersed* coated metal spheres (same as the one whose cross-sectional picture is shown in Fig. 5.2a) in an epoxy matrix. As a reference, the measured amplitude transmission coefficient through a 2.1-cm slab of epoxy is also plotted (open squares connected by thin solid line). The *dashed* and *dot-dashed lines*, respectively, show the calculated transmission amplitudes of a 2.1-cm epoxy slab and a 2.1-cm homogeneous slab of the same density as that of the

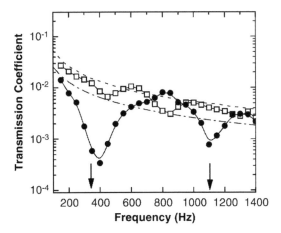

Fig. 5.3 Measured amplitude transmission (solid circles; the solid line is a guide to the eye) through a 2.1-cm slab of composite material containing 48 vol% of randomly dispersed coated lead spheres in an epoxy matrix. As a reference, the measured amplitude transmission through a 2.1-cm slab of epoxy is also plotted (*open squares* connected by a *thin solid line*). The dashed and dot-dashed lines, respectively, show the calculated transmission amplitudes of a 2.1-cm epoxy slab and a 2.1-cm homogeneous slab of the same density as that of the composite material containing the coated spheres. Adapted from [1]

composite material containing the coated spheres. The arrows indicate the dip frequency positions predicted by the multiple-scattering calculation for a mono-layer of hexagonally arranged coated spheres in an epoxy matrix.

In Fig. 5.3, the comparison between the measured results for the composite slab and the mass density predictions shows clearly that the locally resonant sonic materials can break the mass density law at particular low-frequency regimes, thereby exhibiting acoustic metamaterial characteristics.

5.2.2 Theoretical Understanding

In order to gain an understanding of the metamaterial functionality, we have performed finite-element simulations by using the COMSOL Multiphysics. In the simulations, the mass density, Young's modulus, and Poisson's ratio for the lead sphere are $11.6 \times 10^3 \text{kg/m}^3$, $4.08 \times 10^{10}\text{Pa}$, and 0.37, respectively. The mass density, Young's modulus, and Poisson's ratio for the silicone rubber are 1.3×10^3 kg/m^3, $1.18 \times 10^5\text{Pa}$, and 0.469, respectively. Corresponding parameters for epoxy are $1.18 \times 10^3\text{kg/m}^3$, $4.35 \times 10^9\text{Pa}$, and 0.368, respectively. Standard values for air, i.e., $\rho = 1.23\,\text{kg/m}^3$, ambient pressure of 1 atm, and speed of sound in air of $c = 340\,\text{m/s}$, were used. Two types of simulations were performed.

We first calculate the spectrum of transmission coefficients for a plane wave normally incident onto one unit cell along the z-direction. Periodic boundary conditions along the x- and y-directions were used. Radiation boundary conditions

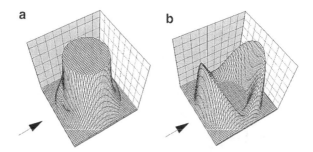

Fig. 5.4 Calculated displacement configurations around the first (**a**) and second (**b**) peak frequencies. The displacement show is for a cross section through the center of one coated sphere, located at the front surface. The *arrows* indicate the direction of the incident wave. Adapted from [1]

were used at the input and output planes of the air domain in the simulations. Two transmission peaks, with frequencies at 606 Hz and 1,576 Hz, were found. We also found two transmission dips, at 374 and 1,339 Hz.

We have also calculated the eigenmodes for one unit cell. Many eigenmodes were found. Out of these, we select the ones that are symmetric with respect to both the x- and y-directions, since otherwise the modes would not couple to the normally incident plane wave. The resulting triply degenerate eigenfrequencies are located at 606 and 1,571 Hz, respectively. They are seen to be *almost identical with the frequencies of the transmission peaks*.

In Fig. 5.4a, we show the calculated displacement configurations around the first peak frequency, where the lead sphere is seen to move as a whole along the direction of wave propagation. Around the second peak, the maximum displacement occurs inside the silicone rubber, as shown in Fig. 5.4b. In Fig. 5.5, we show the calculated strain tensor components ε_{xz} and ε_{yz} at the first and second dip frequencies, respectively. It can be seen that strains occur at the lead–rubber and/or the rubber–epoxy interfaces, which in fact can also be inferred from the displacement configurations as shown in Fig. 5.4. Below we show that the *dip frequencies correspond to anti-resonances* where the dynamic mass density displays a resonance-like behavior.

Figure 5.6 displays the calculated dynamic mass density D_{eff} for one unit cell of the locally resonant sonic material. Around 370 and 1,340 Hz, i.e., the transmission dip frequencies, the dynamic mass density $D_{\text{eff}} = \langle \nabla \cdot \sigma \rangle_z / \langle a_z \rangle$ clearly displays a resonance-like behavior. Thus the transmission peaks correspond with the eigenfrequencies, and the dips in the transmission are associated with anti-resonances at which we have a dynamic mass density resonance profile. In particular, it is shown below that at the anti-resonance frequencies, the average normal displacement of the unit cell surface (in the matrix material) vanishes, hence $\langle a_z \rangle = -\omega^2 \langle u_z \rangle$ goes through a zero and therefore it is easy to see that D_{eff} acquires a resonance-like behavior, with a diverging magnitude at the anti-resonance frequency. In a sense, the mass density law seems to recover its validity–but only if its value replaces the static mass density.

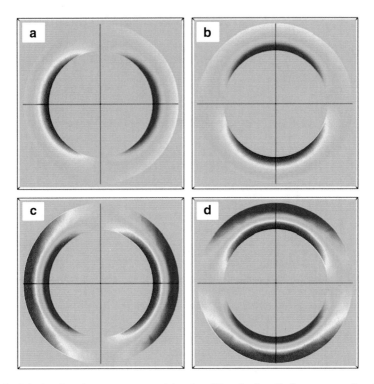

Fig. 5.5 Calculated strain components ε_{xz} (**a**) and ε_{yz} (**b**) at the first dip frequency, and ε_{xz} (**c**) and ε_{yz} (**d**) at the second dip frequency, within the $z = 0$ cross section plane within one unit cell. *Red* and *blue colors* denote positive and negative values of strain components, respectively, and *green* indicates near-zero strain

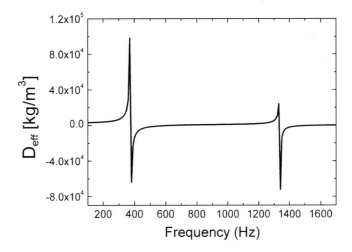

Fig. 5.6 Dynamic effective mass density D_{eff} for one unit cell of the local resonant sonic material as shown in Fig. 5.2. Around the anti-resonance frequencies (transmission dip frequencies), resonant behavior of D_{eff} is evident

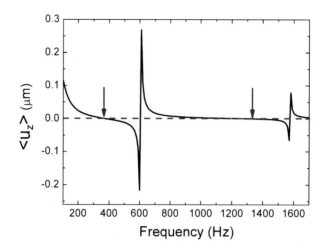

Fig. 5.7 Averaged normal surface displacement $\langle u_z \rangle$ for one unit cell of the locally resonant sonic material when a plane wave is incident along the z-direction. Large $\langle u_z \rangle$ amplitude corresponds with the transmission peak. Around the transmission dip frequency (lower side of the transmission peak frequency), $\langle u_z \rangle$ passes through zero (indicated by the *red arrows*), thereby leading to the divergence of D_{eff} as shown in Fig. 5.6

5.2.3 Physical Underpinning of the Anti-resonances

Mechanical anti-resonances constitute a very common phenomenon [14]. They are also of practical importance in mechanical systems. For example, the change in frequencies of anti-resonances can be an indicator of structural damages [15, 16]; it is also an element that needs to be taken into account in the design and modeling of the cantilever for atomic force microscopes [17–19].

By focusing on the surface normal displacement of the mechanical system, it is possible to appreciate the physical underpinning of this phenomenon. That is, an anti-resonance always occurs between two resonances. At the anti-resonance frequency, the two neighboring resonances are *simultaneously* excited but with the opposite phase, since the resonance response is given by $1/(\omega_i^2 - \omega^2)$, with ω_i denoting the angular frequency of the ith resonance and $\omega_i < \omega < \omega_{i+1}$. As the two eigenfunctions are spatially orthogonal to each other, it is possible to demonstrate that in varying the frequency continuously from ω_i to ω_{i+1}, there must be a point at which the averaged normal surface displacement is zero. In Fig. 5.7, we show the averaged normal surface displacement $\langle u_z \rangle$ at a unit cell when the incident wave is along the z-direction. It can be seen that $\langle u_z \rangle$ passes through zero at around the transmission dip frequencies, and that is the underlying mechanism of the divergence of $D_{\text{eff}} = \langle \nabla \cdot \sigma \rangle_z / (-\omega^2 \langle u_z \rangle)$ in the relevant frequency regime. It therefore follows that the dynamic mass density must have a resonant behavior at anti-resonance, giving rise to total reflection of the acoustic waves. It is also seen that

$\langle u_z \rangle$ exhibits divergent behavior at the eigenmode frequencies where the peak transmissions occur.

The understanding that the dynamic mass density's behavior–as the underlying cause of the anti-resonances–offers the possibility of generalization of this principle to the regime of ultrasound and even optical phonons. However, such experimental manifestations at high frequencies are still to be pursued.

5.3 Membrane-Type Acoustic Metamaterials

The metamaterial functionality of the locally resonant sonic materials operates only in a limited range of frequencies. Such a disadvantage can be overcome if there are membrane-type locally resonant sonic materials since one may be able to stack these membranes, each operative at a different frequency regime, so as to broaden the effective frequency range of the stacked sample.

However, making a membrane-type acoustic metamaterial that can totally reflect the low-frequency sound may seem to be anti-intuitive at first sight because a total-reflecting surface is usually a node, implying no displacement. However, a membrane is generally soft and elastically weak, hence difficult to have zero movement. But what we shall show, both theoretically and experimentally, is that precisely because of its weak elastic moduli, even a small membrane can have multiple low-frequency resonances. As there can be an anti-resonance between two resonances, it follows that the average normal displacement of the membrane vanishes at the anti-resonance frequency, thereby causing a resonant behavior of the dynamic mass density together with a diverging magnitude at the anti-resonance frequency. Total reflection occurs as a result.

It should be noted, however, that even though the average normal displacement is zero, the membrane displacement is *not* everywhere zero. But such nonzero displacement couples only to non-radiating evanescent waves, which can be ignored as far as the far-field transmission and reflection are concerned.

Below we give a detailed account of this simple system.

5.3.1 Sample Construct

In Fig. 5.8, we show our sample to consist of a circular rubber membrane decorated with a small button of varying mass (at the center of the membrane) for the purpose of tuning the eigenfrequencies [20]. These decorated membranes are assembled into a larger plate. The measurement setup, illustrated in the top panel, comprises two Brüel and Kjaer type-4206 impedance tubes with a sample sandwiched in between. The front tube has a loudspeaker at one end to generate a plane wave. There are two sensors in the front tube to sense the incident and reflected waves. The third sensor in the back tube, terminated with a 25-cm-thick anechoic sponge

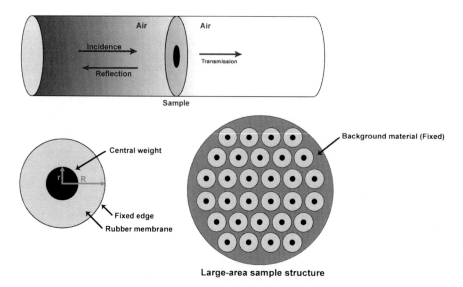

Fig. 5.8 Typical sample structure of the membrane-type acoustic metamaterial (*bottom panels*) and the testing geometry (*upper panel*)

(enough to minimize reflection), senses the transmitted wave. The signals from the three sensors are sufficient to resolve the transmitted and reflected wave amplitudes, in conjunction with their phases.

5.3.2 Vibrational Eigenfunctions and the Anti-resonance Phenomenon

In Fig. 5.9a, c, we show the finite-element COMSOL simulation results on the vibrational eigenmodes of a button-decorated rubber membrane. Here the circular button has a radius of 4.5 mm and a mass of 160 mg, and the rubber membrane is 28 mm in diameter and 0.2 mm in thickness. The mass density, Young's modulus, and Poisson's ratio for the rubber are 980 kg/m^3, 2×10^5 Pa, and 0.49, respectively. A radial pre-stress, on the order of 10^5 Pa, has been applied to the membrane. The two lowest-frequency eigenmodes are shown. It is seen that for the lowest frequency eigenmode, at 250 Hz (Fig. 5.9a), the button and the membrane (on which it is attached) move in unison. However, for the mode at ~1,050 Hz (Fig. 5.9c), the button's oscillation amplitude is small whereas the surrounding rubber's oscillation amplitude is fairly significant. Figure 5.9b shows the profile at the anti-resonance frequency. It should be noted that in contrast to the 3D locally resonant sonic materials (see 5.2), in which the resonance and anti-resonance frequencies are closely grouped together, for the membrane-type acoustic metamaterials the resonance and anti-resonance frequencies are well-separated.

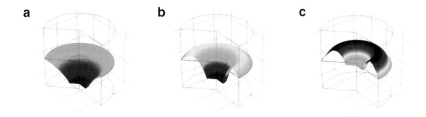

Fig. 5.9 The first eigenmode (**a**) and the second eigenmode (**c**). The profile at the dip frequency is shown in (**b**)

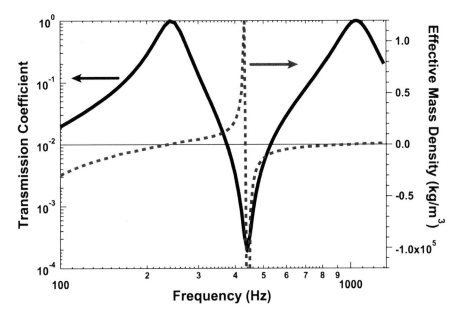

Fig. 5.10 The effective dynamic mass of the membrane-type acoustic metamaterial (*red symbols*, right axis), together with the transmission coefficient (*black solid curve*, left axis), evaluated with an incident wave with pressure modulation amplitude of 1 Pa

In Fig. 5.10, it is shown that each of the transmission peaks corresponds with an eigenmode of the system. Between the two eigenfrequencies, there is clearly a sharp dip in transmission. At this dip frequency (~440 Hz), both eigenmodes are excited, but with opposite phase. Their superposition leads to the mode profile shown in Fig. 5.9b. A closer examination of this transmission dip configuration shows that the averaged normal displacement of the mode is accurately zero. The dynamic mass density, defined as

$$D_{\text{eff}} = -\langle \sigma_{zz} \rangle / (h\langle a_z \rangle) = \langle \sigma_{zz} \rangle / (\omega^2 h \langle w \rangle), \tag{5.6}$$

displays a resonance-like behavior in which D_{eff} has a divergent magnitude precisely at the anti-resonance frequency, as shown in Fig. 5.10. Here σ_{zz} denotes the zz component of the stress tensor, z being the direction normal to the membrane surface, a_z is the acceleration along the z-direction, equal to $-\omega^2 w$ for time-harmonic motions, with w being the normal displacement of the membrane and h being the thickness of the membrane. In accordance with the principle of the mass density law, if one allows the dynamic mass density to play the role of the static mass density, then total reflection should occur. However, a more accurate picture for explaining the total reflection phenomenon is as follows.

5.3.3   Anti-resonance and the Non-radiating Evanescent Mode

Consider the dispersion relation for the acoustic wave in air, $k_\parallel^2 + k_\perp^2 = \omega^2/v^2 = (2\pi/\lambda)^2$, where $\bar{k}_\parallel, k_\perp$ denote the wave vector components parallel or perpendicular to the surface of the membrane, respectively, $v = 340$ m/s is the speed of sound in air, and λ is the wavelength. At the air–membrane interface, we note that the normal displacement (which is usually sub-micron in magnitude and hence small compared to the membrane thickness) pattern of the membrane can be fully described by using 2D Fourier components of $\bar{k}_\parallel$. If we decompose the normal displacement w into an area-averaged component and a component of whatever is left over, i.e., $w = <w> + \delta w$, then it should be clear that their respective Fourier components' magnitudes should have a distribution, illustrated schematically in Fig. 5.11. Here d denotes the lateral size of the membrane. Since d is usually much smaller than the wavelength λ, it follows that for the δw part of the displacement, the overwhelming majority of the $\bar{k}_\parallel$ components will have magnitudes $|k_\parallel| \geq 2\pi/d \gg 2\pi/\lambda$. Hence from the dispersion relation, it follows that the associated $k_\perp^2 < 0$. That is, the δw part of the displacement can only cause evanescent waves. In contrast, for the $\langle w \rangle$ part of the normal displacement, the distribution of the $|k_\parallel|$ must be peaked at zero, owing to its piston-like motion. Thus again from the dispersion relation, the associated $k_\perp^2 \sim (2\pi/\lambda)^2$. It follows that only the average component of the normal displacement can affect far-field transmission. If $\langle w \rangle = 0$, then there can be no far-field transmission. We therefore arrive at the conclusion that total reflection is the necessary consequence of the membrane status at the anti-resonance frequency.

However, even at the anti-resonance frequency, the membrane is not stationary. Figure 5.12 displays the finite-element COMSOL simulation result at the anti-resonance frequency. It indicates evanescent waves being emitted, with a decay length on the order of a millimeter. This fact distinguishes a membrane reflector from its rigid (and heavy) wall counterpart.

In Fig. 5.10, it should be noted that before the first resonance, D_{eff} is negative with a decreasing trend (toward negative infinity) as the frequency approaches zero. This would seem to contradict the common intuition that D_{eff} should reduce to the

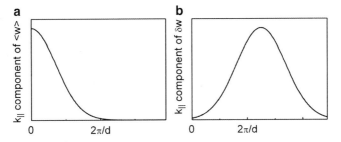

Fig. 5.11 The parallel Fourier components' distribution for (**a**) $\langle w \rangle$ and (**b**) δw components, respectively. For (**b**), the peak of the distribution lies higher than $2\pi/d$ because the feature sizes for the δw component must be smaller than d

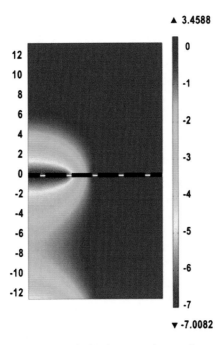

Fig. 5.12 The normal velocity field distribution near the membrane at the transmission dip frequency, where the *black dashed line* denotes the position of membrane plane. The left axis (which is also the symmetry axis) is in units of millimeter, while the velocity is in μm/s (calculated with the same incident wave intensity as that for Fig. 5.10). The wave is incident from the bottom. The decay characteristic near the two sides of the membrane surfaces indicates a decay length of 3 mm

volume-averaged value in the static limit. The fact that it does not do so in the present case is due to two factors. First, the divergent magnitude is a reflection of the boundary condition. Since the boundary of the membrane is fixed, the membrane essentially transfers its load onto the fixed boundary in the long wavelength

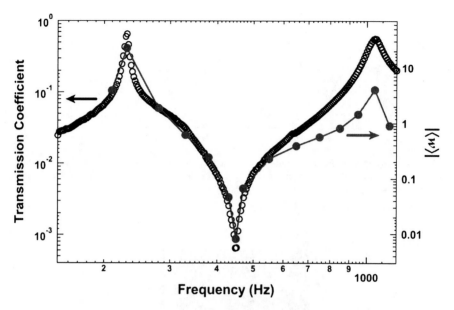

Fig. 5.13 The *black open circles* are the measured transmission coefficient (*left axis*), and the *red solid circles* are the LDV-measured $|\langle w \rangle|$ (*right axis*, arbitrary unit). The *red line* is to guide the eye. A clear correlation is seen

limit. That means the fixed boundary can also be interpreted as a piece of very heavy mass. Second, the negative sign of D_{eff} signifying off-phase response to the external force, is a reflection of Newton's third law—the reaction is opposite to the applied force. Such behavior of D_{eff}, also referred to as the "Drude-type negative mass density" in analogy to free electrons in metal, has been studied in different structures [21, 22].

5.3.4 Experimental Verification

Experimentally, we have used laser Doppler vibrometer (LDV) to directly verify the $\langle w \rangle = 0$ condition at the transmission minimum frequency. The amplitude transmission spectrum of the membrane-type metamaterial system was also measured. Both show very good agreement with the predictions of finite-element COMSOL simulations.

In Fig. 5.13 the correlation between the transmission coefficient and $|\langle w \rangle|$ is clearly demonstrated. In Fig. 5.14, we give a detailed comparison between the measured normal displacement profiles and the COMSOL simulation results on

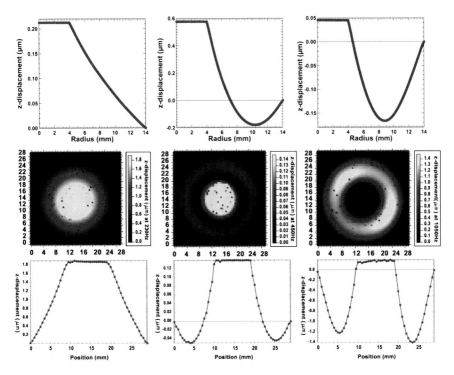

Fig. 5.14 The calculated (*upper panel*) and measured (*middle* and *lower panels*) normal displacement profiles on the two eigenmodes (left and right columns) and the anti-resonance mode (central column). The frequencies of the three profiles are (from left to right) around 230 Hz, 450 Hz, and 1,050 Hz. Displacement profiles are measured with ~0.25 Pa incident wave amplitude. Note that the simulation results (*top panels*) are only half of the experimental profiles (*bottom panels*), since the simulation results are symmetric and therefore the other half need not be shown

the two eigenmodes, together with the profile at the anti-resonance frequency. Very good agreement is seen. In particular, if one uses the experimental profile to calculate the average normal displacement, $\langle w \rangle \cong 0$ is obtained at the anti-resonance point.

In Fig. 5.15, we show a comparison of the theory and experimental transmission spectra, in which the black solid curve denotes the calculated amplitude transmission coefficient and the open circles represent measured data. The dashed red line is the prediction of the mass density law. Excellent agreement is obtained. In particular, the transmission peaks' correspondence with the vibrational eigenmodes, as well as with the transmission dip's amplitude and frequency, all conform to the theory predictions.

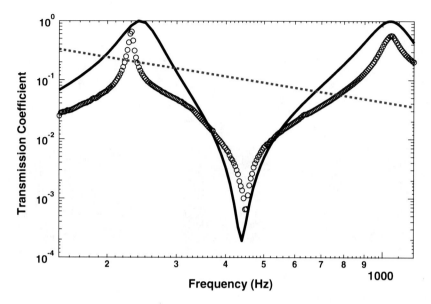

Fig. 5.15 Measured transmission coefficient amplitude (*black open circles*) and the COMSOL simulation results (*black solid curve*). The *red dashed line* is the mass density law prediction

5.3.5 Addition Rule

As stated earlier, one of the purposes of developing the membrane-type metamaterials is to stack them so as to make the stacked sample more effective at a particular frequency as well as to broaden the frequency range of the metamaterial functionality. Here we illustrate the results of such stacking to be indeed in line with what was expected.

An important point about stacking is that the membrane–membrane separation should be larger than the evanescent decay length generated by the δw part of the membrane displacement. Only when this condition is satisfied would the two membranes be regarded as truly independent, in the sense of having no near-field coupling.

We first examine quantitatively the effect of stacking two decorated membranes with the same anti-resonance frequency. In order to contrast with the traditional mass density law, we note that if the thickness of a solid wall is doubled, then the mass density law predicts the transmission amplitude to be halved, i.e.,

$$T \propto \frac{1}{\rho\omega(d+d)} = (0.5)\frac{1}{\rho\omega d}, \tag{5.7}$$

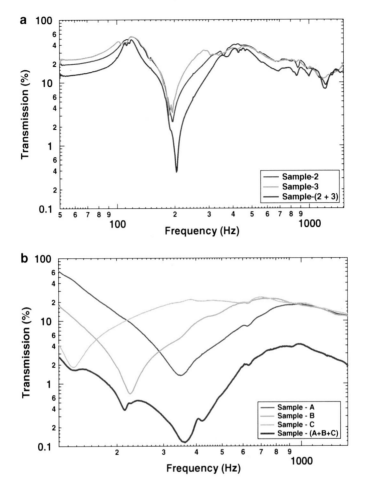

Fig. 5.16 Measured transmission spectra for stacking two membranes operating at almost identical frequencies (**a**) and three membranes operating at different anti-resonance frequencies (**b**)

a 6 dB increase in sound intensity attenuation is expected. In order to achieve 18 dB attenuation, which is the usual desired increment, it follows that the wall thickness has to be increased by a factor of 8! In contrast, for the membrane-type metamaterials, the attenuation rule is given by

$$T \propto \exp\left[-\text{const.}(d+d)\right] = \{\exp\left[-\text{const.}d\right]\}^{2}. \tag{5.8}$$

From the above, it can be seen that in terms of dB, the addition rule for the mass density law is logarithmic in character, whereas it is linearly additive for the membrane-type metamaterials, which is much more effective. In Fig. 5.16a,

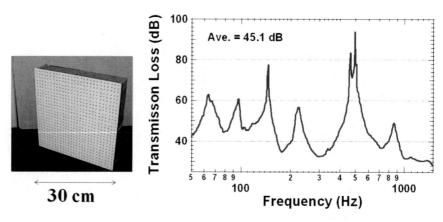

Fig. 5.17 Broadband attenuation sample (*left*) and its measured transmission loss (*right*)

we show the result of stacking two almost identical membrane-type metamaterials. The green and red curves are the transmission spectra of the two membranes, measured individually. The violet curve is the measured result by stacking the two together. At the anti-resonance frequency, almost 49 dB in intensity attenuation has been achieved. That is, stacking two nearly identical membranes shows an enhancement of ~20 dB in attenuation over a single membrane at the anti-resonance frequency.

It should be further noted that the resonant frequency of the first eigenmode is tunable by varying the weight of the central mass, in a manner that is proportional to the *inverse square root* of the central mass. The frequency of the second eigenmode, since its vibrational amplitude is mostly in the membrane, is insensitive to the weight. As the anti-resonance is a superposition of these two eigenmodes, it is viable to tune the anti-resonance frequency by varying the weight of the central mass.

To illustrate that stacking can broaden the frequency range of the membrane-type metamaterial functionality, we have fabricated a panel comprising three membranes operative at different anti-resonance frequencies. In Fig. 5.16b, the individually measured transmission spectra are shown as the red, green, and cyan curves. The transmission spectrum of the stacked sample is shown as the violet curve. The additive character of the panel is clearly seen from the remnant transmission dips of the three membranes.

To achieve broadband attenuation, we have fabricated panels with multiple weights in each unit cell (e.g., four weights in one cell). Multiple weights introduce degenerate eigenmodes, and as a result, the panel's transmission spectrum has many transmission minima. We have further tuned the frequency positions of the anti-resonance dips so that by stacking several panels, a broadband attenuation sample can be achieved. The separation between the neighboring panels is 15 mm, much larger than the evanescent decay length at the transmission dips. This sample (left panel, Fig. 5.17) has a total weight of 15 kg/m^2, and the average transmission loss is 45 dB over the 50–1,500 Hz frequency range (Fig. 5.17, right panel) [23].

5.4 Other Types of Acoustic Metamaterials

Subsequent to the initial demonstration of metamaterial characteristics of the locally resonant sonic materials, there has been a proliferation of other types of acoustic metamaterials during the past decade. This section is devoted to a brief survey of some major achievements in this field, with emphasis on the negativity in bulk modulus.

5.4.1 Negative Effective Bulk Modulus

Elastic constants play an equally important role as the mass density in determining a material's response to elastic/acoustic waves. In the context of elasticity, bulk modulus describes the elastic deformation that leads to a change in volume [24]. Intuitively, such deformation can be understood as a result of hydrostatic pressure with no preferred direction(s). This geometric characteristic of the bulk modulus, which differs from that of mass density, carries over to the consideration of effective bulk modulus (EBM) for acoustic metamaterials.

Multipole expansion is a standard technique that can be used to reveal the geometric character of the response functions. Being omnidirectional, bulk modulus-type response has the highest degree of rotational symmetry. Translated into the language of multipole representation, such response must be dominated by the monopole term [25]. On the other hand, mass density-type response is strongly directional as evidenced by the vibrational modes we analyzed in previous sections. It has the dipole symmetry.

Negativity in bulk modulus means that the medium expands under compression and contracts upon release. Thermodynamics dictates that a system with such a static response characteristic must be unstable. However, negative bulk modulus is possible in the context of dynamic response of an elastic/acoustic system, whereby the material display an out-of-phase response to an AC pressure field. Some theoretical models, such as water with suspending air bubbles [26], have been proposed for the realization of negative EBM.

In terms of experimental realization, there has been only one recipe so far that successfully achieved negative EBM [27]. The structure consists of a fluid channel that is sideway shunted by a series of periodically placed Helmholtz resonators (HRs). Instead of utilizing combinations of several materials, this metamaterial system seeks to produce modulus-type response by shaping the geometry that confines fluid in which sound propagates [28]. Several derivative works also exist on structures that display negative EBM, e.g., HRs in air that are operative in the kHz frequency regime [29, 30], and flute-like structures [31].

Fig. 5.18 A typical
Helmholtz resonator

HR is a well-known acoustic resonance structure that can be analyzed with a spring-and-mass model. An HR is basically a bottle with a large belly and a small opening orifice, connected by a narrow neck. Since the volume of the neck is much smaller than that of the belly, it is a good approximation to consider the fluid in the neck to be incompressible. The fluid in the belly, however, is compressed when the fluid in the neck section moves inward. Once compressed, the fluid pressure in the belly naturally increases, thereby providing a restoring force. Since the wavelength of the sound is generally much larger than the dimension of the entire resonator, the pressure gradient within the cavity can be neglected. From this description of the HR, fluid in the neck serves as the mass and the belly plays the role of a spring. Using this analogy, we obtain the resonance frequency of an HR as $\omega_0^2 = k/m = (\mathrm{d}F/\mathrm{d}x)/m = S^2(\mathrm{d}P/\mathrm{d}V)/m$, with k denoting the spring constant, which can be expressed as the force (F) derivative with respect to displacement (x), and that in turn can be expressed as the pressure (P) derivative with respect to volume (V) times the square of the cross-sectional area S of the neck. By writing $m = \rho S L$, where ρ denotes fluid density and L the length of the neck, we obtain

$$\omega_0 = v\sqrt{S/VL}, \tag{5.9}$$

where $v = \sqrt{(v(\mathrm{d}P/\mathrm{d}v)/\rho)}$ is the speed of sound in the fluid and V is the volume of the resonator chamber (the belly) (Fig. 5.18).

The HRs in [27] were arranged orthogonal to the propagation direction of the sound in the waveguide (Fig. 5.19a). A sound wave can trigger fluid motion in the neck of an HR, and when the excitation frequency approaches the vicinity of the HR eigenfrequency, the EBM response is excited, with a typical frequency dependence of $1/(\omega_0^2 - \omega^2)$. We therefore expect a sign change in the EBM response, arising from the fact that the motion of the fluid column in the neck switches from in-phase to out-of-phase with respect to the external pressure field.

Negative bulk modulus has a similar effect on acoustic wave propagation as the negative mass density—both cause the acoustic waves to be evanescent in character.

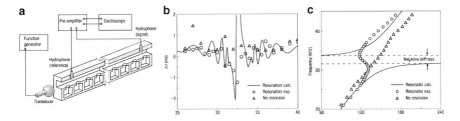

Fig. 5.19 Experimental layout (**a**) and measured results (**b**), (**c**). Negative transit time in (**b**) indicates negative group velocity, as seen in the band structure in (**c**). Figures adapted from [27]

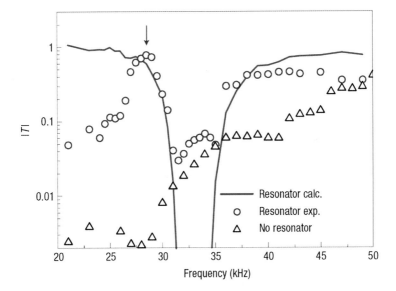

Fig. 5.20 Transmission spectra. A forbidden band is clearly seen around 32 kHz, owing to the HR resonance. The asymmetric peak (*red arrow*) is caused by Fano-like resonance, which is the consequence of interference between continuum channel and resonant channel [27, 32]. Figure adapted from [27]

Accordingly, bandgap was experimentally observed close to the resonant frequency of the metamaterial (Fig. 5.20).

5.4.2 Acoustic Double Negativity

The successful demonstrations of acoustic metamaterials with negative effective parameters naturally lead to the possibility of simultaneous double negativity in the same frequency regime. Early theoretical prediction [25] suggested that monopolar and dipolar resonances of the local scatterers are key to negative EBM and negative effective (dynamic) mass density, respectively. Recipes were conceived for their simultaneous realization [25, 26, 33, 34]. Similar to the electromagnetic case, doubly negative bulk modulus and (dynamic) mass density can lead to negative dispersion, i.e., the so-called left-handed acoustic materials. However, it was not until 2010 that the first success in experimental realization of acoustic double negativity [35] was demonstrated. In their 1D design, periodically arranged elastic membranes were deployed to tune the dipolar resonance [21], with side-opening orifices providing monopolar response [31]. Double-negative transmission band was found in the low-frequency limit. The same group later utilized the same design to demonstrate a reversed Doppler shift of sound within the double-negative band [36].

5.4.3   Focusing and Imaging

With the advent of acoustic metamaterials, a new horizon of possibilities for acoustic wave manipulation has emerged. During the past few years, there has been a proliferation of theoretical/numerical predictions [37–41] for achieving acoustic focusing and superlensing by using acoustic metamaterials. Shu Zhang et al. expanded such concept by building an interconnecting fluid network. Shunted by cavities of different volumes, each unit in the network resembles a Helmholtz resonator. It was experimentally shown that such a network is capable of achieving in-device focusing of ultrasound [42]. Lucian Zigoneanu et al. designed and fabricated flat lens with gradient index of refraction, bringing kHz airborne sound into out-of-device focus [43].

Highly dispersive materials can attain almost flat equi-frequency contours within a certain regime, thereby "canalizes" the propagation of wave [44, 45], achieving imaging effect. Such concept can be adapted to acoustic waves. By arranging locally resonant units in a square lattice, a low-frequency bandgap can emerge, with almost-flat lower band edge. It was numerically shown that the equi-frequency contour is square-like near the band edge and is capable of canalizing even evanescent acoustic wave into propagating modes [46, 47]. X. Ao and C. T. Chan took a step further [47] by incorporating rectangular lattice to introduce anisotropy. And by laying out the lattice in half-cylindrical geometry, a magnifying effect analogous to optical hyperlens [48, 49] was numerically demonstrated.

Anisotropy is at the core of the hyperlens idea. From multipole expansion, waves scattered/emanated from an object can be represented by superposition of modes with different angular momenta. Geometric details of the scatterers are carried in modes with high angular momenta that do not propagate (i.e., evanescent in character). However, for anisotropic materials in which the dielectric constant along one direction is negative, it becomes possible to have hyper-resolution. This is easy to see for a 2D circular geometry in which we have anisotropic dielectric constants $\varepsilon_\theta, \varepsilon_r$

with the condition that $\varepsilon_\theta \varepsilon_r < 0$. Then from the dispersion relation $(k_\theta^2/\varepsilon_r) + (k_r^2/\varepsilon_\theta)$ $= (\omega^2/v^2)$, it is easy to see that both k_θ and k_r can take on very large values, implying high resolution, without violating the dispersion relation. Such a material is denoted a hyperlens [48, 50], which is able to convert evanescent waves with high angular momenta into propagating modes. An acoustic magnifying "hyperlens" was subsequently realized by Jensen Li et al. [51], based not on the negative dielectric constant but rather on the large effective density and the relatively weak bulk modulus, realized by a fan-like structure with alternating fins of brass and air ducts, so that the effective wave speed is low and thereby the relevant wavelength is small. The lens has clearly demonstrated resolution that is less than half of the wavelength (with magnification) in a spatial region that is out of the device.

In the absence of viscous effect, a longitudinal acoustic wave can propagate in ducts (i.e., waveguides) of very small cross section, without the constraint of a cutoff frequency. By exploiting this fact, and with the aid of Fabry–Perot resonances, it was shown theoretically [52] that an "acoustic endoscope" can enhance evanescent waves, therefore open the possibility for sub-diffraction imagining. This idea was subsequently realized [53] with an array of waveguides with deep-subwavelength transverse-scale size.

Besides the approaches discussed above, C. Daraio's group took a different path toward acoustic focusing—nonlinearity in granular materials [54]. They constructed a nonlinear lens by patching granular chains tightly together. Such granular chains can transform an acoustic pulse into solitary waves, whose phase velocity depends on the amplitude. By adjusting the pre-applied static force exerted on each individual chain, the lens was found able to focus sonic pulse into very high intensity.

5.4.4 Cloaking

Acoustic cloaking has attracted theoretical attention in the past few years [55–67]. In particular, researchers have conceived devices by using "transformation acoustics" as a tool. Schemes for the cloaking of acoustic surface waves [68], bending waves on thin plates [69–71] and even fluid flow [72], have been proposed theoretically and studied by numerical simulations.

The experimental breakthrough came from Fang's group [73]. By making analogy between the acoustic wave equation and the telegrapher's equation, they explored the idea of using fluid networks as a platform for realizing acoustic cloaking. The effective mass density and bulk modulus were designed to follow a gradient in the radial direction, such that the ultrasonic wave is bent around the central domain, thereby minimizing the scattering of the object placed inside the domain so as to render it "invisible" to external observers. Experimental demonstration has clearly shown the reduced shadowing effect of the scattering object in the presence of the cloak. Impedance mismatch and the inevitable dissipative loss accounted for the less-than-perfect cloaking effect. Recently, the method of transformation acoustics showed its power in the design and experimental realization of an acoustics "carpet cloak" in air [74].

5.4.5 Acoustic Rectification

Time-reversal symmetry and spatial inversion symmetry are intrinsic to linear acoustic wave equation. Hence, nonreciprocal transmission of wave requires certain extra conditions to break these symmetries. By introducing second harmonics (nonlinear effect) into the wave equation and thereby breaking its time-reversal symmetry, an acoustic one-way mirror was proposed [75]. This was subsequently realized in the ultrasonic regime [76]. More recently, C. Daraio's group used 1D, strongly nonlinear (force-loaded) artificial granular medium to achieve rectification of acoustic waves and proposed prototypes of mechanical logic gates [77]. On the other hand, acoustic "one-way mirror" was also realized using simple 2D phononic crystals with incomplete bandgap [78]. Li et al. incorporated diffraction structures on one end of the phononic crystal to induce spatial modes with different k-vectors, thereby mimicking the condition of oblique incidence to result in transmission for part of the acoustic energy.

5.4.6 Hybrid Elastic Solids

Negativity in the effective mass density and the EBM is a direct outcome of dipolar and monopolar resonances, respectively. A natural question is whether it is possible to have a solid with a unit cell that can display monopole, dipole, and quadrupole resonances [6]. If so, what kind of behavior would such a solid exhibit? A recent publication [79] has proposed a unit cell design that can realize all three resonances, with overlapping resonance frequency regimes. Finite-element calculations found this unique design to simultaneously support dipolar and monopolar/quadrupolar resonances. As a result, two doubly negative bands exist. In one band, with overlapping dipolar and monopolar resonances, only pressure waves can propagate (with negative dispersion) while the shear waves are evanescent. This in effect resembles the acoustic property of a fluid. In the other band, "super-anisotropic behavior" is exhibited—i.e., pressure and shear waves are allowed to propagate only along mutually perpendicular directions. Hence within the frequency range of this band, the material appears to be a rigid solid in one direction but appears fluid-like in the other (Table 5.1).

Table 5.1 Properties of the hybrid elastic solids [79]

Direction	ΓX		ΓM	
Wave type	P-wave	S-wave	P-wave	S-wave
Wave velocities	$\sqrt{\dfrac{\kappa_{\text{eff}} + \mu_{\text{eff}}}{\rho_{\text{eff}}}}$	$\sqrt{\dfrac{c_{44}^{\text{eff}}}{\rho_{\text{eff}}}}$	$\sqrt{\dfrac{\kappa_{\text{eff}} + c_{44}^{\text{eff}}}{\rho_{\text{eff}}}}$	$\sqrt{\dfrac{\mu_{\text{eff}}}{\rho_{\text{eff}}}}$
Lower band $\kappa_{\text{eff}} > 0, \rho_{\text{eff}} < 0$ $\mu_{\text{eff}} \ll 0, c_{44}^{\text{eff}} > 0$	Propagation allowed, double negative in ρ_{eff} and μ_{eff}	Evanescent, negative ρ_{eff}	Evanescent, negative ρ_{eff}	Propagation allowed, double negative in ρ_{eff} and μ_{eff}
Higher band $\kappa_{\text{eff}} < 0, \rho_{\text{eff}} < 0$ $\mu_{\text{eff}} > 0, c_{44}^{\text{eff}} > 0$	Propagation allowed, double negative in ρ_{eff} and κ_{eff}	Evanescent, negative ρ_{eff}	Propagation allowed, double negative in ρ_{eff} and κ_{eff}	Evanescent, negative ρ_{eff}

5.5 Dynamic Mass Density at the Low-Frequency Limit

It is well known that for a time-harmonic wave, the elastic wave equation may be written as

$$\nabla \cdot \mu [\nabla \vec{u} + (\nabla \vec{u})^{\mathrm{T}}] + \nabla(\lambda \nabla \cdot \vec{u}) + D\omega^2 \vec{u} = 0, \qquad (5.10)$$

where D is the mass density, λ and μ are the (spatially varying) Lamé constants, $\vec{u}$ is the displacement vector, and $(\nabla \vec{u})^{\mathrm{T}}$ denotes the transpose of the tensorial quantity $\nabla \vec{u}$. Static effective elastic moduli and mass density are usually defined in the zero-frequency limit, where the limit $\omega \to 0$ is usually taken *first*, so that the mass density term drops out. Thus, the *static* effective moduli are obtained by the homogenization of $\nabla \cdot (\mu \nabla)$ and $\nabla(\lambda \nabla \cdot)$ operators. In contrast, to obtain the dynamic mass density expression, we have to solve the wave equation (5.10) so as to get the low-frequency wave solution and its relevant dispersion relation $\omega(\vec{k})$. The fact that for the fluid–solid composites the two limits are not necessarily the same has already been explained in the introductory Sect. 5.1. Thus the dynamic mass density is obtained from the slope of $\omega(\vec{k})$, i.e., the wave velocity. However, to separate out the elastic constant and mass density information from a single wave speed requires an additional criterion, which turns out to be the different angular momentum channels, as shown below. But at this point, we must first briefly introduce the MST, since our approach in obtaining the $\omega \to 0$ dynamic mass density is simply to examine the low-frequency limit of the MST.

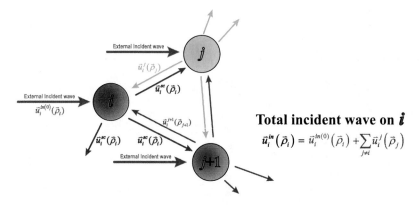

Fig. 5.21 A schematic diagram illustrating the basic idea of the multiple-scattering theory (MST), in which the scattered outgoing wave from any one particular scatterer constitutes part of the incident wave to any other scatterer

5.5.1 Multiple-Scattering Theory

MST represents a solution of the elastic wave equation (5.10) for a periodic composite that accounts fully for *all* the multiple scattering effects between *any* two scatterers, shown schematically in Fig. 5.21, as well as for the inherent vector character of elastic waves [2, 80, 81]. In what follows, we shall attempt to illustrate the basic ideas of the MST by using diagrammatic illustrations. A more detailed mathematical description can be found in Chap. 10.

We shall focus on the case of 2D periodic composites with a fluid matrix, in which MST has a rather simple form, as shown in Fig. 5.22, where $\vec{u}_i^{in}(\vec{\rho}_i) = \sum_n a_n^i \vec{J}_n^i(\vec{\rho}_i)$ and $\vec{u}_i^{sc}(\vec{\rho}_i) = \sum_n b_n^i \vec{H}_n^i(\vec{\rho}_i)$ are the waves incident on, and scattered by the scatterer i, respectively, with $\alpha_1 = \omega\sqrt{D_1/B_1}$ being the wave vector in the fluid matrix. Here D_1 and B_1 denote the mass density and bulk modulus of the matrix, respectively, $\vec{\rho} = (\rho, \varphi)$ is the polar coordinates, and $J_n(x)$ and $H_n(x)$ denote the nth Bessel function and Hankel function of the first kind, respectively.

Since the incident wave on scatterer i comprises the external incident wave $\vec{u}_i^{in(0)}(\vec{\rho}_i)$ plus the scattered waves by all the other scatterers except i (as shown in Fig. 5.21), we have

$$\vec{u}_i^{in}(\vec{\rho}_i) = \vec{u}_i^{in(0)}(\vec{\rho}_i) + \sum_{j\neq i}\sum_{n''} b_{n''}^j \vec{H}_{n''}^j(\vec{\rho}_j), \qquad (5.11)$$

where $\vec{\rho}_i$ and $\vec{\rho}_j$ refer to the position of the same spatial point measured from scatterers i and j, respectively.

In Fig. 5.22, the expansion coefficients $\{a_n\}$ and $\{b_n\}$ are not independent but are in fact related by the so-called T matrix. This is shown in Fig. 5.23, where $T = \{T_{nn'}\}$ is the elastic Mie scattering matrix determined by matching the normal displacement and normal stress component at the fluid–solid interface.

Acoustic wave equation
$$B\nabla(\nabla \cdot \vec{u}) + D\omega^2\vec{u} = 0$$

↓

General Solution in 2D
$$\vec{u}(\vec{\rho}) = \sum_n \left[a_n \vec{J}_n(\vec{\rho}) + b_n \vec{H}_n(\vec{\rho}) \right]$$

$$\begin{cases} \vec{J}_n(\vec{\rho}) = \nabla\left[J_n(\alpha_1\rho)e^{in\phi} \right] \\ \vec{H}_n(\vec{\rho}) = \nabla\left[H_n(\alpha_1\rho)e^{in\phi} \right] \\ \alpha_1 = \omega\sqrt{D_1/B_1} \end{cases}$$

incident (incoming) wave on scatterer i $\vec{u}_i^{in}(\vec{\rho}_i) = \sum_n a_n^i \vec{J}_n^i(\vec{\rho}_i)$

scattered (out-going) wave from i $\vec{u}_i^{sc}(\vec{\rho}_i) = \sum_n b_n^i \vec{H}_n^i(\vec{\rho}_i)$

Fig. 5.22 General solution of the acoustic wave equation for 2D phononic crystals with a fluid matrix. Here J_n denotes the Bessel function of nth order and H_n denotes the nth-order Hankel function of the first kind

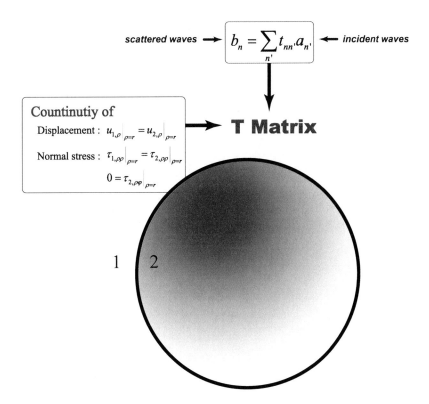

scattered waves → $b_n = \sum_{n'} t_{nn'} a_{n'}$ ← incident waves

Countinutiy of

Displacement : $u_{1,\rho}\big|_{\rho=r} = u_{2,\rho}\big|_{\rho=r}$

Normal stress : $\tau_{1,\rho\rho}\big|_{\rho=r} = \tau_{2,\rho\rho}\big|_{\rho=r}$

$0 = \tau_{2,\rho\phi}\big|_{\rho=r}$

→ **T Matrix**

1 2

Fig. 5.23 T matrix and the boundary conditions. Region 1 denotes the matrix materials and region 2 denotes the solid scatterer

G Matrix (structural constant): lattice summation

$$\bar{H}^{j}_{n'}\left[\bar{\rho}_i - \left(\bar{R}_j - \bar{R}_i\right)\right] = \sum_n G_{n'n}\left(\bar{R}_j - \bar{R}_i\right)\bar{J}^i_n\left(\bar{\rho}_i\right)$$

$$G_{n'n}\left(\bar{k}\right) = (-1)^l\, S_{-l}\left(\alpha_1, \bar{k}\right)$$

$$\longrightarrow \begin{cases} S_l\left(\alpha_1, \bar{k}\right) = -\delta_{l,0} + i S_l^T\left(\alpha_1, \bar{k}\right) \\ S_l^T\left(\alpha_1, \bar{k}\right) J_{l+m}\left(\alpha_1 \eta\right) = -\left[Y_m\left(\alpha_1 \eta\right) + \dfrac{1}{\pi}\sum_{n=1}^{m}\dfrac{(m-n)!}{(n-1)!}\left(\dfrac{2}{\alpha_1 \eta}\right)^{m-2n+2}\right]\delta_{l,0} - i^l\,\dfrac{4}{A}\sum_{k}\left(\dfrac{\alpha_1}{Q_k}\right)^m\dfrac{J_{l+m}\left(Q_k \eta\right)}{Q_k^2 - \alpha_1^2} \end{cases}$$

Fig. 5.24 G matrix and its evaluation method

With the help of addition theorem, it can be proved that

$$\bar{H}^{j}_{n''}(\bar{\rho}_j) = \bar{H}^{j}_{n''}\left(\bar{\rho}_i - (\bar{R}_j - \bar{R}_i)\right) = \sum_n G^{ij}_{n''n}\bar{J}^i_n(\bar{\rho}_i), \tag{5.12}$$

where the G matrix $G^{ij}_{n''n} = G_{n''n}(\bar{R}_j - \bar{R}_i)$ denotes the translation coefficients as shown in Fig. 5.24, with $\phi = \arg(\bar{R}_j - \bar{R}_i)$, $\bar{R}_{i(j)}$ being the position of scatterer $i(j)$. This translation means that the wave scattered by the scatterer j may be expressed in terms of Bessel functions centered at scatterer i. And since the coefficients $\{a_n\}$ and $\{b_n\}$ at scatterer i are related by the T matrix, one can therefore obtain a single matrix equation with $\{a_n\}$ being the variables. Of course, in such a derivation, it is assumed that in a periodic composite every scatterer is the same.

For the purpose of calculating the dispersion relation, we do not need an externally incident wave $\bar{u}^{in(0)}_i(\bar{\rho}_i)$ in (5.2), thus we have

$$\bar{u}^{in}_i(\bar{\rho}_i) = \sum_{j \neq i}\sum_{n''} b^j_{n''}\bar{H}^j_{n''}(\bar{\rho}_j). \tag{5.13}$$

After substituting the expressions for the T matrix and G matrix into this equation and Fourier-transforming the coefficients of $\{a_n\}$, as shown in Fig. 5.25, we arrive at the following secular equation:

$$\det\left|T^{-1}_{nn'} - G_{nn'}(\bar{k})\right| = 0. \tag{5.14}$$

Equation (5.14) is equivalent to (85) in Chap. 10. Written in this particular form, (5.14) is particularly suitable for the low-frequency expansion, as seen below.

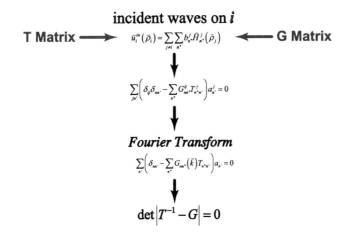

Fig. 5.25 Secular determinant equation of the MST, for the determination of band structures for periodic composites

5.5.2   Dynamic Mass Density at the $\omega \rightarrow 0$ Limit

Equation (5.14) is the secular equation for determining the band structure of a periodic composite. Here we want only the branch at the $\omega \rightarrow 0$ limit, i.e., by letting $\alpha_1 \rightarrow 0$ and retaining the leading-order terms of the secular equation. This is illustrated in Fig. 5.26.

By taking the low-frequency limit and retaining terms to the order of ω^{-2}, both the T^{-1} matrix and the G matrix can be simplified to 3×3 matrices [4, 5]. Therefore, the secular equation in the low-frequency limit is given by

$$
\det \begin{vmatrix}
\dfrac{D_1 + D_2}{D_1 - D_2} + \dfrac{x^2 f}{1 - x^2} & \dfrac{ixf}{1 - x^2} & -\dfrac{f}{1 - x^2} \\[2ex]
-\dfrac{ixf}{1 - x^2} & \dfrac{B_2}{B_2 - B_1} + \dfrac{x^2 f}{1 - x^2} & \dfrac{ixf}{1 - x^2} \\[2ex]
-\dfrac{f}{1 - x^2} & -\dfrac{ixf}{1 - x^2} & \dfrac{D_1 + D_2}{D_1 - D_2} + \dfrac{x^2 f}{1 - x^2}
\end{vmatrix} = 0, \quad (5.15)
$$

in which $f = \pi r_0^2 / A$ is the filling ratio of the solid inclusions, $B_1 = \lambda_1$ and $B_2 = \lambda_2 + \mu_2$ are the bulk moduli of the fluid matrix and solid inclusions, respectively, and $x = V_{\text{eff}} / V_1$ is the variable to be solved in the determinant equation (5.15). By discarding the trivial root, we obtain the effective sound velocity of the composite as

Low-frequency limit, $\alpha_1 \to 0$

$$J_0(\alpha_1 r_0) \to 1 - \frac{1}{4}\alpha_1^2 r_0^2$$

$$J_1(\alpha_1 r_0) \to \frac{1}{2}\alpha_1 r_0 - \frac{1}{16}\alpha_1^3 r_0^3$$

$$J_2(\alpha_1 r_0) \to \frac{1}{8}\alpha_1^2 r_0^2 - \frac{1}{96}\alpha_1^4 r_0^4$$

$$H_0(\alpha_1 r_0) \to 1 - \frac{1}{4}\alpha_1^2 r_0^2 + i\frac{2}{\pi}\ln(\alpha_1 r_0) + i\frac{2}{\pi}(\gamma_0 - \ln 2)$$

$$H_1(\alpha_1 r_0) \to \frac{1}{2}\alpha_1 r_0 - \frac{1}{16}\alpha_1^3 r_0^3 - i\frac{2}{\pi}\frac{1}{\alpha_1 r_0} + i\frac{\gamma_0 - \frac{1}{2}}{\pi}\alpha_1 r_0$$

$$H_2(\alpha_1 r_0) \to \frac{1}{8}\alpha_1^2 r_0^2 - \frac{1}{96}\alpha_1^4 r_0^4 - i\frac{4}{\pi}\frac{1}{\alpha_1^2 r_0^2} - i\frac{1}{\pi}$$

....

$$T^{-1} = -I + \frac{4i}{\pi r_0^2}\frac{1}{\alpha_1^2}\begin{bmatrix} \dfrac{D_1 + D_2}{D_1 - D_2} & 0 & 0 \\ 0 & \dfrac{\lambda_2 + \mu_2}{\lambda_2 + \mu_2 - \lambda_1} & 0 \\ 0 & 0 & \dfrac{D_1 + D_2}{D_1 - D_2} \end{bmatrix}$$

$$G = -I + \frac{4i}{A}\frac{1}{1-x^2}\frac{1}{\alpha_1^2}\begin{bmatrix} -x^2 & xe^{-i\theta_0} & e^{-2i\theta_0} \\ -xe^{i\theta_0} & -x^2 & -xe^{i\theta_0} \\ e^{2i\theta_0} & -xe^{i\theta_0} & -x^2 \end{bmatrix}$$

Fig. 5.26 The T^{-1} matrix and G matrix in the low-frequency limit, where r_0 and A are the radius of solid inclusions and area of unit cell, respectively. λ_1, λ_2, and μ_2 are the Lamé constants, and $\gamma_0 \approx 0.5772$ is the Euler's constant. θ_0 Is the polar angle of wave vector $\vec{k}$, which vanishes in the determinant evaluation of $|T^{-1} - G|$. The variable $x = V_{\text{eff}}/V_1$ is the quantity to be evaluated

$$V_{\text{eff}} = \sqrt{\frac{B_{\text{eff}}}{D_{\text{eff}}}} = \sqrt{\frac{\dfrac{B_2}{B_2 + (B_1 - B_2)f}B_1}{\dfrac{(D_2 + D_1) + (D_2 - D_1)f}{(D_2 + D_1) - (D_2 - D_1)f}D_1}}. \tag{5.16}$$

It is well known that according to the effective medium theory [82], the EBM B_{eff} of the fluid–solid composite is given by

$$\frac{1}{B_{\text{eff}}} = \frac{1-f}{B_1} + \frac{f}{B_2} \tag{5.17}$$

or

$$B_{\text{eff}} = \frac{B_2}{B_2 + (B_1 - B_2)f}B_1. \tag{5.18}$$

It can also be seen from Eq. (5.15) and Fig. 5.26 that the expression for B_{eff}, (5.18), arises from the $n = 0$ angular scattering channel.

By using (5.16) and the effective medium expression for B_{eff} [i.e., (5.18)], we arrive at precisely the Berryman effective mass density in 2D [83, 84]:

$$D_{\text{eff}} = \frac{(D_2 + D_1) + (D_2 - D_1)f}{(D_2 + D_1) - (D_2 - D_1)f} D_1. \tag{5.19}$$

In contrast to the B_{eff} expression, the effective mass density D_{eff} is completely determined by the $n = 1$ angular channel. As pointed out previously, the effective mass density and the EBM represent *separate* but parallel wave scattering channels.

Equation (5.19) is valid for both the square and the hexagonal lattices when the filling fraction of the solid inclusions is not very high. At this leading order of density expansion, both B_{eff} and D_{eff} are noted to be independent of the lattice structure. In particular, they are both relatively accurate for *random* fluid–solid composites as long as the density is not close to the tight-packing limit, and the viscous boundary layer thickness is smaller than the fluid channel width. When the concentration of scatterers becomes larger and larger, it is expected that higher-order angular momentum channels in T^{-1} and G matrices should be included. The effective sound speeds would then be different for the square and the hexagonal lattices, but isotropy still holds.

It is instructive to carry the effective dynamic mass density evaluation to a higher concentration level, by retaining more angular momentum channels in the T^{-1} and G matrices. Through a lengthy derivation, the dynamic mass density is found to be in the form [84–86]

$$D_{\text{eff}} = \frac{(D_2 + D_1) + (D_2 - D_1)(f - g)}{(D_2 + D_1) - (D_2 - D_1)(f + g)} D_1, \tag{5.20}$$

where [87]

$$g = 768 \left(\frac{M_4}{\pi}\right)^2 f^4 \approx 0.3058 f^4 \tag{5.21}$$

for the square lattice, and

$$g = 1,620 \left(\frac{M_6}{\pi^2}\right)^2 f^6 \approx 0.0754 f^6 \tag{5.22}$$

for the hexagonal lattice. Here the lattice sums

$$M_4 = \sum_{h \neq 0} \frac{J_4(K_h a)}{(K_h a)^2} e^{4i\theta_h} \tag{5.23}$$

and

$$M_6 = \sum_{h \neq 0} \frac{J_6(K_h a)}{(K_h a)} e^{6i\theta_h} \tag{5.24}$$

are defined in the reciprocal spaces of the square and hexagonal lattices, respectively, with $\bar{K}_h = (K_h, \theta_h)$ denoting the reciprocal lattice vector in polar coordinates and a being the lattice constant. In contrast, the EBM is still given by (5.18), i.e., the Wood's formula.

Comparing (5.20) with (5.19), we notice that the effective mass density is modified by a correction term, g. When the filling fraction of the inclusions is not very high, g is very small so that it can be safely neglected. When this happens, (5.20) reduces to (5.19), i.e., the dipole solution. However, in case of high concentration of inclusions, Berryman's expression, i.e., (5.19), should be modified to incorporate the influence of higher-order scattering coefficients.

It is worth noting that the correction term g is proportional to f^4 for the square lattice and to f^6 for the hexagonal lattice. Common sense tells us that the correction term should be quadratic in f, but here the correction term g is obviously determined by the symmetry of the square and hexagonal lattices. This point can be easily understood since the coefficients in front of f^4 (square lattice) and f^6 (hexagonal lattice) are respectively the lattice sums M_4 and M_6 defined by (5.23) and (5.24), and they are clearly determined by the lattice symmetry.

If the matrix is made of solid instead of liquid, we can also take the low-frequency limit on the MST in a similar way. But a different effective medium formula for the mass density may be expected since in a solid matrix not only the longitudinal wave but also the transverse waves can propagate. It is well known that in 2D phononic crystals, when the wave vector is confined in the 2D plane (i.e., the x–y plane) perpendicular to the cylinder axis direction (i.e., the z-direction), the elastic waves can be decoupled into an out-of-plane transverse z mode and an in-plane mixed xy mode.

For the transverse z mode, the displacement is perpendicular to the x–y plane and thus easier to deal with. By taking the low-frequency limit and retaining the dominant terms, the $T^{-1} - G$ matrix can also be simplified to a 3×3 matrix [5]:

$$T^{-1} - G = \frac{4i}{\pi r^2} \frac{1}{\beta_1^2} \begin{bmatrix} \dfrac{\mu_2 + \mu_1}{\mu_2 - \mu_1} + f\dfrac{x^2}{1-x^2} & \dfrac{ixf}{1-x^2}e^{-i\theta_0} & -\dfrac{f}{1-x^2}e^{-2i\theta_0} \\[3mm] -\dfrac{ixf}{1-x^2}e^{i\theta_0} & \dfrac{D_1}{D_1 - D_2} + f\dfrac{x^2}{1-x^2} & \dfrac{ixf}{1-x^2}e^{-i\theta_0} \\[3mm] -\dfrac{f}{1-x^2}e^{2i\theta_0} & -\dfrac{ixf}{1-x^2}e^{i\theta_0} & \dfrac{\mu_2 + \mu_1}{\mu_2 - \mu_1} + f\dfrac{x^2}{1-x^2} \end{bmatrix}, \tag{5.25}$$

in which $x = V_{eff}/V_1$ is the quantity to be evaluated. By solving (5.25), we obtain the effective transverse wave velocity of the composite as

$$V_{eff} = \sqrt{\frac{\mu_{eff}}{D_{eff}}} = \sqrt{\frac{\frac{(\mu_2 + \mu_1) + (\mu_2 - \mu_1)f}{(\mu_2 + \mu_1) - (\mu_2 - \mu_1)f}\mu_1}{(1-f)D_1 + fD_2}}. \qquad (5.26)$$

It can be recognized from (5.26) that the effective shear modulus μ_{eff}, determined by the $n = 1$ angular channel [see (5.25)], is given by

$$\mu_{eff} = \frac{(\mu_2 + \mu_1) + (\mu_2 - \mu_1)f}{(\mu_2 + \mu_1) - (\mu_2 - \mu_1)f}\mu_1. \qquad (5.27)$$

It is interesting to point out that (5.27) has the same form as (5.19), and this similarity is due to the fact that both μ_{eff} and D_{eff} arise from the $n = 1$ angular channel scattering.

According to (5.26) and the effective shear modulus expression for μ_{eff}, i.e., (5.27), we arrive at the volume-averaged mass density expression for the transverse z mode:

$$D_{eff} = \rho_{eff} = (1-f)D_1 + fD_2, \qquad (5.28)$$

which is distinct from the fluid-matrix case. Here the effective mass density for the solid-matrix composite is determined by the $n = 0$ angular channel. Equation (5.28) for the solid matrix case is noted to be identical to that found by Berryman [83] through a different approach.

If we let $\mu_1 \to 0$, then according to (5.27), we have $\mu_{eff} \to 0$. That is, when the solid matrix is gradually reduced to the limit of zero shear modulus, the whole composite would also act like a zero-shear modulus system, i.e., the composite behaves like a fluid. However, it is important to note that even in this limit, the volume-averaged density formula, i.e., (5.28), still holds. Therefore, by first taking the $\omega \to 0$ limit and then the $\mu_1 \to 0$ limit, we arrive at the volume-averaged mass density expression. However, reversing the order of taking the two limits leads to the expression given by (5.19). Therefore, the order of taking the two limits *cannot* be interchanged, as explained in the introductory Sect. 5.1.

5.5.3 Comparison with Experimental Data

Cervera et al. have measured the sound velocity in a 2D phononic crystal composed of hexagonal array of aluminum cylinders in air [86]. Here the frequency of sound is 600 Hz, and the wavelength of sound in air, 57 cm, is much larger than either the cylinder diameter or the lattice constant. The wavelengths of sound in Al, for both longitudinal and transverse waves, are even larger. The use of effective medium theory is thus justified. The viscosity and mass density of air at normal temperature are 1.827×10^{-5}Pa s and 1.292 kg/m^3, respectively. At the experimental frequency of ~600 Hz, the viscous boundary layer thickness $\ell_{\mathrm{vis}} = \sqrt{\eta/\rho_{\mathrm{air}}\omega} = 6.12 \times 10^{-3}$ cm is much smaller than either the cylinder diameter, the lattice constant, or the fluid channel width ℓ. Thus the condition $\ell \gg \ell_{\mathrm{vis}}$ is valid.

In the experiment, the maximum filling ratio of Al cylinders is about 0.36, shown as open triangles in Fig. 5.27, where it can be seen that there is nearly an order of magnitude discrepancy between the experimentally measured velocity with that predicted by using the volume-averaged mass density and the EBM B_{eff} given by (5.18). In contrast, when the dynamic effective mass density, (5.19), is used, excellent agreement is seen.

For higher filling ratio of Al cylinders, we have used COMSOL Multiphysics, a finite-element solver, to perform a band-structure calculation for the same periodic system. From the band structure, i.e., the dispersion relation, one can compute the effective wave speed by using $c = \omega/k$ in the low-frequency limit. The corresponding results are plotted in Fig. 5.27 in green circles. They are seen to be in excellent agreement with (5.20), as shown with red solid curve, where the correction term g is included.

In Fig. 5.28a, we show the numerically calculated displacement field intensities for the relevant experiment. It can be seen that the displacement field is nearly zero inside the cylinders, hence it is almost impossible to have the condition for the validity of volume-averaged density formula. However, when the impedance mismatch is relatively moderate, e.g., when the mass density contrast is small, then the effective dynamic mass density reduces the volume-averaged mass density, which means that the static mass density is a special case of the dynamic mass density. For comparison with Fig. 5.28a, we have also plotted the displacement field intensities for the poly(methyl methacrylate) (PMMA)–water system in Fig. 5.28b, in which the wavefield homogeneity is very evident. As our derivation of the dynamic mass density is obtained by taking the long wavelength limit of the scattering wave field solutions, it is not surprising that such formula inherently accounts for the wavefield inhomogeneities as they exist in reality. As explained in Sect. 5.1, the *relative motion* between the components of a composite is the basic reason leading to the difference between the static and dynamic mass densities, and such relative motion is evident when the impedance mismatch is large and $\ell \gg \ell_{\mathrm{vis}}$.

In a solid-matrix composite, the presence of a nonzero shear modulus for the matrix component means that in the long wavelength limit, uniform motion of the matrix and the inclusions is guaranteed. As a result, the dynamic mass density for the solid-matrix composites is always the volume-averaged value. When one further takes the limit of $\mu_1 \to 0$ in that case, only the relative ratio of the longitudinal wavelength to the transverse (shear) wavelength is altered, which is the reason that the effective mass density expression still remains the same as the static mass density.

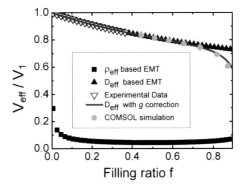

Fig. 5.27 The effective sound velocities calculated with the effective bulk modulus given by Wood's formula with volume-averaged mass density (*solid squares*) and with the mass density given by (5.19) (*solid triangle*). Experimentally measured effective sound velocity is shown as open triangles. While the volume-averaged mass density gives results very far removed from the experiment, the mass density given by (5.19) is shown to yield almost perfect agreement with measured results when the filling ratio of the Al cylinders is not very high. When the filling ratio is larger than 0.6, however, the correction term g should be included [see (5.20)], with the prediction shown by the *red solid curve*. It can be seen that the prediction of (5.20) agrees very well with the finite-element simulation results, shown as *green dots*, even when the filling ratio is close to the tight-packing limit

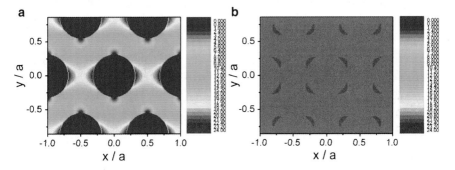

Fig. 5.28 (**a**) MST-calculated displacement field intensities in a 2D hexagonal lattice of Al cylinders in air, with the relevant experimental parameter values as stated in the text. *Blue* indicates low field intensity, and *yellow* indicates high field intensity. The wave vector is along the y-direction, with a being the lattice constant. It is seen that the wave amplitude is nearly zero inside the Al cylinders. Decreasing the frequency further does not alter this fact. (**b**) The same for PMMA cylinders in water. Wave field is seen to be much more homogeneous than that in (**a**). Figure adapted from [4]

5.6 Concluding Remarks

Acoustics has been one of the oldest topics of scientific investigation. Its robust revival during the past two decades has been a most gratifying experience for many researchers in this area. The purpose of this chapter is to give a vignette on some of the more recent developments. In particular, we present an overview on the different ramifications of the dynamic mass density issue that includes both the acoustic metamaterials manifestations and the effective mass density of fluid–solid composite in the zero frequency limit. The connection with the anti-resonance behavior is emphasized and clarified, especially with respect to the membrane-type acoustic metamaterials. A brief review of other types of acoustic metamaterials is also included.

In contrast to electromagnetic metamaterials, the role of dissipation is minimal for acoustic metamaterials—at least in the low-frequency limit. However, since the presence of dissipation is inevitable, its consideration, while still in the incipient stage at present, may become more important in the future. Another issue is the role of evanescent waves, which can be expected to play an increasingly important part in transformational acoustics, just as in the case of electromagnetic metamaterials. However, unlike the electromagnetic case, the elasticity of solid composites has more parameters and therefore can be expected to display a much richer variety of behaviors. An example along this direction is the recent work on hybrid elastic solids [79].

Potential applications of acoustic metamaterials would undoubtedly be a consideration for the future developments in this area. Pursuit of such a worthy goal may not only open up new topics for basic research, but can also impact those disciplines that are traditionally related to acoustics—such as architecture, noise pollution, medical ultrasound, acoustic imaging, etc. Cross-disciplinary pollination can imply exciting potential possibilities.

Acknowledgments Research works in dynamic mass density and membrane-type acoustic metamaterials have been supported by Hong Kong Research Grant Council grants HKUST6145/99P, HKUST6143/00P, HKUST 605405, and HKUST604207.

References

1. Z. Liu, X. Zhang, Y. Mao, Y.Y. Zhu, Z. Yang, C.T. Chan, P. Sheng, Locally resonant sonic materials. Science **289**, 1734 (2000). It should be noted that in this initial publication the metamaterial characteristic was mis-attributed to negative elastic modulus. This was subsequently corrected in Ref. 2
2. Z. Liu, C. Chan, P. Sheng, Analytical model of phononic crystal with local resonances. Phys. Rev. B **71**, 014103 (2005)
3. P. Sheng, X. Zhang, Z. Liu, C. Chan, Locally resonant sonic materials. Physica B Condens. Matter **338**, 201 (2003)
4. J. Mei, Z. Liu, W. Wen, P. Sheng, Effective mass density of fluid-solid composites. Phys. Rev. Lett. **96**, 24301 (2006)
5. J. Mei, Z. Liu, W. Wen, P. Sheng, Effective dynamic mass density of composites. Phys. Rev. B **76**, 134205 (2007)
6. Y. Wu, Y. Lai, Z.Q. Zhang, Effective medium theory for elastic metamaterials in two dimensions. Phys. Rev. B **76**, 205313 (2007)
7. G.W. Milton, J.R. Willis, On modifications of Newton's second law and linear continuum elastodynamics. Proc. R. Soc. A **463**, 855 (2007)

8. S. Yao, X. Zhou, G. Hu, Experimental study on negative effective mass in a 1d mass–spring system. New J. Phys. **10**, 043020 (2008)

9. D.L. Johnson, Equivalence between fourth sound in liquid He Ii at low temperatures and the Biot slow wave in consolidated porous media. Appl. Phys. Lett. **37**, 1065 (1980)

10. M.A. Biot, Theory of propagation of elastic waves in a fluid-saturated porous solid. I. Low-frequency range. J. Acoust. Soc. Am. **28**, 168 (1956)

11. T.J. Plona, Observation of a second bulk compressional wave in a porous medium at ultrasonic frequencies. Appl. Phys. Lett. **36**, 259 (1980)

12. P. Sheng, M.Y. Zhou, Dynamic permeability in porous media. Phys. Rev. Lett. **61**, 1591 (1988)

13. D. Weaire, Existence of a gap in the electronic density of states of a tetrahedrally bonded solid of arbitrary structure. Phys. Rev. Lett. **26**, 1541 (1971)

14. D.J. Ewins, *Modal Testing: Theory, Practice and Application* (Research Studies Press. Ltd., Hertfordshire, 2000)

15. M. Dilena, A. Morassi, The use of antiresonances for crack detection in beams. J. Sound Vib. **276**, 195 (2004)

16. M. Dilena, A. Morassi, Damage detection in discrete vibrating systems. J. Sound Vib. **289**, 830 (2006)

17. G. Schitter, K.J. Astrom, B.E. DeMartini, P.J. Thurner, K.L. Turner, P.K. Hansma, Design and modeling of a high-speed Afm-scanner. IEEE Trans. Control. Syst. Technol. **15**, 906 (2007)

18. G. Schitter, F. Allgöwer, A. Stemmer, A new control strategy for high-speed atomic force microscopy. Nanotechnology **15**, 108 (2004)

19. G. Schitter, P.J. Thurner, P.K. Hansma, Design and input-shaping control of a novel scanner for high-speed atomic force microscopy. Mechatronics **18**, 282 (2008)

20. Z. Yang, J. Mei, M. Yang, N. Chan, P. Sheng, Membrane-type acoustic metamaterial with negative dynamic mass. Phys. Rev. Lett. **101**, 204301 (2008)

21. S.H. Lee, C.M. Park, Y.M. Seo, Z.G. Wang, C.K. Kim, Acoustic metamaterial with negative density. Phys. Lett. A **373**, 4464 (2009)

22. S. Yao, X. Zhou, G. Hu, Investigation of the negative-mass behaviors occurring below a cut-off frequency. New J. Phys. **12**, 103025 (2010)

23. Z. Yang, H. Dai, N. Chan, G. Ma, P. Sheng, Acoustic metamaterial panels for sound attenuation in the 50–1000 Hz regime. Appl. Phys. Lett. **96**, 041906 (2010)

24. L. Landau, E. Lifshitz, *Theory of Elasticity*, 3rd edn. (Elmsford, New York, 1982)

25. J. Li, C. Chan, Double-negative acoustic metamaterial. Phys. Rev. E **70**, 055602 (2004)

26. Y. Ding, Z. Liu, C. Qiu, J. Shi, Metamaterial with simultaneously negative bulk modulus and mass density. Phys. Rev. Lett. **99**, 93904 (2007)

27. N. Fang, D. Xi, J. Xu, M. Ambati, W. Srituravanich, C. Sun, X. Zhang, Ultrasonic metamaterials with negative modulus. Nat. Mater. **5**, 452 (2006)

28. L. Fok, M. Ambati, X. Zhang, Acoustic metamaterials. MRS Bull. **33**, 931 (2008)

29. C. Ding, L. Hao, X. Zhao, Two-dimensional acoustic metamaterial with negative modulus. J. Appl. Phys. **108**, 074911 (2010)

30. C.L. Ding, X.P. Zhao, Multi-band and broadband acoustic metamaterial with resonant structures. J. Phys. D Appl. Phys. **44**, 215402 (2011)

31. S.H. Lee, C.M. Park, Y.M. Seo, Z.G. Wang, C.K. Kim, Acoustic metamaterial with negative modulus. J. Phys. Condens. Matter **21**, 175704 (2009)

32. C. Goffaux, J. Sánchez-Dehesa, A.L. Yeyati, P. Lambin, A. Khelif, J. Vasseur, B. Djafari-Rouhani, Evidence of fano-like interference phenomena in locally resonant materials. Phys. Rev. Lett. **88**, 225502 (2002)

33. Y. Cheng, J. Xu, X. Liu, One-dimensional structured ultrasonic metamaterials with simultaneously negative dynamic density and modulus. Phys. Rev. B **77**, 045134 (2008)

34. X. Hu, K.M. Ho, C. Chan, J. Zi, Homogenization of acoustic metamaterials of Helmholtz resonators in fluid. Phys. Rev. B **77**, 172301 (2008)

35. S.H. Lee, C.M. Park, Y.M. Seo, Z.G. Wang, C.K. Kim, Composite acoustic medium with simultaneously negative density and modulus. Phys. Rev. Lett. **104**, 54301 (2010)
36. S.H. Lee, C.M. Park, Y.M. Seo, C.K. Kim, Reversed Doppler effect in double negative metamaterials. Phys. Rev. B **81**, 241102 (2010)
37. M. Ambati, N. Fang, C. Sun, X. Zhang, Surface resonant states and superlensing in acoustic metamaterials. Phys. Rev. B **75**, 195447 (2007)
38. S. Guenneau, A. Movchan, G. Pétursson, S. Anantha Ramakrishna, Acoustic metamaterials for sound focusing and confinement. New J. Phys. **9**, 399 (2007)
39. K. Deng, Y. Ding, Z. He, H. Zhao, J. Shi, Z. Liu, Theoretical study of subwavelength imaging by acoustic metamaterial slabs. J. Appl. Phys. **105**, 124909 (2009)
40. X. Zhou, G. Hu, Superlensing effect of an anisotropic metamaterial slab with near-zero dynamic mass. Appl. Phys. Lett. **98**, 263510 (2011)
41. J. Li, Z. Liu, C. Qiu, Negative refraction imaging of acoustic waves by a two-dimensional three-component phononic crystal. Phys. Rev. B **73**, 054302 (2006)
42. S. Zhang, L. Yin, N. Fang, Focusing ultrasound with an acoustic metamaterial network. Phys. Rev. Lett. **102**, 194301 (2009)
43. L. Zigoneanu, B.I. Popa, S.A. Cummer, Design and measurements of a broadband two-dimensional acoustic lens. Phys. Rev. B **84**, 024305 (2011)
44. P.A. Belov, C.R. Simovski, P. Ikonen, Canalization of subwavelength images by electromagnetic crystals. Phys. Rev. B **71**, 193105 (2005)
45. Y. Jin, S. He, Canalization for subwavelength focusing by a slab of dielectric photonic crystal. Phys. Rev. B **75**, 195126 (2007)
46. Z. He, F. Cai, Y. Ding, Z. Liu, Subwavelength imaging of acoustic waves by a canalization mechanism in a two-dimensional phononic crystal. Appl. Phys. Lett. **93**, 233503 (2008)
47. X. Ao, C. Chan, Far-field image magnification for acoustic waves using anisotropic acoustic metamaterials. Phys. Rev. E **77**, 025601 (2008)
48. Z. Jacob, L.V. Alekseyev, E. Narimanov, Optical hyperlens: far-field imaging beyond the diffraction limit. Opt. Express **14**, 8247 (2006)
49. Z. Liu, H. Lee, Y. Xiong, C. Sun, X. Zhang, Far-field optical hyperlens magnifying sub-diffraction-limited objects. Science **315**, 1686 (2007)
50. D. Smith, D. Schurig, Electromagnetic wave propagation in media with indefinite permittivity and permeability tensors. Phys. Rev. Lett. **90**, 77405 (2003)
51. J. Li, L. Fok, X. Yin, G. Bartal, X. Zhang, Experimental demonstration of an acoustic magnifying hyperlens. Nat. Mater. **8**, 931 (2009)
52. F. Liu, F. Cai, S. Peng, R. Hao, M. Ke, Z. Liu, Parallel acoustic near-field microscope: a steel slab with a periodic array of slits. Phys. Rev. E **80**, 026603 (2009)
53. J. Zhu, J. Christensen, J. Jung, L. Martin-Moreno, X. Yin, L. Fok, X. Zhang, F. Garcia-Vidal, A holey-structured metamaterial for acoustic deep-subwavelength imaging. Nat. Phys. **7**, 52 (2010)
54. A. Spadoni, C. Daraio, Generation and control of sound bullets with a nonlinear acoustic lens. Proc. Natl. Acad. Sci. **107**, 7230 (2010)
55. G.W. Milton, M. Briane, J.R. Willis, On cloaking for elasticity and physical equations with a transformation invariant form. New J. Phys. **8**, 248 (2006)
56. H. Chen, C. Chan, Acoustic cloaking in three dimensions using acoustic metamaterials. Appl. Phys. Lett. **91**, 183518 (2007)
57. S.A. Cummer, D. Schurig, One path to acoustic cloaking. New J. Phys. **9**, 45 (2007)
58. Y. Cheng, F. Yang, J.Y. Xu, X.J. Liu, A multilayer structured acoustic cloak with homogeneous isotropic materials. Appl. Phys. Lett. **92**, 151913 (2008)
59. J. Pendry, J. Li, An acoustic metafluid: realizing a broadband acoustic cloak. New J. Phys. **10**, 115032 (2008)
60. D. Torrent, J. Sánchez-Dehesa, Acoustic cloaking in two dimensions: a feasible approach. New J. Phys. **10**, 063015 (2008)

61. S.A. Cummer, B.I. Popa, D. Schurig, D.R. Smith, J. Pendry, M. Rahm, A. Starr, Scattering theory derivation of a 3d acoustic cloaking shell. Phys. Rev. Lett. **100**, 24301 (2008)
62. A.N. Norris, Acoustic cloaking theory. Proc. R. Soc. A **464**, 2411 (2008)
63. Y. Urzhumov, F. Ghezzo, J. Hunt, D.R. Smith, Acoustic cloaking transformations from attainable material properties. New J. Phys. **12**, 073014 (2010)
64. Y. Bobrovnitskii, Impedance acoustic cloaking. New J. Phys. **12**, 043049 (2010)
65. Z. Liang, J. Li, Bending a periodically layered structure for transformation acoustics. Appl. Phys. Lett. **98**, 241914 (2011)
66. X. Zhu, B. Liang, W. Kan, X. Zou, J. Cheng, Acoustic cloaking by a superlens with single-negative materials. Phys. Rev. Lett. **106**, 14301 (2011)
67. H. Chen, C. Chan, Acoustic cloaking and transformation acoustics. J. Phys. D Appl. Phys. **43**, 113001 (2010)
68. M. Farhat, S. Enoch, S. Guenneau, A. Movchan, Broadband cylindrical acoustic cloak for linear surface waves in a fluid. Phys. Rev. Lett. **101**, 134501 (2008)
69. M. Farhat, S. Guenneau, S. Enoch, A.B.. Movchan, Cloaking bending waves propagating in thin elastic plates. Phys. Rev. B **79**, 033102 (2009)
70. M. Farhat, S. Guenneau, S. Enoch, Ultrabroadband elastic cloaking in thin plates. Phys. Rev. Lett. **103**, 24301 (2009)
71. A. Norris, A. Shuvalov, Elastic cloaking theory. Wave Motion **48**, 525–538 (2011)
72. Y.A. Urzhumov, D.R. Smith, Fluid flow control with transformation media. Phys. Rev. Lett. **107**, 074501 (2011)
73. S. Zhang, C. Xia, N. Fang, Broadband acoustic cloak for ultrasound waves. Phys. Rev. Lett. **106**, 24301 (2011)
74. B.I. Popa, L. Zigoneanu, S.A. Cummer, Experimental acoustic ground cloak in air. Phys. Rev. Lett. **106**, 253901 (2011)
75. B. Liang, B. Yuan, J. Cheng, Acoustic diode: rectification of acoustic energy flux in one-dimensional systems. Phys. Rev. Lett. **103**, 104301 (2009)
76. B. Liang, X. Guo, J. Tu, D. Zhang, J. Cheng, An acoustic rectifier. Nat. Mater. **9**, 989 (2010)
77. N. Boechler, G. Theocharis, C. Daraio, Bifurcation-based acoustic switching and rectification. Nat. Mater. **10**, 665 (2011)
78. X.F. Li, X. Ni, L. Feng, M.H. Lu, C. He, Y.F. Chen, Tunable unidirectional sound propagation through a sonic-crystal-based acoustic diode. Phys. Rev. Lett. **106**, 84301 (2011)
79. Y. Lai, Y. Wu, P. Sheng, Z.Q. Zhang. Hybrid elastic solids. Nat. Mater. **10**, 620 (2011)
80. J. Mei, Z. Liu, J. Shi, D. Tian, Theory for elastic wave scattering by a two-dimensional periodical array of cylinders: an ideal approach for band-structure calculations. Phys. Rev. B **67**, 245107 (2003)
81. J. Mei, Z. Liu, C. Qiu, Multiple-scattering theory for out-of-plane propagation of elastic waves in two-dimensional phononic crystals. J. Phys. Condens. Matter **17**, 3735 (2005)
82. P. Sheng, *Introduction to Wave Scattering, Localization and Mesoscopic Phenomena*, vol. 88, Springer Series in Materials Science (Springer, 2006)
83. J.G. Berryman, Long wavelength propagation in composite elastic media I. Spherical inclusions. J. Acoust. Soc. Am. **68**, 1809 (1980)
84. L.M. Schwartz, D.L. Johnson, Long-wavelength acoustic propagation in ordered and disordered suspensions. Phys. Rev. B **30**, 4302 (1984)
85. R. Lakes, T. Lee, A. Bersie, Y. Wang, Extreme damping in composite materials with negative-stiffness inclusions. Nature **410**, 565 (2001)
86. F. Cervera, L. Sanchis, J. Sanchez-Perez, R. Martinez-Sala, C. Rubio, F. Meseguer, C. Lopez, D. Caballero, J. Sanchez-Dehesa, Refractive acoustic devices for airborne sound. Phys. Rev. Lett. **88**, 23902 (2001)
87. J. Mei, Y. Wu, Z. Liu, Effective medium of periodic fluid-solid composites, Eur. Phys. Lett. **98**, 54001 (2012)

Chapter 6
Damped Phononic Crystals and Acoustic Metamaterials

Mahmoud I. Hussein and Michael J. Frazier

Abstract The objective of this chapter is to introduce the topic of *damping* in the context of both its modeling and its effects in phononic crystals and acoustic metamaterials. First, we provide a brief discussion on the modeling of damping in structural dynamic systems in general with a focus on viscous and viscoelastic types of damping (Sect. 6.2) and follow with a non-exhaustive literature review of prior work that examined periodic phononic materials with damping (Sect. 6.3). In Sect. 6.4, we consider damped 1D diatomic phononic crystals and acoustic metamaterials as example problems (keeping our attention on 1D systems for ease of exposition as in previous chapters). We introduce the generalized form of Bloch's theorem, which is needed to account for both temporal and spatial attenuation of the elastic waves resulting from the presence of damping. We also describe the transformation of the governing equations of motion to a state-space representation to facilitate the treatment of the damping term that arises in the emerging eigenvalue problem. Finally, the effects of dissipation (based on the two types of damping models considered) on the frequency and damping ratio band structures are demonstrated by solving the equations developed for a particular choice of material parameters.

6.1 Introduction

Damping is an innate property of materials and structures. Its consideration in the study of wave propagation is important because of its association with energy dissipation. We can concisely classify the sources of damping in phononic crystals and acoustic metamaterials into three categories, depending on the type and

M.I. Hussein (✉) • M.J. Frazier
Department of Aerospace Engineering Sciences, University of Colorado Boulder, Boulder, CO 80309, USA
e-mail: mih@colorado.edu; michael.frazier@colorado.edu

P.A. Deymier (ed.), *Acoustic Metamaterials and Phononic Crystals*,
Springer Series in Solid-State Sciences 173, DOI 10.1007/978-3-642-31232-8_6,
© Springer-Verlag Berlin Heidelberg 2013

configuration of the unit cell. These are (1) bulk material-level dissipation stemming from deformation processes (e.g., dissipation due to friction between internal crystal planes that slip past each other during deformation); (2) dissipation arising from the presence of interfaces or joints between different components (e.g., lattice structures consisting of interconnected beam elements [1]); and (3) dissipation associated with the presence of a fluid within the periodic structure or in contact with it. In general, the mechanical deformations that take place at the bulk material level, or similarly at interfaces or joints, involve microscopic processes that are not thermodynamically reversible [2]. These processes account for the dissipation of the oscillation energy in a manner that fundamentally alters the macroscopic dynamical characteristics including the shape of the frequency band structure. Similar yet qualitatively different effects occur due to viscous dissipation in the presence of a fluid. While the representation of the inertial and elastic properties of a vibrating structure is adequately accounted for by the usual "mass" and "stiffness" matrices, finding an appropriate damping model to describe observed experimental behavior can be a daunting task. This is primarily due to the difficulty in identifying which state variables the damping forces depend on and in formulating the best functional representation once a set of state variables is determined [3, 4].

6.2 Modeling of Material Damping

Due to the diversity and complexity of dissipative mechanisms, the development of a universal damping model stands as a major challenge. A rather simple model proposed by Rayleigh [5] is the *viscous damping* model in which the instantaneous generalized velocities, $\dot{\mathbf{u}}$, are the only relevant state variables in the calculation of the damping force vector $\mathbf{f}_d$ [4]. Using $\mathbf{C}$ to denote the damping matrix, this relationship is given by

$$\mathbf{f}_d(t) = \mathbf{C}\dot{\mathbf{u}}(t). \tag{6.1}$$

While this description may be suitable when accounting for dissipation associated with the presence of a standard viscous medium (e.g., a Newtonian fluid), a physically realistic model of material damping will generally depend on a wider assortment of state variables. Such a model would represent nonviscous damping, of which *viscoelastic damping* is the most common type.

In treating viscoelasticity, it is suitable to use Boltzmann's hereditary theory whereby the damping force depends upon the past history of motion via a convolution integral over a kernel function $G(t)$:

$$\mathbf{f}_d(t) = \mathbf{C}\int_0^t G(t-\tau)\,\dot{\mathbf{u}}(\tau)\mathrm{d}\tau. \tag{6.2}$$

The kernel function $G(t)$ may take several forms while recognizing that in the limit where $G(t-\tau) = \delta(t-\tau)$, the familiar viscous damping model of (6.1) is recovered [4]. Fundamentally, any form is valid if it guarantees a positive rate of energy

Fig. 6.1 Maxwell model

dissipation. Thus there are numerous possibilities. For example, in Fig. 6.1, the Maxwell model for viscoelastic damping is illustrated; it consists of a linear spring and a viscous dashpot in a series configuration. The spring accounts for the fraction of mechanical energy that is stored during loading, while the viscous dashpot accounts for the remainder that is lost (not stored) from the system. The dashpot also adds a time dependence to the model as the rate of deformation becomes a factor. In this arrangement, the spring and dashpot experience the same axial force, $F = ku = c\dot{u}$. In addition, the total displacement has contributions from both elements, that is, $u = u_s + u_d$, where the subscripts "s" and "d" denote the spring and dashpot, respectively. Differentiating u with respect to time gives $\dot{u} = \dot{u}_s + \dot{u}_d$, which, by recalling the aforementioned equality of force within each element, can be written in the following form:

$$\dot{u} = \frac{\dot{F}}{k} + \frac{F}{c}. \tag{6.3}$$

Assuming an initial displacement $u(0) = F(0)/k$, we can integrate (6.3) with respect to time to obtain the displacement function, $u(t)$. Corresponding to an elongation $u(t) = H(t)$, where $H(t)$ is the unit step-function, a relaxation response function may be expressed as $h(t) = G(t)$. Based on the Maxwell model of Fig. 6.1, the kernel function is [6]

$$G(t) = ke^{-\frac{k}{c}t}H(t). \tag{6.4}$$

If the spring constant $k \to \infty$ in the Maxwell model, then elasticity, the mechanism of storing energy, is lost, and only the dissipative viscous mechanism remains. This is immediately apparent in (6.3), where $\dot{F}/k \to 0$ thus leading to the omission of the force–displacement portion of the Maxwell element. Returning to (6.2), we will use a more general form of the kernel function, $G(t) = \mu_1 e^{-\mu_2 t}H(t)$, where the constants $\mu_{1,2}$ are called relaxation parameters and may be determined from experiment.

6.3 Elastic Wave Propagation in Damped Periodic Media

There are several studies in the literature that consider the treatment of damping in the context of periodic phononic materials. Many of these focus on simulating finite periodic structures (e.g., [7–11]), which is different from carrying out a unit cell analysis. The latter approach has the advantage that it allows us to elucidate the broad effects of damping on the band structure characteristics. It is, therefore, more comprehensive because it provides information that can be relevant to a range of finite-structure simulation scenarios.

In unit cell analysis, the dynamics of a periodic material [(e.g., atomic-scale crystalline materials, phononic (or photonic) crystals, periodic acoustic (or electro-magnetic) metamaterials] is fully characterized by the application of Bloch's theorem [12] on a single representative unit cell. As discussed in earlier chapters, this theorem states that the wave field in a periodic medium is also periodic, except that its periodicity is determined by the frequency versus wavevector dispersion relation. The form of the displacement response in a non-dissipative phononic material following Bloch's theorem is given by

$$\mathbf{u}(\mathbf{x}, \boldsymbol{\kappa}; t) = \tilde{\mathbf{u}}(\mathbf{x}, \boldsymbol{\kappa}) e^{i(\boldsymbol{\kappa} \cdot \mathbf{x} - \omega t)}, \tag{6.5}$$

where $\mathbf{u} = \{u_x, u_y, u_z\}$ is the displacement field, $\tilde{\mathbf{u}}$ is the displacement Bloch function, $\mathbf{x} = \{x, y, z\}$ is the position vector, $\boldsymbol{\kappa} = \{\kappa_x, \kappa_y, \kappa_z\}$ is the wavevector, $i = \sqrt{-1}$, and ω and t denote real-valued temporal frequency and time, respectively.

Among the earlier studies that adopted a unit cell approach is the paper by Mead [13], which presented 1D discrete mass–spring–dashpot models and solved for the dispersion under structural damping (i.e., damping exhibiting velocity-independent forces). Similarly, Mukherjee and Lee [14], Castanier and Pierre [10], Zhang et al. [15], and Merheb et al. [16] provided dispersion relations using a complex elastic modulus (or a convolution integral expression in the case of [16]), and Langley [17] presented a corresponding analysis using a complex inertial term to account for the damping. In these studies, damping has therefore been incorporated in either the stiffness or mass matrix in the governing equations. Representing damping directly in the form of viscous or viscoelastic forces (as discussed in Sect. 6.2) represents another avenue, but requires the incorporation of an additional state variable—velocity—in the governing equations of motion. Naturally, this leads to an eigenvalue problem with a non-traditional format. Several studies considered this problem for different types of configurations (i.e., concerning the geometry, boundary conditions, and constitutive material behavior), for example, [18].

A critical limitation to using (6.5) is that it assumes that a spatially propagating wave does not attenuate in time [19, 20]. The allowance of a temporal loss factor was adopted by Mukherjee and Lee [14, 21] in their investigations of damped periodic composites (although limited to structural damping). In recent publications [22–24], the possibility of temporal attenuation has been incor-porated in the Bloch formulation for viscous damping, and it was shown that such treatment is consistent with results emanating from a free vibration analysis of corresponding finite periodic structures (whose theory of damping is well established [4]). It was observed that the band gaps in the damped unit cell dispersion match with the damped natural frequency gaps in the finite periodic structures only when the temporal component of Bloch's theorem is generalized to include a complex root λ, that is,

$$\mathbf{u}(\mathbf{x}, \boldsymbol{\kappa}; t) = \tilde{\mathbf{u}}(\mathbf{x}, \boldsymbol{\kappa}) e^{i(\boldsymbol{\kappa} \cdot \mathbf{x})} e^{\lambda t}. \tag{6.6}$$

Here it should be noted that for the undamped case, $\lambda = -i\omega$ and (6.5) is recovered. While this form was successfully applied in [22–24], the models considered were limited to simple Kelvin–Voigt viscous damping models. In the next section, we present formulations for analyzing damped one-dimensional diatomic phononic crystals and acoustic metamaterials on the basis of the generalized Bloch's theorem and for both viscous and viscoelastic damping as described in Sect. 6.2.

6.4 Damped One-Dimensional Diatomic Phononic Crystals and Acoustic Metamaterials

By considering lumped masses, springs, and damping elements, in this section we construct a simple 1D model of a damped diatomic phononic crystal (represented by a "mass-and-mass" configuration as shown in Fig. 6.2a) and similarly a simple ID model of a damped diatomic acoustic metamaterial (represented by a "mass-in-mass" configuration as shown in Fig. 6.2b).

Considering unit cell periodicity, the set of homogeneous differential equations describing the motion of each mass in the phononic crystal model is obtained as follows [23]:

$$m_1 \ddot{u}_1^j + (c_1 + c_2)\dot{u}_1^j - c_2 \dot{u}_2^j - c_1 \dot{u}_2^{j-1} + (k_1 + k_2)u_1^j - k_2 u_2^j - k_1 u_2^{j-1} = 0, \quad (6.7)$$

$$m_2 \ddot{u}_2^j + (c_1 + c_2)\dot{u}_2^j - c_2 \dot{u}_1^j - c_1 \dot{u}_1^{j+1} + (k_1 + k_2)u_2^j - k_2 u_1^j - k_1 u_1^{j+1} = 0, \quad (6.8)$$

where u_α^j is the displacement of mass α in an arbitrary jth unit cell. In general, a unit cell and its neighbors may be identified by $j + n$, where $n = 0, -1, 1$ denote the present, previous, and subsequent unit cells, respectively. Similarly, for the acoustic metamaterial model, the equations of motion corresponding to the two masses are [24]

$$m_1 \ddot{u}_1^j + c_1 \left(2\dot{u}_1^j - \dot{u}_1^{j-1} - \dot{u}_1^{j+1} \right) + c_2 \left(\dot{u}_1^j - \dot{u}_2^j \right) + k_1 \left(2u_1^j - u_1^{j-1} - u_1^{j+1} \right)$$
$$+ k_2 \left(u_1^j - u_2^j \right) = 0, \quad (6.9)$$

$$m_2 \ddot{u}_2^j + k_2 \left(u_2^j - u_1^j \right) + c_2 \left(\dot{u}_2^j - \dot{u}_1^j \right) = 0. \quad (6.10)$$

6.4.1 Generalized Bloch's theorem and State-Space Transformation

Generalized Bloch's theorem
Writing the generalized Bloch's theorem of (6.6) in discrete format for the models shown in Fig. 6.2 involves the product of the spatial function $A(x, \kappa) = e^{i(\kappa x + n\kappa a)}$

Fig. 6.2 Lumped parameter
unit cell model of (**a**)
phononic crystal (mass-and-
mass) and (**b**) acoustic
metamaterial (mass-in-mass)

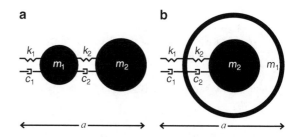

and a temporal function, which takes the form of $\bar{U}_\alpha(t) = \tilde{U}_\alpha e^{\lambda t}$. Here, $\tilde{U}_\alpha$ represents the complex wave amplitude, the variable a denotes the lattice constant, and $\kappa = \kappa_x$ for brevity. Thus, the displacement function of mass α in the $(j+n)$th unit cell is given by

$$u_\alpha^{j+n}(x, \kappa; t) = A(x, \kappa)\bar{U}_\alpha(t) = \tilde{U}_\alpha e^{i(\kappa x + n\kappa a) + \lambda t}. \tag{6.11}$$

If we apply this form of the Bloch wave solution to (6.7) and (6.8) for the phononic crystal, we obtain the following complex eigenvalue problem:

$$\lambda^2 m_1 \tilde{U}_1 + \lambda(c_1 + c_2)\tilde{U}_1 - \lambda c_2 \tilde{U}_2 - \lambda c_1 \tilde{U}_2 e^{-i\kappa a} + (k_1 + k_2)\tilde{U}_1 - k_2 \tilde{U}_2 - k_1 \tilde{U}_2 e^{-i\kappa a} = 0, \tag{6.12a}$$

$$\lambda^2 m_2 \tilde{U}_2 + \lambda(c_1 + c_2)\tilde{U}_2 - \lambda c_2 \tilde{U}_1 - \lambda c_1 \tilde{U}_1 e^{i\kappa a} + (k_1 + k_2)\tilde{U}_2 - k_2 \tilde{U}_1 - k_1 \tilde{U}_1 e^{i\kappa a} = 0, \tag{6.12b}$$

which in matrix form is represented as

$$\begin{bmatrix} \lambda^2 m_1 + \lambda(c_1 + c_2) + k_1 + k_2 & -\lambda(c_1 e^{-i\kappa a} + c_2) - (e^{-i\kappa a} k_1 + k_2) \\ -\lambda(c_1 e^{i\kappa a} + c_2) - (e^{i\kappa a} k_1 + k_2) & \lambda^2 m_2 + \lambda(c_1 + c_2) + k_1 + k_2 \end{bmatrix} \begin{bmatrix} \tilde{U}_1 \\ \tilde{U}_2 \end{bmatrix} = \begin{bmatrix} 0 \\ 0 \end{bmatrix}. \tag{6.12c}$$

Equation (6.12c) can be segregated in the following manner:

$$\left[\lambda^2 \mathbf{M} + \lambda \mathbf{C}(\kappa) + \mathbf{K}(\kappa)\right]\tilde{\mathbf{U}} = \mathbf{0}. \tag{6.13}$$

Thus, we identify the mass matrix $\mathbf{M}$, damping matrix $\mathbf{C}(\kappa)$, and stiffness matrix $\mathbf{K}(\kappa)$ as

$$\mathbf{M} = \begin{bmatrix} m_1 & 0 \\ 0 & m_2 \end{bmatrix}, \tag{6.14}$$

$$\mathbf{C}(\kappa) = \begin{bmatrix} c_1 + c_2 & -(c_1 e^{-i\kappa a} + c_2) \\ -(c_1 e^{i\kappa a} + c_2) & c_1 + c_2 \end{bmatrix}, \tag{6.15}$$

$$\mathbf{K}(\kappa) = \begin{bmatrix} k_1 + k_2 & -(k_1 e^{-i\kappa a} + k_2) \\ -(k_1 e^{i\kappa a} + k_2) & k_1 + k_2 \end{bmatrix}. \tag{6.16}$$

Applying (6.11) to (6.9) and (6.10) for the acoustic metamaterial yields the following complex eigenvalue problem:

$$\lambda^2 m_1 \tilde{U}_1 + 2\lambda c_1(1 - \cos\kappa a)\tilde{U}_1 + \lambda c_2(\tilde{U}_1 - \tilde{U}_2) + 2k_1(1 - \cos\kappa a)\tilde{U}_1 + k_2(\tilde{U}_1 - \tilde{U}_2) = 0, \tag{6.17a}$$

$$\lambda^2 m_2 \tilde{U}_2 + \lambda c_2(\tilde{U}_2 - \tilde{U}_1) + k_2(\tilde{U}_2 - \tilde{U}_1) = 0, \tag{6.17b}$$

which in matrix form can be represented as

$$\begin{bmatrix} \lambda^2 m_1 + 2\lambda c_1(1 - \cos\kappa a) + \lambda c_2 + 2k_1(1 - \cos\kappa a) + k_2 & -(\lambda c_2 + k_2) \\ -(k_2 + \lambda c_2) & \lambda^2 m_2 + \lambda c_2 + k_2 \end{bmatrix} \begin{bmatrix} \tilde{U}_1 \\ \tilde{U}_2 \end{bmatrix} = \begin{bmatrix} 0 \\ 0 \end{bmatrix}. \tag{6.17c}$$

Again, by segregating the coefficients, the mass matrix will be as given in (6.14) and the damping and stiffness matrices will be identified and written as follows:

$$\mathbf{C}(\kappa) = \begin{bmatrix} 2c_1(1 - \cos\kappa a) + c_2 & -c_2 \\ -c_2 & c_2 \end{bmatrix}, \tag{6.18}$$

$$\mathbf{K}(\kappa) = \begin{bmatrix} 2k_1(1 - \cos\kappa a) + k_2 & -k_2 \\ -k_2 & k_2 \end{bmatrix}. \tag{6.19}$$

Finally, if we define the set of material parameters r_m, r_c, and r_k as follows:

$$r_m = m_2/m_1, \qquad r_c = c_2/c_1, \qquad r_k = k_2/k_1, \tag{6.20}$$

then we may write the system matrices for each of the two models in a more convenient form. Thus the system matrices for the phononic crystal become

$$\mathbf{M} = m_2 \begin{bmatrix} 1/r_m & 0 \\ 0 & 1 \end{bmatrix} = m_2 \mathbf{M}_r, \tag{6.21}$$

$$\mathbf{C}(\kappa) = c_2 \begin{bmatrix} 1 + 1/r_c & -(1 + e^{-i\kappa a}/r_c) \\ -(1 + e^{i\kappa a}/r_c) & 1 + 1/r_c \end{bmatrix} = c_2 \mathbf{C}_r(\kappa), \tag{6.22}$$

$$\mathbf{K}(\kappa) = k_2 \begin{bmatrix} 1 + 1/r_k & -\left(1 + e^{-i\kappa a}/r_k\right) \\ -\left(1 + e^{i\kappa a}/r_k\right) & 1 + 1/r_k \end{bmatrix} = k_2 \mathbf{K}_r(\kappa), \qquad (6.23)$$

and for the acoustic metamaterial,

$$\mathbf{C}(\kappa) = c_2 \begin{bmatrix} 2(1 - \cos \kappa a)/r_c + 1 & -1 \\ -1 & 1 \end{bmatrix} = c_2 \mathbf{C}_r(\kappa), \qquad (6.24)$$

$$\mathbf{K}(\kappa) = k_2 \begin{bmatrix} 2(1 - \cos \kappa a)/r_k + 1 & -1 \\ -1 & 1 \end{bmatrix} = k_2 \mathbf{K}_r(\kappa), \qquad (6.25)$$

and $\mathbf{M}$ is the same as in (6.21). It should be noted that in general the matrices $\mathbf{C}_r(\kappa)$ and $\mathbf{K}_r(\kappa)$ are unitary matrices.

State-space transformation

For general damping, viscous or nonviscous, the equations of motion cannot be uncoupled by using an alternate set of coordinates (as done, for example, in [22], which treated proportional Rayleigh damping using Bloch modal analysis). To determine the complex eigenvalues, λ_s, $s = 1, 2$, we develop a Bloch state-space formulation for each of the damping types. The formulation is based on a transformation of variables of the form:

$$\bar{\mathbf{Y}} = \begin{bmatrix} \dot{\bar{\mathbf{U}}} \\ \bar{\mathbf{U}} \end{bmatrix}. \qquad (6.26)$$

6.4.1.1 Viscous Damping

For general viscous damping, the Block state-space formulation is as follows [23]:

$$\begin{bmatrix} \mathbf{0} & m_2\mathbf{M}_r \\ m_2\mathbf{M}_r & c_2\mathbf{C}_r(\kappa) \end{bmatrix} \dot{\bar{\mathbf{Y}}} + \begin{bmatrix} -m_2\mathbf{M}_r & \mathbf{0} \\ \mathbf{0} & k_2\mathbf{K}_r(\kappa) \end{bmatrix} \bar{\mathbf{Y}} = \mathbf{0}. \qquad (6.27)$$

Now we write the solution as $\bar{\mathbf{Y}} = \tilde{\mathbf{Y}}e^{\gamma t}$. It is at this point that we introduce, for convenience, two additional material parameters, $\bar{\omega} = \sqrt{k_2/m_2}$ and $\beta = c_2/m_2$. Thus, (6.27) becomes

$$\left(\begin{bmatrix} \mathbf{0} & \mathbf{M}_r \\ \mathbf{M}_r & \beta\mathbf{C}_r(\kappa) \end{bmatrix} \gamma + \begin{bmatrix} -\mathbf{M}_r & \mathbf{0} \\ \mathbf{0} & \bar{\omega}^2\mathbf{K}_r(\kappa) \end{bmatrix} \right) \tilde{\mathbf{Y}} = \mathbf{0}. \qquad (6.28)$$

Equation (6.28), which is a first-order representation of the original second-order eigenvalue problem, has two complex conjugate pairs of eigenvalues, γ, and eigenvectors, $\tilde{\mathbf{Y}}$. Given their orthogonality with respect to the system matrices, the eigenvectors decouple the equations into four modal equations with complex roots γ_{s^*}, $s^* = 1, \ldots, 4$ appearing in complex conjugate pairs and thus effectively representing two single-degree of freedom systems. Thus we can write

$$\gamma_s = -\xi_s \omega_s \pm i \omega_{d_s} = -\xi_s \omega_s \pm i \omega_s \sqrt{1 - \xi_s^2}, \quad s = 1, 2. \qquad (6.29)$$

If we focus our attention on only the first eigenvalue in each complex conjugate pair, then we extract the wavenumber-dependent damped frequency as:

$$\omega_{ds}(\kappa) = \text{Im}[\gamma_s(\kappa)], \quad s = 1, 2, \qquad (6.30)$$

and the corresponding wavenumber-dependent damping ratio:

$$\xi_s(\kappa) = -\frac{\text{Re}[\gamma_s(\kappa)]}{\text{Abs}[\gamma_s(\kappa)]}, \quad s = 1, 2. \qquad (6.31)$$

Note that $\omega_s(\kappa) = \text{Abs}[\gamma_s(\kappa)]$ and is referred to as the resonant frequency. For the special case of proportional viscous damping, the resonant frequency is equal to the undamped frequency.

6.4.1.2 Viscoelastic Damping

In this section, we replace the viscous damping elements in Fig. 6.2 with Maxwell elements and apply the Bloch state-space approach to the viscoelastic case by introducing a set of internal variables. We develop the state-space matrices using an approach proposed by Wagner and Adhikari [25] for finite structural dynamics systems; here we extend it to the analysis of the unit cell problem. The approach is specific to the case in which the constants $\mu_{1,2}$ in (6.4) are equal (i.e., $\mu = \mu_1 = \mu_2$).

According to [25], we define an internal variable $\bar{\mathbf{V}}(t)$ as follows:

$$\bar{\mathbf{V}}(t) = \int_0^t \mu e^{-\mu(t-\tau)} \dot{\bar{\mathbf{U}}}(\tau) d\tau. \qquad (6.32)$$

The Leibniz integral rule gives the following formula for differentiation of a definite integral whose limits are functions of the differential variable:

$$\frac{\partial}{\partial t} \int_{a(t)}^{b(t)} f(x, t) dt = \int_{a(t)}^{b(t)} \frac{\partial}{\partial t} f(x, t) dt + f[b(t), t] \cdot \frac{\partial}{\partial t} b(t) - f[a(t), t] \frac{\partial}{\partial t} a(t). \qquad (6.33)$$

Applying this rule to (6.32), we obtain

$$\dot{\mathbf{V}}(t) = \int_0^t -\mu^2 e^{-\mu(t-\tau)}\dot{\mathbf{U}}(\tau)d\tau + \mu\dot{\mathbf{U}}(t) = \mu\left[\dot{\mathbf{U}}(t) - \bar{\mathbf{V}}(t)\right], \qquad (6.34)$$

which we may rewrite as

$$\dot{\bar{\mathbf{V}}}(t) + \mu\bar{\mathbf{V}}(t) = \mu\dot{\mathbf{U}}(t). \qquad (6.35)$$

According to (6.2) and (6.4), the system of equations for a viscoelastically damped chain is

$$\mathbf{M}\ddot{\mathbf{U}} + \mathbf{C}(\kappa)\int_0^t \mu e^{-\mu(t-\tau)}\dot{\mathbf{U}}(\tau)d\tau + \mathbf{K}(\kappa)\bar{\mathbf{U}} = 0. \qquad (6.36)$$

Utilizing (6.21)–(6.25) and the definitions $\bar{\omega}$ and β, we get:

$$\mathbf{M}_r\ddot{\bar{\mathbf{U}}} + \beta\mathbf{C}_r(\kappa)\int_0^t \mu e^{-\mu(t-\tau)}\dot{\mathbf{U}}d\tau + \bar{\omega}^2\mathbf{K}_r(\kappa)\bar{\mathbf{U}} = \mathbf{0}. \qquad (6.37)$$

With (6.32), this becomes

$$\mathbf{M}_r\ddot{\bar{\mathbf{U}}} + \beta\mathbf{C}_r(\kappa)\bar{\mathbf{V}} + \bar{\omega}^2\mathbf{K}_r(\kappa)\bar{\mathbf{U}} = \mathbf{0}. \qquad (6.38)$$

Next, we solve for $\bar{\mathbf{V}}(t)$ in (6.35) and substitute the result into (6.38):

$$\mathbf{M}_r\ddot{\bar{\mathbf{U}}} + \beta\mathbf{C}_r(\kappa)\left[\dot{\bar{\mathbf{U}}} - \frac{1}{\mu}\dot{\bar{\mathbf{V}}}\right] + \bar{\omega}^2\mathbf{K}_r(\kappa)\bar{\mathbf{U}} = \mathbf{0}. \qquad (6.39)$$

At this point, incorporating (6.39) into a state-space matrix equation format will result in non-square matrices. To produce square and block-symmetric state-space matrices, we formulate another equation. Premultiplying (6.35) by $\mathbf{C}_r(\kappa)$ and dividing by μ^2 yields

$$-\frac{1}{\mu}\mathbf{C}_r(\kappa)\dot{\bar{\mathbf{U}}} + \frac{1}{\mu^2}\mathbf{C}_r(\kappa)\dot{\bar{\mathbf{V}}} + \frac{1}{\mu}\mathbf{C}_r(\kappa)\bar{\mathbf{V}} = \mathbf{0}. \qquad (6.40)$$

In the first-order state-space form, (6.39) and (6.40) become

$$\begin{bmatrix} \mathbf{0} & \mathbf{M}_r & \mathbf{0} \\ \mathbf{M}_r & \beta\mathbf{C}_r(\kappa) & -\dfrac{\beta}{\mu}\mathbf{C}_r(\kappa) \\ \mathbf{0} & -\dfrac{\beta}{\mu}\mathbf{C}_r(\kappa) & \dfrac{\beta}{\mu^2}\mathbf{C}_r(\kappa) \end{bmatrix}\dot{\mathbf{Z}} + \begin{bmatrix} -\mathbf{M}_r & \mathbf{0} & \mathbf{0} \\ \mathbf{0} & \bar{\omega}^2\mathbf{K}_r(\kappa) & \mathbf{0} \\ \mathbf{0} & \mathbf{0} & \dfrac{\beta}{\mu}\mathbf{C}_r(\kappa) \end{bmatrix}\bar{\mathbf{Z}} = \mathbf{0}, \quad (6.41)$$

where $\bar{\mathbf{Z}} = \begin{bmatrix} \dot{\bar{\mathbf{U}}} & \bar{\mathbf{U}} & \bar{\mathbf{V}} \end{bmatrix}^{\mathrm{T}}$. We assume the solution $\bar{\mathbf{Z}} = \tilde{\mathbf{Z}}e^{\gamma t}$ and subsequently develop the eigenvalue problem:

$$\left(\begin{bmatrix} \mathbf{0} & \mathbf{M}_r & \mathbf{0} \\ \mathbf{M}_r & \beta\mathbf{C}_r(\kappa) & -\frac{\beta}{\mu}\mathbf{C}_r(\kappa) \\ \mathbf{0} & -\frac{\beta}{\mu}\mathbf{C}_r(\kappa) & \frac{\beta}{\mu^2}\mathbf{C}_r(\kappa) \end{bmatrix} \gamma + \begin{bmatrix} -\mathbf{M}_r & \mathbf{0} & \mathbf{0} \\ \mathbf{0} & \bar{\omega}^2\mathbf{K}_r(\kappa) & \mathbf{0} \\ \mathbf{0} & \mathbf{0} & \frac{\beta}{\mu}\mathbf{C}_r(\kappa) \end{bmatrix} \right)\tilde{\mathbf{Z}} = \mathbf{0}.$$

(6.42)

Upon obtaining the eigenvalues γ_{s^*}, now mathematically a set of six values, $s^* = 1,\ldots,6$, we can extract the two sets of roots appearing as complex conjugate pairs (the two remaining roots are spurious) that physically represent the modes of damped wave propagation, exactly as defined in (6.30) and (6.31).

As implied in the above formulation, and supported by the definition of the Maxwell element [Fig. 6.1 and (6.3)], in the limit $\mu \to \infty$, the viscous Bloch state-space formulation of (6.28) is recovered. That is, high values of μ represent more viscous behavior (less dependence on the past history), while low values of μ represent more viscoelastic behavior (more dependence on the past history).

6.4.2 Damped Bragg Scattering and Local Resonance

For both the phononic crystal and the acoustic metamaterial models in Fig. 6.2, we generate dispersion curves using, for demonstration, a specific set of material parameters: $r_m = 9$, $r_c = 0.5$, $r_k = 1$, and $\bar{\omega} = 149.07$ rad/s. The parameter β is varied to show the dependence on the damping intensity. In Figs. 6.3–6.6, plots corresponding to the phononic crystal (mass-and-mass model) appear on the left while those pertaining to the acoustic metamaterial (mass-in-mass model) appear on the right. While these results are dependent on values chosen for the parameters r_m, r_c, and r_k, they provide a basic insight into the effects of damping on the elastodynamic behavior of phononic crystals and acoustic metamaterials.

6.4.2.1 Viscous Damping

Here we show the frequency (Fig. 6.3) and damping ratio (Fig. 6.4) band structures for the case of viscous damping, obtained by solving (6.28). We observe in Fig. 6.3 that as the damping intensity $\beta/\bar{\omega}$ is increased, the optical branch for both the phononic crystal and the acoustic metamaterial drops while the acoustic branch experiences little change—this in turn leads to a reduction in the size of the band gap. For the phononic crystal, the descent of the optical branch takes place at slightly faster rates at low wavenumbers compared to high wavenumbers, whereas for the acoustic metamaterial, significantly higher drop rates take place at high

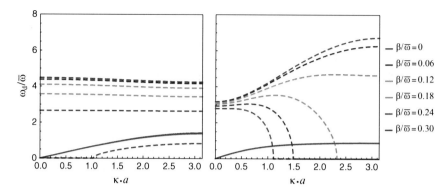

Fig. 6.3 Frequency band structure for viscous damping case: phononic crystal (*left*) and acoustic metamaterial (*right*)

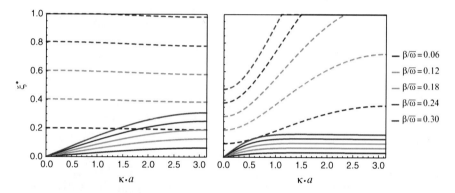

Fig. 6.4 Damping ratio band structure for viscous damping case: phononic crystal (*left*) and acoustic metamaterial (*right*)

wavenumbers compared to low wavenumbers. The damping ratio band diagram in Fig. 6.4 follows a corresponding trend with an indication that the effect of damping is slightly more significant at low wavenumbers for the phononic crystal and noticeably more significant at high wavenumbers for the acoustic metamaterial. We also observe that at low wavenumbers, the damping ratio values for the phononic crystal exceed those of the acoustic metamaterial for a given damping intensity $\beta/\bar{\omega}$ and vice versa at high wavenumbers. With regard to the damping ratios of the acoustic branch modes, these are higher for the phononic crystal compared to the acoustic metamaterial.

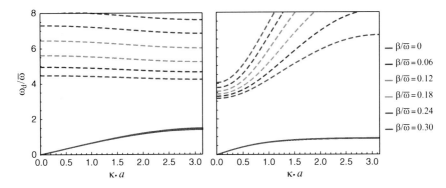

Fig. 6.5 Frequency band structure for viscoelastic damping case: phononic crystal (*left*) and acoustic metamaterial (*right*)

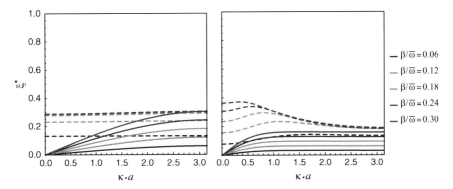

Fig. 6.6 Damping ratio band structure for viscoelastic damping case: phononic crystal (*left*) and acoustic metamaterial (*right*)

6.4.2.2 Viscoelastic damping

As noted in Sect. 6.4.1.2, viscoelastic behavior is better represented by "low" μ and, in contrast, viscous behavior is better represented by "high" μ. What qualifies as a low/high value of μ depends on the other parameters in the damping matrix $\mathbf{C}(\kappa)$. If μ is such that $\frac{1}{\mu}\mathbf{C}(\kappa) \approx \mathbf{0}$ compared to $\mathbf{C}(\kappa)$ (i.e., $\mathbf{C}(\kappa) + \frac{1}{\mu}\mathbf{C}(\kappa) \approx \mathbf{C}(\kappa)$), then dominantly viscous behavior can be expected. For our models, we find that $\mu = 10^8$ reflects viscous behavior well. To represent the viscoelastic behavior, we set $\mu = 10^3$ and show the solution of (6.42) in Figs. 6.5 and 6.6.

Unlike the viscous damping case, we notice that in this model of viscoelastic damping, the optical dispersion branches for both the phononic crystal and the acoustic metamaterial rise with the damping intensity. Subsequently, this increases the size of the band gap—a desirable feature for many applications such as vibration isolation and frequency sensing. The sensitivity of the changes in the optical branches again seems to be highest in the acoustic metamaterial at high wavenumbers. The damping ratio band structures shown in Fig. 6.6 indicate less sensitivity overall to damping intensity compared to the viscous damping case. This is because a portion of the energy associated with the viscoelastic forces is stored in the spring element (as discussed in Sect. 6.2) and not dissipated.

Recall that both the phononic crystal and the acoustic metamaterial models considered are based on the same ratios of mass, stiffness, and damping (see Fig. 6.2 and 6.20). Yet the effects of damping are different. This is a manifestation of the fundamental difference in the wave propagation mechanism of Bragg scattering (that takes place in phononic crystals) and that of local resonance (that takes place in acoustic metamaterials).

References

1. A.S. Phani, J. Woodhouse, N.A. Fleck, Wave propagation in two-dimensional periodic lattices. J. Acoust. Soc. Am. **119**, 1995–2005 (2006)
2. A.S. Nowick, B.S. Berry, *Anelastic Relaxation in Crystalline Solids* (Academic, New York, 1972)
3. C.W. Bert, Material damping: an introductory review of mathematic measures and experimental techniques. J. Sound Vib. **29**, 129–153 (1973)
4. J. Woodhouse, Linear damping models for structural vibration. J. Sound Vib. **215**, 547–569 (1998)
5. J.W.S. Rayleigh, *Theory of Sound* (Macmillan and Co., London, 1878)
6. Y.C. Fung, P. Tong, *Classical and Computational Solid Mechanics* (World Scientific Publishing Co., Singapore, 2001)
7. Y. Yong, Y.K. Lin, Propagation of decaying waves in periodic and piecewise periodic structures of finite length. J. Sound Vib. **129**, 99–118 (1989)
8. I.E. Psarobas, Viscoelastic response of sonic band-gap materials. Phys. Rev. B **64**, 012303 (2001)
9. P.W. Mauriz, M.S. Vasconcelos, E.L. Albuquerque, Acoustic phonon power spectra in a periodic superlattice. Phys. Status Solidi B **243**, 1205–1211 (2006)
10. M.P. Castanier, C. Pierre, Individual and interactive mechanisms for localization and dissipation in a mono-coupled nearly-periodic structure. J. Sound Vib. **168**, 479–505 (1993)
11. Y. Liu, D. Yu, H. Zhao, J. Wen, X. Wen, Theoretical study of 2D PC with viscoelasticity based on fractional derivative models. J. Phys. D Appl. Phys. **41**, 065503 (2008)
12. F. Bloch, Über die quantenmechanik der electron in kristallgittern. Z. Phys. **52**, 555–600 (1928)
13. D.J. Mead, A general theory of harmonic wave propagation in linear periodic systems with multiple coupling. J. Sound Vib. **27**, 235–260 (1973)
14. S. Mukherjee, E.H. Lee, Dispersion relations and mode shapes for waves in laminated viscoelastic composites by finite difference methods. Comput. Struct. **5**, 279–285 (1975)
15. X. Zhang, Z. Liu, J. Mei, Y. Liu, Acoustic band gaps for a 2D periodic array of solid cylinders in viscous liquid. J. Phys. Condens. Mater. **15**, 8207–8212 (2003)
16. B. Merheb, P.A. Deymier, M. Jain, M. Aloshyna-Lesuffleur, S. Mohanty, A. Baker, R.W. Greger, Elastic and viscoelastic effects in rubber-air acoustic band gap structures: a theoretical and experimental study. J. Appl. Phys. **104**, 064913 (2008)
17. R.S. Langley, On the forced response of one-dimensional periodic structures: vibration localization by damping. J. Sound Vib. **178**, 411–428 (1994)
18. E. Tassilly, Propagation of bending waves in a periodic beam. Int. J. Eng. Sci. **25**, 85–94 (1987)

19. R.P. Moiseyenko, V. Laude, Material loss influence on the complex band structure and group velocity in phononic crystals. Phys. Rev. B **83**, 064301 (2011)
20. F. Farzbod, M.J. Leamy, Analysis of Bloch's method in structures with energy dissipation. J. Vib. Acoust. **133**, 051010 (2011)
21. R. Sprik, G.H. Wegdam, Acoustic band gaps in composites of solids and viscous liquids. Solid State Commun. **106**, 77–81 (1998)
22. M.I. Hussein, Theory of damped Bloch waves in elastic media. Phys. Rev. B **80**, 212301 (2009)
23. M.I. Hussein, M.J. Frazier, Band structures of phononic crystals with general damping. J. Appl. Phys. **108**, 093506 (2010)
24. M. J. Frazier, M. I. Hussein, Dissipative Effects in Acoustic Metamaterials, in *Proceedings of Phononics 2011*, Paper Phononics-2011-0172, Santa Fe, New Mexico, USA, May 29–June 2, 2011, pp. 84–85
25. N. Wagner, S. Adhikari, Symmetric state-space method for a class of nonviscously damped systems. AIAA J. **41**, 951–956 (2003)

Chapter 7
Nonlinear Periodic Phononic Structures and Granular Crystals

G. Theocharis, N. Boechler, and C. Daraio

Abstract This chapter describes the dynamic behavior of nonlinear periodic phononic structures, along with how such structures can be utilized to affect the propagation of mechanical waves. Granular crystals are one type of nonlinear periodic phononic structure and are the focus of this chapter. The chapter begins with a brief history of nonlinear lattices and an introduction to granular crystals. This is followed by a summary of past and recent work on one-dimensional (1D) and two-dimensional (2D) granular crystals, which is categorized according to the crystals' periodicity and dynamical regime. The chapter is concluded with a commentary by the authors, which discusses several possible future directions relating to granular crystals and other nonlinear periodic phononic structures. Throughout this chapter, a richness of nonlinear dynamic effects that occur in granular crystals is revealed, including a plethora of phenomena with no linear analog such as solitary waves, discrete breathers, tunable frequency band gaps, bifurcations, and chaos. Furthermore, in addition to the description of fundamental nonlinear phenomena, the authors describe how such phenomena can enable novel engineering devices and be applied to other nonlinear periodic systems.

7.1 Introduction

7.1.1 Nonlinearity in Periodic Phononic Structures

The effect of structural periodicity on wave propagation has been studied in a wide array of fields. This includes vibrations in spring-mass systems, electrons in crystalline lattices, light waves in photonic periodic structures, cold atoms in optical lattices, and plasmons in networks of Josephson junctions or metal surfaces [1–4].

G. Theocharis (✉) • N. Boechler • C. Daraio
Engineering and Applied Science, California Institute of Technology, Pasadena, CA 91125, USA
e-mail: Georgios.Theocharis@univ-lemans.fr

P.A. Deymier (ed.), *Acoustic Metamaterials and Phononic Crystals*,
Springer Series in Solid-State Sciences 173, DOI 10.1007/978-3-642-31232-8_7,
© Springer-Verlag Berlin Heidelberg 2013

The preceding chapters of this book have considered, in particular, the effect of structural discreteness and periodicity on the propagation of phonons, sound, and other mechanical waves. Phononic crystals and acoustic metamaterials are examples of materials designed for this purpose. By studying the linear response of these systems, many common properties are revealed, such as the existence of band gaps. However, as the amplitude of the wave excitation is increased, the response of the material can become nonlinear and the wave propagation becomes more complex. As a result, the study of nonlinearity in periodic structures has revealed unique phenomena with no analogs in linear theory. Such phenomena include nonlinear resonances, bifurcations, chaos, self-trapping, and intrinsic localization. Nonlinear devices thus have potential for novel applications such as frequency conversion, energy harvesting, and switching, among others.

Although the role of nonlinearity has been extensively studied in non-phononic periodic structures and metamaterials, such as photonic periodic structures, optical metamaterials, and atomic Bose-Einstein Condensates in optical lattices [5], there are thus-far few examples of nonlinear phononic crystals or nonlinear acoustic metamaterials. Potential sources of nonlinearity in phononic/acoustic materials can be categorized into (1) intrinsic and (2) extrinsic. The former derives from nonlinearities in the material constitutive response (i.e., interatomic forces, nonlinear elasticity, plasticity, or ferroelasticity) [6]. The latter derives from the geometry or topology of the fundamental building blocks (i.e., contact forces between particles [7], deformation of micro-nano mechanical oscillators [8], or the nonlinearity related to geometrical instabilities [9]).

Homogenous materials with nonlinear elastic [6] or nonlinear acoustic responses [10] have long been studied. Nonlinear bulk and surface waves, resulting from the interplay between the intrinsic nonlinearity and geometrical dispersion, have also been studied and observed in solids [11–13]. However, until recently, this research has not been combined with the new capabilities of linear phononic crystals and acoustic metamaterials, as described in the previous chapters of this book. The far most studied nonlinear periodic phononic structures within the sonic regime (0–20 kHz) are granular crystals. Granular crystals are arrays of elastic particles in contact [14] whose nonlinearity results from the geometry of adjacent particles. In addition to granular crystals, some of the few studied examples of nonlinear periodic phononic structures are as follows. In the ultrasonic regime (greater than 1 MHz), nonlinear energy localization has been observed in micromechanical oscillator arrays [15]. Moreover, recent work by Liang B et al. suggested theoretically [16] and later demonstrated experimentally [17] the ability to use nonlinear acoustic materials, e.g., a contrast agent micro-bubble suspension, coupled to a linear superlattice to obtain acoustic rectification. Finally, at much higher frequencies (greater than 1 GHz), several studies have explored mechanical wave propagation in periodic nonlinear structures, focusing on high-amplitude stress wave and thermal phonon propagation. Several studies by Maris and collaborators investigated the propagation of high-amplitude picosecond pulses in crystalline solids, which are a type of naturally occurring nonlinear periodic structure [18, 19]. With respect to the propagation of high-frequency thermal phonons, many studies have focused on the use of nonlinear lattices for thermal rectification. The earliest of these studies were conducted by Terraneo et al. in 2002 [20] and then by Li et al. in 2004 [21]. Later, an experimental study by Chang et al., 2006, also demonstrated thermal rectification using mass-loaded carbon and boron-nitride nanotubes, and attributed the rectification to nonlinear processes [22]. Based on these studies, several following works have extended this concept further to suggest that nonlinear lattices could be used as thermal transistors [23], logic gates [24], and memory [25]. Several computational studies have also investigated and suggested multiple device concepts for thermal rectification building blocks, including carbon nanocones [26] and graphene ribbons [27].

One of the most common ways to model the behavior of granular crystals, and many other types of nonlinear periodic structures, is to describe them as nonlinear lattices. The study of nonlinear lattices can thus offer many potential lessons and insights into the behavior of nonlinear periodic phononic structures. As such, the section directly following gives a brief history of the major types of nonlinear lattices. The review of nonlinear lattices is then followed by an introduction to granular crystals, which is one of the most widely studied types of nonlinear periodic phononic structures and is the subject matter that comprises the focus of this chapter.

7.1.2 Nonlinear Lattices

Since the first computational experiments in nonlinear mass-spring lattices by Fermi, Pasta, and Ulam in 1955 [28], there has been a wealth of interest in the dynamics of nonlinear lattices. Using one of the first modern computers, Fermi, Pasta, and Ulam (FPU) studied a system where the restoring (spring) force between two adjacent masses was nonlinearly related to the relative displacement between masses, and investigated how long would it take for long-wavelength oscillations to transfer their energy (thermalize) into an equilibrium distribution between all the modes of the system. Instead of the predicted thermalization, they found that over the course of the simulation, most of the energy had returned to the mode with which they had initialized the system in coherent form [29].

This discovery initiated whole fields of research relating to the study of nonlinear waves in discrete lattices [30–32]. This includes many different types of nonlinear lattices inspired by physical systems (in addition to the FPU lattice), and the study of physical phenomena occurring in them. As described in the review by Kevrekidis, P. G. [32], three of the most commonly studied types of nonlinear lattices are the discrete nonlinear Schrödinger (DNLS), the Klein-Gordon (KG), and the FPU lattices. The 1D forms of these lattice equations are as follows:

The DNLS can be written as

$$j\dot{u}_i = -\epsilon(u_{i+1} + u_{i-1}) - |u_i|^2 u_i,$$

the KG as

$$\ddot{u}_i = \epsilon(u_{i+1} - 2u_i + u_{i-1}) - V'(u_i),$$

and the FPU as

$$\ddot{u}_i = V'(u_{i+1} - u_i) - V'(u_i - u_{i-1}),$$

where u_i is the dynamical variable of interest at site i, ϵ is a coupling parameter (constant), $j = \sqrt{-1}$, and V is a nonlinear potential function [32]. The DNLS equation has been used to describe nonlinear waveguide arrays and Bose-Einstein condensates, among others [32]. Additionally, under small-amplitude assumptions, it is interesting to note that the DNLS can be derived from either the KG or the FPU lattices [33]. The KG system has been used to model systems of coupled pendula, electrical systems, and metamaterials with split ring resonators, among others [32]. In contrast to the KG system, the FPU has no onsite potential term, and instead involves a nonlinear potential based on nearest neighbor interactions (nonlinear springs). The FPU system has been used to describe the behavior of crystalline solids and structures, including granular crystals.

Studies of these lattices have helped to predict and understand the existence of localized nonlinear coherent structures and other nonlinear phenomena in many naturally occurring and artificial nonlinear (not necessarily discrete) systems. Two examples of nonlinear coherent structures, which are particularly applicable to the study of granular crystals are solitary waves and discrete breathers. Solitary waves were first observed by Russel in a shallow water-filled canal in 1844 [34]. Since then they were shown to be a solution of the Korteweg-de Vries (KdV), a nonlinear partial differential equation, and have been discovered in myriad systems and discrete nonlinear lattices of all the above types [35, 36]. Discrete breathers are a type of intrinsic (not tied to any structural disorder) localized mode, and have been the subject of many theoretical and experimental investigations [33, 36, 37]. Discrete breathers have been demonstrated in charge-transfer solids, superconducting Josephson junctions, photonic crystals, biopolymers, micromechanical cantilever arrays, and more [33]. In addition to nonlinear localized structures, the presence of nonlinearity in dynamical lattices makes available an array of useful phenomena including quasiperiodic and chaotic states, sub- and superharmonic generation, bifurcations, the breaking of time-reversal symmetry, and frequency conversion [38–43].

7.1.3 Introduction to Granular Crystals

Granular crystals, which can be defined as ordered aggregates of elastic particles in contact with each other, are a type of nonlinear periodic phononic structure (Fig. 7.1). Their nonlinearity emerges from two characteristics: (1) the geometry

Fig. 7.1 Granular crystals in one, two, and three dimensions composed by metallic particles confined by supporting walls or confined in a matrix [The three-dimensional image has been adapted from (Daraio, C.; Nesterenko, V.F.; Jin, S.; *"Strongly Nonlinear Waves in 3D Phononic Crystals"* APS – Shock Compression of Condensed Matter, 197–200, American Institute of Physics, Conference Proceedings, Portland (OR), 2003)]

of the particles is such that the force at the contact between neighboring elements is nonlinearly related to the displacement of the particle centers, as can be described by Hertzian contact [7] and (2) in an uncompressed state, granular crystals cannot support tensile loads, effectively creating an asymmetric potential between neighboring elements. An unusual feature of granular crystals that results from these nonlinearities is the negligible linear range of the interaction forces between neighboring particles in the vicinity of a zero compression force. This results in nonexistent linear sound speed in the uncompressed material. This phenomena has led to the term "sonic vacuum," which describes a medium where the traditional wave equation does not support a characteristic speed of sound [14].

The study of granular crystals emerged in 1983 with work by Nesterenko that showed analytically, numerically [44], and later experimentally [45], the concept of "sonic vacuum" and the formation and propagation of highly nonlinear solitary waves in one-dimensional granular crystals. Granular crystals have since been shown to support many other unique dynamic phenomena. This wide array of phenomena supported by granular crystals is enabled by a tunable nonlinear response that encompasses linear, weakly nonlinear, and highly nonlinear behaviors, and can be controlled by essentially linearizing the system through the application of a variable static load [14, 46–48].

In their linear and weakly nonlinear dynamic regime, granular crystals have shown the ability to support tunable acoustic band gaps [49, 50] and discrete breathers [51, 52]. In the strongly nonlinear regime, they have been shown to support compact solitary waves [44–46, 48], nonlinear normal modes [53] anomalous reflections [54], and energy-trapping phenomena when interacting with defects and interfaces [55]. Because of this array-rich dynamics, which has been confirmed by theory, numerical simulations, and simple experiments, granular crystals have become one of the most studied examples of nonlinear lattices. Granular crystals have also been proposed and designed for use in numerous engineering applications including tunable vibration filters [50, 56], optimal shock protectors [57], nondestructive evaluation devices [58], acoustic lenses [59], and acoustic rectifiers [60].

As previously described, the nonlinearity of the interaction law results from the Hertzian contact between particles with elliptical contact area [7, 61, 62]. The Hertzian contact relates the contact force $F_{i,i+1}$ between two particles (i and $i+1$) to the relative displacement $\Delta_{i,i+1}$ of their particle centers, as shown in the following equation:

$$F_{i,i+1} = A_{i,i+1} [\Delta_{i,i+1}]_+^{n_{i,i+1}}.$$

Values inside the bracket $[s]_+$ only take positive values, which denotes the tensionless characteristic of the system (i.e., there is no force between the particles when they are separated). For $\Delta_{i,i+1} = 0$ the particles are just touching, $\Delta_{i,i+1} > 0$ the particles are in compression, and $\Delta_{i,i+1} < 0$ the particles are separated. This tensionless characteristic is one part of the nonlinearity of the Hertzian contact.

For two spheres (or a sphere and a cylinder), the coefficient $A_{i,i+1}$ in the Hertz relationship is defined as:

$$A_{i,i+1} = \frac{4 E_i E_{i+1} \sqrt{\dfrac{R_i R_{i+1}}{R_i + R_{i+1}}}}{3 E_{i+1}(1 - v_i^2) + 3 E_i(1 - v_{i+1}^2)}, \quad n_{i,i+1} = \frac{3}{2}, \tag{7.1}$$

where E_i, v_i, and R_i are the elastic modulus, the Poisson's ratio, and the radius of the i-th particle, respectively. The $n_{i,i+1} = 3/2$ comes from the geometry of the contact between two linearly elastic particles with elliptical contact area, as can be seen in [61]. In addition to assuming that the contact area is elliptical and that both particles remain linearly elastic, the derivation of Hertzian contact assumes that [61]: (1) the contact area is small compared to the dimensions of the particle, (2) the contact surface is frictionless with only normal forces between them, and (3) the motion between the particles is slow enough that the material responds quasi-statically. Variation of the contact geometry will result in a variation of the interaction law stiffness and/or nonlinearity, and ultimately in a variation of the acoustic properties of the crystals. Several recent works have studied this variation theoretically, numerically, and experimentally, by exploring the dynamic response of chains of particles composed of grains with different geometries [63–65] (see Sect. 7.3.5).

The remainder of this chapter describes past and recent work in one-dimensional (1D) and two-dimensional (2D) granular crystals, categorized according to periodicity and dynamical regime.

7.2   One-Dimensional Granular Crystals

The dynamic properties of one-dimensional (1D) granular crystals have been extensively studied, using analytical, numerical, and experimental methods. The following sections describe some of the most interesting physical phenomena supported by these nonlinear systems.

If the stiffness of the contact between two adjacent particles is very low compared to the bulk stiffness of the particles composing the crystal and the contact area small compared to the particle size, 1D granular crystals can be modeled as a system of nonlinear springs and point masses (FPU-like nonlinear lattices). Another perspective from which to approach this same idea is that the characteristic (resonant) frequencies of the particles themselves must be very high compared to the modal frequencies of the granular crystal involving the rigid body-like motion of the particles in the system. Neglecting any dissipation, a statically compressed 1D array of elastic granules can be described by the following system of coupled nonlinear differential equations:

$$m_i \ddot{u}_i = A_{i-1,i} [\delta_{0,i-1,i} + u_{i-1} - u_i]_+^{n_{i-1,i}} - A_{i,i+1} [\delta_{0,i,i+1} + u_i - u_{i+1}]_+^{n_{i,i+1}}. \qquad (7.2)$$

For spherical particles we recall that $n_{i,i+1} = \frac{3}{2}$ and $A_{i,i+1}$ is defined as in (7.1). Here, the static overlap $\delta_{0,i,i+1} = \left(\frac{F_0}{A_{i,i+1}}\right)^{2/3}$, and F_0 is the homogeneous static compression force. m_i is the mass of the ith particle and u_i is the dynamic displacement of the ith particle from its equilibrium position in the initially statically compressed chain. The bracket $[s]_+$ of (7.2) takes the value s if $s > 0$ and the value 0 if $s \leq 0$, which signifies that adjacent beads are not in contact. Within this framework, the dynamic of the system can be tuned to encompass linear, weakly nonlinear, and strongly nonlinear regimes of dynamic behavior, as will be demonstrated for the mono-atomic case in the following section.

7.3   One-Dimensional Monoatomic Granular Crystals

This section focuses on the nonlinear dynamic behavior of a statically compressed 1D monoatomic granular crystal (all particles are the same). A granular crystal composed of identical elastic spherical granules is considered, as shown in Fig. 7.2. For this crystal, $R_i = R$, $m_i = m = \frac{4}{3}\pi R^3 \rho_0$, and $A_{i,i+1}$ of (7.2) is reduced to $A_{i,i+1} = A = \frac{E\sqrt{2R}}{3(1-v^2)}$, where m is the mass of the sphere, E and ρ_0 are the Young's modulus and density of particle material, R is the particle radius, and v is the Poisson's ratio. Moreover, it is assumed that the chain is subjected to constant static force F_0 applied to both ends, resulting in an initial displacement δ_0 between neighboring particle centers, $\delta_{0,i,i+1} = \left(\frac{F_0}{A}\right)^{2/3} = \delta_0$. The particle equations of motion, shown in (7.2) thus reduce to:

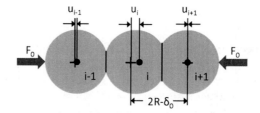

Fig. 7.2 One-dimensional monoatomic crystal compressed by a static force F_0. The *crosses* represent the initial positions of the particle centers in a statically compressed chain while the *black circles* denote the current positions [14]

$$m\ddot{u}_i = A[\delta_0 + u_{i-1} - u_i]_+^{3/2} - A[\delta_0 + u_i - u_{i+1}]_+^{3/2}, \qquad (7.3)$$

where u_i is the displacement of the ith bead from its equilibrium position in the *initially compressed chain*, as shown in Fig. 7.2, and $i \in \{2, \cdots, N-1\}$.

7.3.1   Near-Linear Regime

To approximate the fully nonlinear equations of motion shown in (7.3), a power series expansion of the forces can be taken. For dynamical displacements with amplitude *much less* than the static overlap, i.e. $\frac{|u_{i-1} - u_i|}{\delta_{0,i}} \ll 1$, one can keep only the harmonic term of the expansion. In this case, the granular crystal can be considered as a linear lattice with spring constant $K_2 = \frac{3}{2} A \delta_0^{1/2}$, where the equations of motion are reduced to:

$$m_i \ddot{u}_i = K_2(u_{i-1} - u_i) - K_2(u_i - u_{i+1}). \qquad (7.4)$$

The spectral band of the ensuing linear chain (see Chap. 2 for more details) has an upper cutoff frequency of $\omega_m = \sqrt{4K_2/m}$. As a consequence of the nonlinear relation $F_0 \propto \delta_0^{3/2}$, for the case of the spherical granules, the cutoff frequency (as well as the sound velocity of the 1D monoatomic granular crystal) scales as $F_0^{1/6}$. These results have been confirmed experimentally [47, 66].

7.3.2   Weakly Nonlinear Regime

If the dynamic displacements have small amplitudes $\frac{|u_{i-1} - u_i|}{\delta_{0,i}} < 1$ relative to those due to static load, a power series expansion of the forces (up to quartic displacement terms) can be calculated to yield the $K_2 - K_3 - K_4$ model:

$$m\ddot{u}_i = \sum_{k=2}^{4} K_k[(u_{i+1} - u_i)^{k-1} - (u_i - u_{i-1})^{k-1}], \qquad (7.5)$$

where $K_2 = \frac{3}{2} A \delta_0^{1/2}$, $\quad K_3 = -\frac{3}{8} A \delta_0^{-1/2}$, $\quad K_4 = -\frac{3}{48} A \delta_0^{-3/2}$.

This model is an example of the celebrated FPU model. Many theoretical studies have focused on the dynamical properties of this type of nonlinear lattice, revealing the existence of coherent nonlinear structures such as nonlinear periodic waves, solitary waves [67], and discrete breathers [68].

Seeking traveling waves with a characteristic spatial size L that is much larger than the inter-particle distance $a = 2R - \delta_0$, one can apply the so-called long-wavelength or continuum approximation.

Using the replacement:

$$u_i = u(x), \; u_{i\pm1} \approx u \pm au_x + \frac{1}{2}a^2u_{xx} \pm \frac{1}{6}a^3u_{xxx} + \frac{1}{24}a^4u_{4x}, \qquad (7.6)$$

equation (7.5) is transformed into the nonlinear Boussinesq equation and into the Korteweg-de Vries equation (see for example [69]). In (7.6), $u_x = \partial u/\partial x$, and the number of subscripts x denotes the order of the derivative of u. Nesterenko applied this method (taking into account only up to the K_3 term) to a strongly compressed granular chain, and derived the following KdV equation [14]:

$$\xi_t + c_0\xi_{xx} + \gamma\xi_{xxx} + \frac{\sigma}{2c_0}\xi\xi_x = 0, \; \xi = -u_x$$
$$c_0^2 = \frac{A\delta_0^{1/2}6R^2}{m}, \; \gamma = \frac{c_0R^2}{6}, \; \sigma = \frac{c_0^2R}{\delta_0}. \qquad (7.7)$$

In (7.7), $\xi = -u_x$, $\xi_x = \partial\xi/\partial t$, and c_0 is the linear sound speed. The solutions of (7.7) are well known, and include nonlinear *periodic waves* and *solitary waves*.

On the other hand, to investigate how quasi-monochromatic plane waves or narrow-band packets evolve by nonlinear effects, one can derive another well-known nonlinear wave equation—the Nonlinear Schrödinger (NLS) equation. This equation predicts many nonlinear phenomena, including second harmonic generation, modulation instability, and the existence of bright and dark solitons [35]. The derivation of the NLS from (7.5) is possible using the method of multiple scales combined with a quasi-discreteness approximation (see [70] for an application of this method to a generic FPU lattice in the form of (7.5) and [71] for a recent application of this method to a monoatomic strongly compressed granular crystal).

Another generic feature of nonlinear lattices is the existence of nonlinear localized modes called discrete breathers (DBs). DBs have been studied extensively in monoatomic FPU chains [68]. One of the mechanisms for the generation of such nonlinear localized modes is the modulational instability (MI) of a plane wave at the band edge. A detailed analysis of this instability (bifurcation) shows that the MI of the upper cutoff mode manifests itself when the coefficients of the FPU lattice are such that $3K_2K_4 - 4K_3^2 > 0$ (see Sect. 4.3 of [33] and references therein). In the monoatomic granular crystal setting, one can show that this inequality does not hold. This is an indication that small-amplitude DBs bifurcating from the upper band mode do not exist in monoatomic granular crystals. However, the existence of dark discrete breathers or large-amplitude DBs remains an interesting open question.

Another interesting weakly nonlinear effect, self-demodulation, was studied by Tournat and collaborators in compressed 1D granular crystals [72]. In this work, they explored how, in a nonlinear medium, two primary frequencies can mix to form a propagating wave with frequency that is the difference of the two primaries.

7.3.3 Highly Nonlinear Regime: Long-Wavelength Approach

A very interesting, non-classical wave behavior appears if the granular material is weakly compressed and the particle displacements are larger than the initial relative displacement δ_0 resulting from the static compression. This regime is termed the highly nonlinear regime. Most of the studies in 1D monoatomic granular crystals have been devoted to this dynamical regime. This section summarizes the basic steps of the long-wavelength method that Nesterenko applied [14]. A review of alternate analytical approaches and experimental observations will also be presented in the following sections.

Including δ_0 in displacement u_i which is calculated from the particle positions in the uncompressed system (see Fig. 7.3 and [14] for more details), (7.3) becomes:

$$m\ddot{u}_i = A[u_{i-1} - u_i]_+^{3/2} - A[u_i - u_{i+1}]_+^{3/2} \tag{7.8}$$

In the long-wavelength approximation, the displacements u_{i-1}, u_{i+1} can be expanded in a power series according to a small parameter $\varepsilon = a/L$ up to the fourth order [see (7.6)]. By substituting (7.6) into (7.8), and conducting some additional calculation, a new wave equation is obtained [14]:

$$u_{tt} = -c^2 \left\{ (-u_x)^{3/2} + \frac{a^2}{12} \left[\left((-u_x)^{3/2} \right)_{xx} - \frac{3}{8} \left((-u_x)^{-1/2} \right) u_{xx}^2 \right] \right\}_x, \quad -u_x > 0,$$

$$c^2 = \frac{2E}{\pi \rho_0 (1 - v^2)}. \tag{7.9}$$

Despite the complex nature of the presented strongly nonlinear wave equation, the stationary solutions of (7.9), such as nonlinear periodic and solitary waves, can be found in the form $u(x - Vt)$ [14]. The waveform of a periodic wave with speed $V = V_p$ is given by the following expression:

$$\xi = \left(\frac{5V_p^2}{4c^2} \right)^2 \cos^4 \left(\frac{\sqrt{10}}{5a} x \right). \tag{7.10}$$

The dependence of the speed of the periodic wave, V_p, on the minimal and maximum strains is presented in [14].

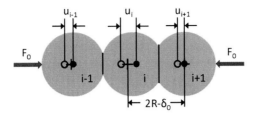

Fig. 7.3 Weakly compressed chain of particles. The *crosses* represent the initial positions of the particles in the statically compressed chain, the *black circles* correspond to the current positions of spheres, and the *open circles* the initial positions of the spheres in the uncompressed [14]

The solitary shape (for the case when the initial prestrain ξ_0 approaches 0) is one hump of the periodic solution of (7.10), with a *finite wave length* equal to about *five particle diameters*. This solitary wave is a supersonic one, similarly to the KdV soliton, but differs from the KdV soliton in other fundamental properties. A unique feature of this solitary wave is the independence of its width on amplitude. Accordingly, this property is quite different from the property of weakly nonlinear KdV solitary wave. Here, the speed of the solitary wave V_s has a nonlinear dependence on maximum strain ξ_m (and particle velocity v_m):

$$V_s = \frac{2}{\sqrt{5}} c\, \xi_m^{1/4} = \left(\frac{16}{25}\right)^{1/5} c^{4/5} v_m^{1/5} = \left(\frac{8E}{5\pi\rho_0(1-v^2)}\right)^{2/5} v_m^{1/5}. \qquad (7.11)$$

This result shows that the speed of the strongly nonlinear solitary wave V_s does not depend on particle size in the granular material. At the same time it does depend on the elastic properties of the particles (E and v) and their density. The presented theoretical results allow us to design strongly nonlinear granular materials with exceptionally low velocity of signal propagation. Simple estimation based on (7.11) shows that it is possible to create materials with nonlinear impulse speed in the interval 10–100 m/s.

7.3.4 Review of Alternate Strongly Nonlinear Wave Theoretical Approaches

The solitary wave solution (*or a soliton with compact support*, known also as *compacton* [73]) presented in the previous section describes well the solitary wave that an impulsive excitation generates in a weakly compressed or uncompressed granular crystal. This was verified in simulations and experiments by different authors (see references below). The rigorous proof of the existence of solitary waves in a monoatomic granular crystal composed of spherical particles was done by MacKay [74], who applied the existence theorem for solitary waves on lattices by Friesecke and Wattis [75]. Ji and Hong extended the proof given by MacKay to the general case of an arbitrary power-law type contact force [76].

An analytical solution of the form tanh(f_n) for stationary waves in discrete chains, where f_n is represented by a series, was also presented by Sen and Manciu [77]. Their result is very close to a soliton obtained by the long-wavelength approximation. Chatterjee studied the asymptotic description of the tail of the soliton in an uncompressed chain and he revealed that it has a double exponential decay [78]. He also presented a new asymptotic solution for the full solitary wave, which is closer to the results of numerical simulations than the approximate solution given by Nesterenko. A quite different analytic approach for the study of pulse propagation in granular crystals was developed by Lindenberg and collaborators [79, 80]. This method uses the binary collision approximation to reduce the problem of propagation to collisions involving only two granules at a time.

English and Pego [81] studied the shape of the solitary wave that propagates in a 1D granular chain without precompression ($\delta_0 = 0$). Their method is based on a reformulation of the equations of motion using the difference coordinates $r_i = u_{i-1} - u_i$ such that:

$$m\ddot{r}_i = A\left[[r_{i+1}]_+^{3/2} - 2[r_i]_+^{3/2} + [r_{i-1}]_+^{3/2}\right]. \qquad (7.12)$$

Seeking for traveling wave solutions, $r_i = r(x) \equiv r(i - ct)$, one obtains the following advanced delay equation:

$$r''(x) = \frac{A}{mc^2}\left[r^{3/2}(x-1) - 2r^{3/2}(x) + r^{3/2}(x+1)\right]. \qquad (7.13)$$

By rewriting this equation in an equivalent integral form and studying its asymptotic behavior, they proved that the solitary wave decays super-exponentially. Moreover, they applied an iterative method for the computation of the *numerically exact shape of the solitary waves*.

Later, Ahnert and Pikovksy [82] applied a different type of quasi-continuum approximation by expanding, up to fourth order, the difference coordinate r_i instead of the displacement u_i. Substituting these expansions in (7.12), they obtained a strongly nonlinear partial differential equation (see (7.6) in [82]) that supports a solitary wave with compact form. The analytic solution has the same form as Nesterenko's solution, but with slightly different amplitude and width constants. Moreover, they presented an accurate numerical method for the numerical solution of the advanced delay equation (7.13) and they compared the numerically obtained solutions with those of approximated PDEs.

Recently, Starosvetsky and Vakakis [83], working directly on the nonlinear lattice equations with no precompression, developed semi-analytical approaches for computing different families of nonlinear traveling waves. These waves involve both separation and compression between adjacent particles and therefore they cannot be resolved using quasi-continuum approximations. In addition, they showed that these wave families converge to the solitary wave in a certain asymptotic limit. They also solved the reduced advanced delay equation numerically, and applied the method of Pade approximations.

7.3.5 Review of Experiments with Strongly Nonlinear Solitary Waves

The quantitative agreement of analytical and numerical predictions with experiments, regarding solitary waves in *strongly nonlinear* granular crystals, was first found by Lazaridi and Nesterenko [45]. They observed for the first time the rapid decomposition of the initial impulse excitation into multiple solitary waves in a distance comparable to the solitary wave width, (Fig. 7.4). Since then there have been several experimental studies and observations of solitary waves and other strongly nonlinear phenomena in multiple settings. Optical observations of strongly nonlinear waves in arrays of photoelastic disks excited by a local explosive loading were reported by Zhu, Shukla, and Sadd [84]. Coste, Falcon, and Fauve [46], and Coste and Gilles [47] conducted a very detailed quantitative study of the speed and shape of solitary waves at different amplitudes. They reported a negligible decay of the solitary wave in chains composed of 50 particles, and they concluded that the solitary waves shape observed in experiments were in very good agreement with the predictions obtained from theoretical solutions such as (7.10).

The relatively low speed of the solitary waves detected by Coste et al. [46] is very unusual for solid materials. In an uncompressed granular system, according to (7.11), the minimum propagation speed of a solitary wave can be close to zero if the amplitude of the disturbance is approaching zero ("sonic vacuum") [13]. Using polymeric and composite particles [85, 86], for example, one can design granular crystals with a solitary wave speed corresponding to a signal in the interval of 10–100 m/s, an order of magnitude less than that previously observed in experiments by Coste et al. [45]. In addition to the experimental observation of solitary waves, many works relating to highly nonlinear phenomena in granular crystals have followed. Nesterenko et al. showed experimentally the presence of anomalous reflections when highly nonlinear waves interact with interfaces [54], effectively demonstrating for the first time the concept of an acoustic diode. Daraio et al. [48], described in detail the ability to tune the dynamic response of granular crystals by controlling the static precompression and the dynamic excitation applied to the system. Job et al. [87], investigated the behavior of solitary waves interacting with a boundary, showing for the first time the sensitivity of solitary waves to the mechanical properties of an adjacent medium. Thorough experimental, numerical, and theoretical descriptions of the formation, propagation [88], and collision of solitary wave trains were published a few years later [89].

Recently, several experimental works described wave propagation in granular crystals composed of particles with elliptical and cylindrical geometries [63, 64]. Chains composed of ellipsoidal or cylindrical particles were shown to support the

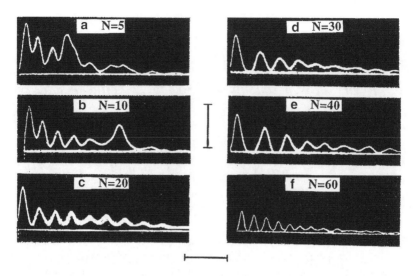

Fig. 7.4 Evolution of a soliton train excited experimentally by a striker impact ($M_s = 10\ m$, $u_s = 0.5$ m/s), after a propagation distance of N particles: (**a**) $N = 5$, (**b**) $N = 10$, (**c**) $N = 20$, (**d**) $N = 30$, (**e**) $N = 40$, (**f**) $N = 60$. The vertical scale corresponds to a force of 80 N, the horizontal scale to a total time of 50 μs (**a–e**) and 100 μs (**f**) [14]. Figure reproduced from [Nesterenko, V.F., Dynamics of Heterogeneous Materials. 2001, Chapter 1, pp. 70, NY: Springer-Verlag] with permission from the publisher

formation and propagation of highly nonlinear solitary waves similar to the solitary waves observed in chains of spherical particles. These systems were also found to be highly dependent not only on the particles' geometry but also on the orientation angles between particles in the chain. This dependence on orientation angle between beads provides an additional free parameter to design acoustic materials with unprecedented transmission properties.

Experimental studies have also described the dynamic response of chains composed of spherical steel particles coated with a soft polymeric material. These studies showed that this type of system also supports the formation and propagation of highly nonlinear solitary waves [85]. However, one interesting property of these systems is that the contact interaction between thin-coated spheres does not follow the classical Hertzian interaction between two solid spheres [90]. The dynamic response of chains composed of coated spheres is governed by a quadratic power law dependence between the contact force, F, and the displacement, δ, instead of the Hertzian, non-integer power of 3/2. This new nonlinear contact interaction dramatically changes the dynamics of solitary wave propagation compared to its counterpart in chains of solid spheres. Here, the spatial width of the wave becomes shorter (3.14 particles size instead of 5), the wave speed (V_s) is relatively slower, and its dependence on force amplitude (F_m) is also different ($V_s \sim F_m^{1/4}$ instead of $V_s \sim F_m^{1/6}$).

Studies of chains of hollow spherical particles also presented interesting nonlinear acoustic phenomena. Highly nonlinear solitary waves were observed to propagate through the system, but the wave properties were different from the highly nonlinear solitary waves in the chains of solid spheres. The spatial width of the solitary wave in chain of hollow spheres was approximately 8 particles (larger than 5 particles, which is the characteristic length of a solitary wave forming in a chain of uniform, solid spheres). The wave speed was found to be proportional to the force amplitude to the power 1/11 [65]. It was shown that such behavior resulted from the unique contact interaction between thin hollow spheres, which for the range of wave amplitude studied, could be approximated by a power-law type relation ($F = k\delta^n$). In this case, the exponent n was found to be smaller than the value 3/2 as in the classical Hertzian interaction between solid spheres. The contact stiffness k and the exponent n were also found to be dependent on the thickness of the hollow sphere's shell. This dependence of the dynamic behavior of granular crystals on the coating and/or shell thickness of spherical particles provides yet another free parameter to employ in tuning the dynamics of nonlinear acoustic crystals.

7.4 One-Dimensional Diatomic Granular Crystals

By increasing the degree of periodicity, from a homogenous monoatomic granular crystal to a diatomic granular crystal composed of alternating particles, additional interesting phenomena can be accessed. This section describes some of those phenomena characteristic of 1D diatomic granular crystals, including tunable band gaps, discrete breathers (DBs), and highly nonlinear solitary waves with widths up to ten particles.

An example of a 1D diatomic granular crystal is illustrated in the bottom of Fig. 7.1. The equation of motion for the general 1D granular crystal, shown in (7.2), can be reduced to the 1D diatomic crystal model, as follows:

$$m_i \ddot{u}_i = A[\delta_0 + u_{i-1} - u_i]_+^{3/2} - A[\delta_0 + u_i - u_{i+1}]_+^{3/2}, \tag{7.14}$$

where the subscript i is the index of the ith particle, the particle masses are $m_{2i-1} = m$ and $m_{2i} = M$. By convention, M is taken to be the larger of the two masses and m to be the smaller of the two masses. Because all contacts (aside from any boundaries) are the same, there is a single Hertzian contact coefficient A and static overlap δ_0 that are used to represent the system, which have been defined in the previous sections. Within this framework, as before, the dynamic response of the system can be tuned to encompass linear, weakly nonlinear, and strongly nonlinear regimes of dynamic behavior. Also as before, the $K_2-K_3-K_4$ model can be applied in the weakly nonlinear regime, and the K_2 linearized model in the linear regime.

7.4.1 Near-Linear Regime: Localized Surface Modes

For dynamical displacements with amplitude much less than the static overlap ($|u_{i+1} - u_i| \ll \delta_0$), the nonlinear K_3 and K_4 terms can be neglected, and the linear dispersion relation of the system can be easily computed. This results in an effectively linear diatomic system of springs and point masses, as was presented in Chap. 2, but with a tunable stiffness K_2.

Several previous studies explored the existence of band gaps in highly compressed granular crystals. Initially, studies focused on 1D, two-particle unit cell, arrays of glued [91], welded [92], and elastically compressed spherical particles [49, 51, 56]. These studies demonstrated tunable vibration spectra with two bands of propagation (called the acoustic and optical bands) separated by a band gap in the diatomic case. Boechler et al. [50] later extended this work by investigating the response of one-dimensional diatomic granular crystals with three-particle unit cells, and showing their tunability based on variations of the particles geometry and on the applied static load. In contrast to diatomic granular crystals with two-particle unit cells, the three-particle unit cell granular crystal was shown to contain up to three distinct pass bands and two finite band gaps.

In addition to acoustic and optical band modes, the diatomic semi-infinite harmonic granular crystal also supports a gap mode, provided the crystal has a light particle at the surface and free boundary conditions. This mode is localized at the surface (i.e., at the first particle), and its displacements have the following form [93]:

$$u_{2i+1} = B(-1)^i \left(\frac{m}{M}\right)^i e^{j\omega_s t} \tag{7.15}$$

$$u_{2i+2} = B(-1)^{i+1} \left(\frac{m}{M}\right)^{i+1} e^{j\omega_s t},$$

with particle number $i \geq 0$, frequency $\omega_s = \sqrt{K_2(1/m + 1/M)}$ is in the gap of the linear spectrum, and B is an arbitrary constant. This particular mode with frequency in the band gap, that is localized around the surface, proves to have a nonlinear counterpart and to be very closely related to the DB in the strongly discrete regime, as will be described in the following section.

7.4.2 Weakly Nonlinear Regime: Discrete Breathers

By increasing the relative amplitude of the dynamic to static displacements ($|u_{i+1} - u_i| < \delta_0$), and thus entering the weakly nonlinear regime, a type of intrinsic localized mode called a discrete breather (DB) can be supported by the system. DBs have been widely studied in the realm of nonlinear lattices, as previously described [33]. They are nonlinear modes that have frequency within the gap of the linear spectrum and are localized in space. As such, discrete breathers have practical importance as a mechanism to localize vibrational energy in frequency and space without the introduction of any extrinsic disorder.

DBs were rigorously proven to exist in diatomic FPU-type lattices, with alternating heavy and light masses, by Mackay in 1997 [94]. Furthermore, several studies also investigated the specific case of DBs located in the gap between the acoustic and optical bands of anharmonic diatomic lattices [95–97].

A recent study by Theocharis et al. systematically studied the existence and stability of DBs in diatomic granular crystals [52]. Studies in other diatomic anharmonic lattices have shown the existence of up to two types of DBs. The study by Theocharis et al. demonstrated that both types of DBs can arise in granular chains. They examined both of these two families of discrete gap breathers, and studied their existence, stability, and structure throughout the gap of the linear spectrum. The first family was an unstable DB that is centered on a heavy particle and characterized by a symmetric spatial energy profile, and the second family is a potentially stable DB that is centered on a light particle, and is characterized by an asymmetric spatial energy profile.

Although the FPU and granular crystal lattices are analogous in many respects, there exists an important difference because of the additional nonlinearity caused by the tensionless characteristic of the granular crystal lattice. Accordingly, Theocharis et al., contrasted discrete breathers in anharmonic FPU-type diatomic chains with those in diatomic granular crystals, and found that for the case when the DB was very narrow (highly discrete), the asymmetric nature of the latter interaction potential led to a form of hybrid bulk-surface localized solutions (see Fig. 7.5). Figure 7.5 shows the two families of DB solutions at times $t = T$ and $t = T/2$ (where T is the periodic of the DB), and the profile of a linear surface mode. This similarity between the shapes of the two modes suggests that the temporary creation of a new interior surface, allowed by tensionless characteristic of the system, has contributed to a modified type of intrinsic localized mode.

The existence of DBs in diatomic granular crystals was experimentally proven in a recent study by Boechler et al. [51]. In this study, the authors utilized the modulational instability (MI) of the lower optical mode to generate DBs in an 80 particle diatomic granular crystal. In the weakly nonlinear regime, granular crystals can be showed to be subject to MI when $K_3^2/K_2K_4 < 3/4$. To excite the MI, they drove the granular crystal from one boundary at the lower optical mode frequency, at high amplitude. Upon reaching a critical amplitude for the MI to occur, the anharmonic lattice vibration decayed into a localized DB.

Figure 7.6 shows, as per Boechler et al. [51], an experimental observation of a DB, generated in an 80 particle diatomic granular crystal. This example shows how the interplay of nonlinearity and discreteness/periodicity leads to the localization of vibrational energy within a narrow spatial regime (around the 14th particle from the boundary), at a specific frequency within the gap of the linear spectrum ($f_b = 8.31$ kHz). In panels (a) and (b), far from the center of the DB, a periodic response at the driving frequency ($f_d = 8.9$ kHz) can be seen. Alternatively, in panels (c) and (d), near the center of the DB, a quasiperiodic response appears, which is characterized by the driving frequency and frequency of the generated DB. This spatial localization is further clarified in the spatial profile of the energy distribution shown in panel (e).

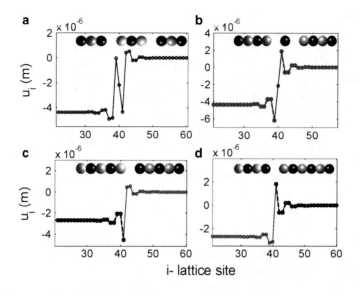

Fig. 7.5 *Top panels*: Spatial profile of a DB in the heavy mass-centered symmetric family at times (a) $t = 0$ and (b) $t = T/2$. *Bottom panels*: As with the top panels, but for the light mass centered asymmetric family of DB solutions. The *dashed curves* correspond to the spatial profile of the surface mode obtained using (7.15). In each panel, a visualization of particle positions is included, along with the corresponding spatial gap openings, for the corresponding time and DB solution. Copyright (2010) by The American Physical Society [52]

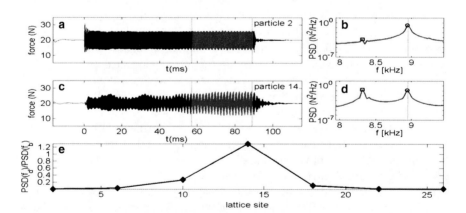

Fig. 7.6 Experimental observation of a DB, in an 80 particle granular crystal, at $f_b = 8.31$ kHz. (a), (c) Force at particle 2 and 14, respectively. (b), (d) Power spectral density (PSD) for the highlighted time regions in (a), (c) of the same color. *Square (circular) markers* denote the DB (driving) frequency and PSD amplitude. (e) The ratio of the PSD amplitude at the discrete breather frequency divided by the PSD amplitude of the driving frequency as a function of sensor location. The *vertical dashed line* in (b) and (d) denotes the lower cutoff frequency of the optical band, and the *vertical dashed lines* in (a) and (c) denote the time region for the PSD calculation. Image reproduced from [98]

7.4.3 Highly Nonlinear Regime: Strongly Nonlinear Solitary Waves

In this section, the effects of increased periodicity on the propagation of solitary waves in 1D diatomic granular crystals with no static load is described. Solitary waves in such systems were first studied and described by Nesterenko in 2001 [14]. He found that by assuming the mass of one particle type to be much larger than the mass of the other ($m_1/m_2 >> 1$) and by applying the long wavelength approximation, the resulting wave equation supports a solitary wave solution with a characteristic spatial width of ~10 particles. This demonstrates how an increase in periodicity (or redistribution of the monoatomic particle masses to two neighboring particles) can result in wider solitary wave.

Later, Porter et al. [99] applied the long-wavelength approximation to diatomic granular crystals with arbitrary mass ratios by postulating a "consistency condition" between the displacements of the two particles in the unit cell. They showed that the diatomic chain supports a finite-width soliton-like solution, and they obtained an analytical expression for the width of the solution as a function on the mass ratio. This expression generalizes the previously known limiting cases, namely, $m_1/m_2 = 1$ (monoatomic) and $m_1/m_2 >> 1$ (diatomic with ~10 particle length solitary wave width). In the same study, Porter et al. compared these analytical predictions with simulations and experiments and found good agreement.

Recently, Vakakis et al. [100] presented an extensive numerical and theoretical study of solitary waves in diatomic chains. They showed that in a diatomic granular crystal, scattering at the interfaces of the dissimilar light and heavy beads will typically cause a slow disintegration of the traveling wave and the formation of small amplitude oscillating tails. However, they also found that for specific discrete values of the mass ratio between heavy and light particles, the system supports solitary waves which travel without distortion. These discrete values of the mass ratio correspond to the case where the light beads always stay in contact with adjacent heavy beads. For this case, the entire energy of the main pulse is conserved and transferred without loss to the next heavy bead. These solutions can be considered analogous to the propagation of solitary waves in monoatomic granular crystals, in that their velocity profiles decay to zero. Finally, they also observed that the diatomic family of solitary waves propagates faster than the corresponding solitary waves in monoatomic systems.

7.5 One-Dimensional Monoatomic Granular Crystals with Defects

By placing one or more defects into an otherwise perfectly periodic mono-atomic granular crystal, disorder can be introduced into the system. The presence of disorder, and its interplay with the nonlinearity of the system, causes interesting and useful phenomena throughout the granular crystal's range of dynamic regimes. In contrast to the case of increasing periodicity, introducing disorder adds new ways to break the spatial-symmetry of the granular crystal. In combination with the ability of nonlinear systems to break the time-reversal symmetry of the dynamic response, the introduction of spatially asymmetric disorder can be particularly useful. In the following section several recent studies are described relating to defects in monoatomic granular crystals, including: tunability of defect modes in the linear regime [101], localized nonlinear defect modes and spontaneous symmetry breaking in the weakly nonlinear regime [102], the interplay of solitary waves with defects in the highly nonlinear regime [103], and tunable bifurcation-based acoustic rectification in a driven granular crystal [60].

7.5.1 Near-Linear Regime: Tunable Defect Modes

A strongly compressed (with respect to the dynamic displacements) homogenous granular crystal with light-mass defects will contain exponentially localized modes with frequencies above the acoustic band of the granular crystal, localized around the defect sites. The frequency of these localized defect modes is tunable with changes in static load, similar to the tunability of the linear dispersion relation of a periodic granular crystal.

The existence and tunability of defect modes localized around one and two light-mass defects in a strongly compressed 1D otherwise homogenous granular crystal was investigated first numerically and analytically by Theocharis et al. [101], and then experimentally by Man et al. [101]. In the work by Man et al., they placed one and two light-mass defects near the edge of a 20 stainless-steel particle granular crystal, applied white-noise excitation from the edge of the crystal, and measured the frequency of the defect modes localized in the vicinity of the defects as a function of defect size and relative defect position. The observed defect mode frequencies were compared with eigen-analysis of the linearized 20 particle granular crystal (as described in Theocharis et al. [101]), and analytical expressions based on few-site considerations [100]. They showed that, for a sufficiently small single light mass defect in an otherwise homogenous granular crystal, the frequency of the defect mode can be approximated as [101]:

f_{3bead}

$$= \frac{1}{2\pi} \sqrt{\frac{2K_{Rr}M + K_{RR}m + K_{Rr}m + \sqrt{-8K_{Rr}K_{RR}mM + [2K_{Rr}M + (K_{RR} + K_{Rr})m]^2}}{2mM}}.$$

(7.16)

This expression is obtained by solving the eigenvalue problem of the three-particle system in the vicinity of the defect (large particle–defect particle–large particle). Here M is the mass of the homogenous particles, m is the mass of the defect particle, f_{3bead} is the frequency of the localized defect mode, $K_{RR} = 3/2$ $A_{RR}^{2/3} F_0^{1/3}$ is the linearized stiffness between two large particles, and $K_{Rr} = 3/2$ $A_{Rr}^{2/3} F_0^{1/3}$ is the linear stiffness of the contact between a defect-particle and a large particle. A_{Rr} and A_{RR} are the Hertz contact coefficients between the respective particles. From this expression, it is clear how the defect modes are tunable with static load, geometry, and material properties.

In both studies [101, 102], it was found that when two defects were placed sufficiently far from each other (outside the localization length of each individual defect mode), the granular crystal presented two isolated linear defect modes with frequencies of a single-defect mode. The further the distance between the defects, the closer the modes are to isolated ones with near-identical frequencies. However, when the defects are brought sufficiently close together (within the localization length of a single-defect mode) each defect was found to affect the other. This caused the formation of a symmetric and anti-symmetric pair of defect modes, with two new separate frequencies, involving both defects.

7.5.2 Weakly Nonlinear Regime: Nonlinear Localized Modes and Symmetry Breaking

If the amplitudes of the dynamic displacements are increased, relative to the static overlap, and thus the nonlinearity of the dynamic response is also increased, the nonlinear localized defect modes depart from their linear counterparts and new phenomena are introduced. In addition to exploring the near-linear behavior of one-dimensional, strongly compressed granular crystals with one or two light-mass defects, Theocharis et al. investigated the behavior of defects in the weakly nonlinear regime [102]. As previously described, by analyzing the problem's linear limit, they identified the system eigen frequencies and the linear defect modes. Using continuation techniques, they found localized nonlinear defect mode solutions that bifurcate from their linear counterparts and studied their linear stability in detail by computing the Floquet multipliers of the nonlinear periodic solutions.

For the case of a single light-mass defect, it was found that the inherent nonlinearity of the system leads to *long-lived* localized breathing oscillations, which form robust nonlinear localized modes. Their frequency depends not only on the static load, the geometry, and the material properties of the granular crystal and defect particle, but also on the amplitude of the oscillations. Because of the type of the nonlinearity in the system, the defect mode's frequency decreases with increasing dynamic amplitude (and nonlinearity). These are examples of two ways where nonlinearity can be used to tune the frequency of a localized mode: by changing the static load, and thus the stiffness of the contacts, or by changing the relative amplitude of the dynamic displacements to the static overlap.

For the case of two defects, nonlinearity can create further interesting phenomenology when the defects are sufficiently close. A particularly intriguing example is the case of next-nearest neighbor defects, where the two defects are separated by one large particle. This resembles the situation of a "double well" potential, which has been studied systematically in various settings, including nonlinear optics [104] and atomic physics [105, 106]. In these settings, it has been predicted analytically (via a two-mode reduction), manifested numerically, and observed experimentally that beyond a certain nonlinearity threshold, a pitchfork bifurcation arises that causes the *spontaneous symmetry breaking* of the relevant configurations, and results in asymmetric nonlinear modes. The investigations of this phenomena in granular crystals, by Theocharis et al., indicate that this phenomenology is more *generic*. Figure 7.7 shows the bifurcation of the antisymmetric linear defect mode as a function of the defect mode frequency and relative force between the defect sites, for a next-nearest neighbor configuration. As the antisymmetric defect mode (Fig. 7.7, inset A1) becomes progressively more nonlinear (and decreases in frequency), at a critical point, the mode becomes unstable via a pitchfork-like bifurcation. This bifurcation signals the emergence of two asymmetric modes (Fig. 7.7, insets A2 and A3), which are mirror images of each other and predominantly centered on one of the two defect sites.

The case of the bifurcation of the antisymmetric two-defect mode is a good example of how, through the addition of nonlinearity, sharp transitions can be created between two acutely different states, the spatial symmetry of the dynamic response broken, and new mechanisms accessed to control the distribution and frequency of vibrational energy.

7.5.3 Highly Nonlinear Regime: Transient Localized Modes

By increasing the nonlinearity of the dynamic response further, the interaction of traveling waves with defects in a nonlinear system can be explored. The interaction of highly nonlinear solitary waves with a mass defect placed in a 1D, unloaded granular crystal has been investigated analytically and computationally first by Sen [107–109] and then by Hascoet, in 2000 [110]. This work was later followed by a more in depth numerical and experimental study by Job, in 2009 [103]. Two different physical pictures emerge whether one considers a light or a heavy impurity mass. The scatter of the solitary wave with a light impurity yields transient oscillations of the defect which leads to the emission of lower amplitude solitary waves in both directions [110]. In contrast, a heavy-mass defect is shifted by the solitary wave, a solitary wave is reflected back, and the transmitted wave loses its soliton characteristics and is fragmented into smaller waves of decreasing amplitude [110]. In the work by Job, it was shown that the interaction with a light-mass defect will also lead to the transient excitation of a localized mode [103]. They described how the slow-timescale local compression caused by the solitary wave around the defect site can act analogously to the linearizing static compression described in

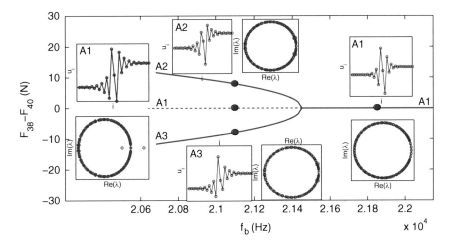

Fig. 7.7 Pitchfork bifurcation illustrated by force differential between two next-nearest neighbor defects, as a function of the mode frequency. This shows the transition from a single antisymmetric mode to two (mirror-symmetric between them) asymmetric modes after the onset of the symmetry breaking bifurcation. *Insets*: spatial profiles and locations of Floquet multipliers λ in the complex plane of solutions for different frequencies

the previous sections, and create an oscillating localized defect mode [103]. Starosvetsky et al. also analyzed analytically and numerically the interaction of the solitary wave with light mass defects. They used reduced models that take into account only the interaction of the defect mass with its neighboring particles [111].

7.5.4 Driven-Damped Granular Crystals: Quasiperiodicity, Chaos, and Acoustic Rectification

In the previous sections, the existence of linear and nonlinear localized modes surrounding defects in an otherwise homogenous granular crystal was discussed. The transient interaction of traveling solitary waves with defects was also explored. Neither of these cases involved a high-amplitude continuous driving force nor damping. Studying cases with damping and continuous driving is useful for both real-world applications and devices, and involves interesting new phenomena.

In 2011, Boechler et al. [60] studied experimentally and computationally the case of a 1D statically compressed granular crystal that contains a light-mass defect close to one end, and is subject to a harmonic driving force (see left panel of Fig. 7.8). As described in the previous section, a light mass defect will create a localized mode with frequency above the acoustic band of the homogenous part of the granular crystal. Boechler et al. selected the frequency of the driving force to be close to the defect mode frequency. Because the driving force has frequency above the acoustic band of the homogenous granular crystal, the signal cannot propagate

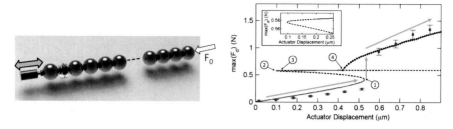

Fig. 7.8 (*Left*) Schematic diagram of a 1D granular crystal designed for acoustic rectification and switching. (*Right*) Bifurcation diagram. *Right panel* reproduced from [60] with permission from the publisher

through the crystal at that frequency. However, at sufficiently high amplitudes, and only from the boundary that is close to the defect, a jump phenomenon occurs from periodic to quasiperiodic and then chaotic states, where the energy of the driver is redistributed to different frequencies that can transmit through the system. This example illustrates how the combination of nonlinearity, periodicity, driving, and asymmetric disorder can create new material and device capabilities. In this case, this combination allowed energy to propagate predominantly in one direction.

To understand the nature of the bifurcations, and the jump to the quasiperiodic and chaotic states that allowed the asymmetric acoustic energy transmission, Boechler et al. conducted parametric continuation using the Newton-Raphson (NR) method in phase space [33] and numerical integration of the fully nonlinear equations of motion that describe the granular crystal. Dissipation was taken into account by using linear damping (see more about dissipative effects in the next section). Applying NR, they followed the periodic family of solutions of the driven system as a function of driving amplitude and studied its linear stability. Right panel of Fig. 7.8 shows the maximum dynamic force amplitude (four particles from the actuator) for each solution as a function of the driving amplitude. The stable (unstable) periodic solutions are denoted with solid blue (dashed black) lines. At turning points 1,2, stable and unstable periodic solutions collide and mutually annihilate (saddle-center bifurcation [40]). At points 3,4, the periodic solution changes stability and a new two-frequency stable quasiperiodic state emerges (Naimark-Sacker bifurcation [38]). Following this bifurcation picture, they observed in their experimental setup and numerical simulations that with increasing amplitude, a progression of the system response that followed the low-amplitude stable periodic solution up to point 1, where the system jumps past the unstable periodic solution to the high-amplitude stable quasiperiodic state. Further increase of the driver's amplitude led to a continued cascade of double period bifurcations and resulted in the merging of distinct frequency peaks, the formation of continuous bands, and chaotic dynamics. As the quasiperiodic and chaotic states redistribute energy from the driver to frequencies within the transmitting band, it is the existence of these states which enables the previously described acoustic rectification.

7.6 Dissipative Granular Crystals

Most of the studies to date involving granular crystals ignore dissipative effects. However, it is clear from the experiments that in many settings dissipation is strong and should be included. The sources of dissipation in granular crystals are many, including friction, plasticity, viscoelasticity, and viscous drag, among others. In the past few years there have been a number of analytical and numerical studies that have introduced dissipative terms into the equations of motion.

In [112], the authors studied the effects of two dissipative mechanisms on pulse propagation in nonlinear chains. The first was an intrinsic mechanism—an incomplete restitution mechanism that resulted in partial trapping of the impulse energy in the internal modes of the grain. The second mechanism was extrinsic—a velocity-dependent friction $f = -\gamma \dot{u}_i$. In both cases, they showed that the decay of the energy was well approximated by an exponential function. The attenuation of traveling pulses in 1D unloaded granular crystals due to on-site linear damping $f = -\gamma \dot{u}_i$ was also analyzed in [113]. They found an overall exponential decay of the energy, which depends on the exponent of the interaction potential, and causes the pulse to slow down as it propagates. They also showed that the shape and the width of the pulse remained unchanged.

Job and his collaborators studied the interaction of a solitary wave with boundaries in a 1D granular crystal, considering two dissipative mechanisms: internal viscoelasticity and solid friction of the beads due to their weight on the track aligning the granular crystal [114]. Viscoelastic dissipation was taken into account by considering a dissipative force at the contact of the two beads in the form $f = \eta A \partial_t ([u_{i-1} - u_i]_+^{3/2})$ [114], where η includes unknown coefficients due to internal friction of the material. Solid friction was included by considering a force $f = \mu m g$. These dissipative terms were also shown to produce broader solitary waves.

In [115] viscous dissipation, depending on the relative velocity between neighboring particles, was included in the model as $f = p(\dot{u}_{i-1} - 2\dot{u}_i + \dot{u}_{i+1})$, where p is the viscosity coefficient. The authors investigated its influence on the shape of a steady shock wave. Using this type of viscous dissipation, in [116], they solved the following system of nondimensional equations:

$$\ddot{u}_i = [p(\dot{u}_{i+1} - \dot{u}_i) - (u_i - u_{i+1})^n]\theta(u_i - u_{i+1}) - [p(\dot{u}_{i-1} - \dot{u}_i) + (u_{i-1} - u_i)^n]\theta(u_{i-1} - u_i),$$

where θ is the Heaviside function. They found that the inclusion of this relative velocity-dependent viscous damping may yield interesting effects such as the creation of secondary pulses. A different approach was presented in [117], where the authors provided a quantitative characterization of dissipative effects for solitary wave propagation in 1D granular crystals. They incorporated a phenomenological nonlinear dissipation that depends on the particle's relative velocities. By using optimization schemes and experiments, they calculated a common dissipation exponent with a material-dependent prefactor.

Most of the above studies concern the attenuation of propagating pulses generated by an impulsive excitation. Recent experiments in 1D compressed granular crystals, subjected to continuous harmonic driving at one end, also revealed a strong attenuation of the signal [60]. To account for the dissipation in these experimental settings, a linear on-site damping term $f = -\gamma \dot{u}_i$ with a damping coefficient γ was selected to match the experimental results.

7.7 Two-Dimensional Granular Crystals

Given the richness of the nonlinear dynamic phenomena found in one-dimensional systems, higher dimensional nonlinear systems are expected to present a plethora of new dynamic effects. For example, two- and three-dimensional nonlinear systems are expected to present additional families of wave modes not realizable in the 1D case; new types of solitary waves propagating in the axial and lateral directions (particularly interesting for wave energy redirection and wave guiding); complex nonlinear resonance interactions occurring between spatially extended modes and localized waves; and enhanced possibilities for acoustic wave energy localization and trapping across spatial or temporal scales.

The dynamic properties of 2D granular crystals have only been partially characterized. In particular, experimental efforts are few, although such systems are expected to present a variety of novel dynamic phenomena. Several authors have previously proposed models to characterize the dynamic response of two-dimensional, ordered granular media. For example, [118] described a model for a square lattice of elastically interacting particles, which included relative particle rotation. Tournat et al. [119] proposed a theoretical model to describe out-of-plane elastic waves in a monolayer granular membrane consisting of a hexagonal lattice of particles. Their model was the first one to include shear and bending rigidity at the contact between particles, and to calculate dispersion relations that accounted for these effects.

The simplest example of a highly nonlinear 2D granular crystal consists of a uniform, uncompressed square packing of elastic particles in contact with each other. When this system is excited on one side by a uniform, planar waveform, its response is expected to be quasi-one-dimensional [14] and the response of the system can be characterized by a "curtain solution" derived similarly to (7.10). The first experimental characterization of the dynamic behavior of a square packing of particles was provided in [84], using photoelastic elliptical disks, excited by an explosive charge. The same study characterized the stress wave propagation in arrays of elliptical disks of various geometrical packings, and concluded that it is the contact normals and the vector-connecting particles' centers of mass that primarily influence wave propagation characteristics such as load transfer path and load attenuation. Discrete element numerical models (DEM) were also used to analyze the dynamics of similar systems [120].

The formation and propagation of solitary waves in 2D square granular crystals was reported and studied quantitatively for the first time by Leonard et al. [121] using triaxial accelerometers embedded within selected particles in the crystals. A larger number of studies also explored the dynamic behavior of hexagonal packing under different (near-linear to highly nonlinear) loading conditions [84, 120, 122–128].One of the major difficulties in the experimental realization of acoustic materials based on two-dimensional nonlinear granular lattices is the sensitivity of such systems to the presence of variation in the particles' geometry. In the ideal configuration, all particles have an equal number of contacts and equal equilibrium forces. The presence of small defects in experiments, however, can lead to the loss of contact between particles or to the local compression in the surrounding particles. Such loss of contacts or local compression ultimately results in a disordered energy transfer between the particles. A few past works studied the effects of imperfections in two-dimensional granular crystals and their role in the stress wave propagation [122–125, 129]. While Hertzian behavior predicts a 1/6 power-law between maximum force and wave speed [47], it was found that the presence of defects tends to increase the wave propagation speed to a 1/4 power law relationship, effectively inducing deviations from the theoretical Hertzian behavior. This deviation from Hertzian behavior was observed only for granular crystals with low precompression. Increasing the precompression applied on hexagonal arrays was seen to cause a transition to a fully Hertzian behavior [122–125]. More recently, Leonard et al. experimentally characterized the dynamic response of regular 2D square granular crystals, and showed that variation in the packing geometry/composition (Fig. 7.9, left) can dramatically vary the directionality of wave propagation [130].

Two-dimensional arrays of particles have also been shown to form tunable acoustic lenses (Fig. 7.9, right) that support the formation of concentrated acoustic pulses at the focal point ("sound bullets", [59]). The ability to redirect nonlinear acoustic pulses in two-dimensional systems has also been studied by looking at pulse splitting and recombination in y-shaped granular networks [131–133]. These works showed theoretically, numerically, and experimentally the ability to bend and split incident pulses, and redirect mechanical energy as a function of the branch geometry.

Additional work on the dynamic behavior of ordered two-dimensional granular crystals is needed to fully understand the dynamic response of such systems, and to characterize how nonlinear wave formation and propagation depends on the underlying particle arrangement. Variations of the excitation type (impulsive or harmonic forcing) are expected to lead to the discovery of interesting new acoustic/dynamic phenomena including wave guiding, trapping, filtering, and localized breathing modes.

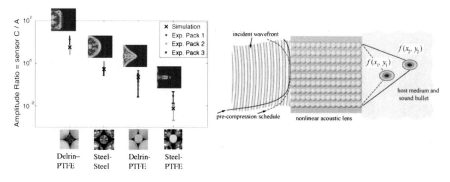

Fig. 7.9 (*Left*) Dynamic response of two-dimensional granular crystals formed by square-centered packings of different material cylinders and spheres (see *inset*). Variation of the materials configuration leads to dramatic changes of the wave propagation front, as shown from experiments and numerical simulations [130]. (*Right*) Design concept of a tunable, nonlinear acoustic lens obtained with a two-dimensional array of particle chains. The formation of the focal spot (i.e., the "sound bullet") is evident on the host medium on the right [59]. Images reproduced from [130] and [59]

7.8 Future Directions and Conclusions

The preceding chapters of this book have demonstrated how structural periodicity can be utilized to create new materials with unprecedented physical properties. In such materials the individual building blocks are assembled in carefully designed structures, where by working together, they cause the bulk material to present properties greater than those of the individual components. This general concept of obtaining "materials by design" is not new, and has been a long-term quest for chemists and material scientists alike. For instance, chemists have long been trying to engineer crystals and molecules by arranging atoms in specific lattices and geometries, to obtain a specific bulk property. However, by extending this concept past molecules and crystal grains, to specially designed structural building blocks—from the nano to macroscales—a whole new field of possibilities is enabled.

One of the main benefits of such designed materials is that they enable new technological capabilities. New materials with multifunctional properties can be designed, which have both structural and dynamic functionalities. Perhaps more importantly, by creating materials with previously unseen properties, new devices and applications are enabled. Furthermore, as such materials are "designed" by construction, and they can be easily tailored for use in specifically targeted applications.

The range of possible bulk responses from such designed materials depends in part on the complexity of the interaction between the fundamental building blocks. As described in the previous chapters, the design of these periodic structures has historically been based on linear interactions. The presence of nonlinearity in these systems gives added advantages through complexity. This chapter predominately focused on nonlinear dynamic phenomena in granular crystal systems, where the nonlinearity was caused by the geometric inter-particle interactions between elastic particles. As was described, nonlinear dynamics enables the existence of new useful dynamic phenomena and coherent structures. This includes solitary waves, discrete breathers, bifurcations, quasiperiodicity, and chaos, among others. Nonlinearity also enables a dramatic tunability of the material responses, by providing an unprecedented sensitivity to variations of materials and external parameters. However, because of the inherent complexity of nonlinear systems, which enables such useful phenomena, analyzing and predicting the behavior of such systems is also more difficult. In the future, the development of new predictive theoretical and computational tools will be necessary to further guide the development, design, and testing of nonlinear periodic phononic structures.

Some particular future areas of interest, with respect to the study of nonlinear periodic phononic structures, include, but are not limited to the following. As nonlinearity has been applied to spring-mass-like systems, in granular crystals, nonlinearity could also further be applied to the study of nonlinear metamaterials, nonlinear resonant structures, or phononic crystals with nonlinearly elastic components. The study of hybrid linear-nonlinear systems, could lead to the observation of new dynamic phenomena such as the amplitude-dependent filtering of acoustic signals [134].

New material systems where nonlinear material responses interplay with active building blocks or other multi-physical effects is another area that could lead to the discovery of unprecedented material responses. The ability to couple multi-physical effects in periodic structures can also lead to the creation of tunable multifunctional and energy-harvesting devices, such as opto-mechanical sensors [135], or phoXonic systems [136]. For example, the generation of nonlinear modes in nonlinear acoustic crystals could be used as a mechanism for frequency conversion, or the presence of nonlinear localized modes could be exploited for energy localization and harvesting.

Because of the similarity of acoustic and elastic wave propagation to phonon propagation, the effects studied here could also be extended to smaller scales involving heat propagation. For instance, as described in this chapter for acoustics, nonlinear periodic structures have been utilized to create tunable rectifiers based on the onset of bifurcation instabilities. This type of device could provide new ways to control the flow of acoustic energy, enable acoustic logic devices, and be used in novel energy-harvesting systems [60]. However, these same ideas could be scaled down to create new ways to control heat propagation, and enable materials with direction-dependent thermal conductivities or thermal logic devices.

Furthermore, the newly explored phenomena, which occur in granular crystals and other nonlinear periodic phononic structures, should be further explored for their potential in engineering applications. The ability to engineer the dispersion relation through nonlinearities could be implemented in tunable vibration filtering devices and in noise and vibration-insulating systems. Compact solitary waves with robust properties and large amplitudes could find use in biomedical devices with improved resolution and signal-to-noise ratio [59], or in the nondestructive evaluation of materials [58].

The study of nonlinearity in engineered materials like phononic crystals and metamaterials is still at an early stage of development. By understanding the fundamental properties of nonlinear acoustic crystals, nonlinear phononic systems, and nonlinear resonant structures, new physical phenomena can be discovered and lead to a new ability to design and implement materials and devices.

References

1. F. Duang, J. Guojin, *Introduction to Condensed Matter Physics*, vol. 1 (World Scientific, Singapore, 2005)
2. O. Morsch, M. Oberthaler, Dynamics of Bose-Einstein condensates in optical lattices. Rev. Mod. Phys. **78**(1), 179–215 (2006)
3. P. Markos, C.M. Soukoulis, *Wave Propagation: From Electrons to Photonic Crystals and Left-Handed Materials* (Princeton University Press, Princeton, NJ, 2008)
4. L. Brillouin, *Wave Propagation in Periodic Structures* (McGraw-Hill, New York, 1953)
5. *Nonlinearities in Periodic Structures and Metamaterials*, ed. by C. Denz, S. Flach, Y.S. Kivshar. Springer Series in Optical Sciences, vol. 150 (Springer, Berlin, 2010)
6. Y.C. Fung, P. Tong, *Classical and Computational Solid Mechanics* (World Scientific Publishing, Singapore, 2001)
7. H. Hertz, Journal fur Die Reine und Angewandie Mathematic **92**, 156–171 (1881)
8. L. Lifshitz, M.C. Cross, *Nonlinear Dynamics of Nanomechanical and Micromechanical Resonators*, in *Review of Nonlinear Dynamics and Complexity*, ed. by H.G. Schuster (2008)
9. K. Bertoldi, M.C. Boyce, Mechanically triggered transformations of phononic band gaps in periodic elastomeric structures. Phys. Rev. B **77**(5), 052105 (2008)
10. F.M. Hamilton, D.T. Blackstock, *Nonlinear Acoustics: Theory and Applications* (Academic, New York, 1997)
11. A.M. Samsonov, *Strain Solitons and How to Construct Them* (Chapman & Hall/CRC, Boca Raton, 2001)
12. G.A. Maugin, *Nonlinear Waves in Elastic Crystals* (Oxford University Press, New York, NY, 1999)
13. P. Hess, Surface acoustic waves in materials science. Phys. Today **55**(3), 42–47 (2002)
14. V.F. Nesterenko, *Dynamics of Heterogeneous Materials* (Springer, New York, 2001)
15. M. Sato, B.E. Hubbard, A.J. Sievers, Colloquium: nonlinear energy localization and its manipulation in micromechanical oscillator arrays. Rev. Mod. Phys. **78**(1), 137–157 (2006)
16. B. Liang, B. Yuan, J.C. Cheng, Acoustic diode: rectification of acoustic energy flux in one-dimensional systems. Phys. Rev. Lett. **103**(10), 104301 (2009)
17. B. Liang et al., An acoustic rectifier. Nat. Mater. **9**(12), 989–992 (2010)
18. H.Y. Hao, H.J. Maris, Experiments with acoustic solitons in crystalline solids. Phys. Rev. B **64**(6), 064302 (2001)
19. H.J. Maris, S. Tamura, Propagation of acoustic phonon solitons in nonmetallic crystals. Phys. Rev. B **84**(2), 024301 (2011)
20. M. Terraneo, M. Peyrard, G. Casati, Controlling the energy flow in nonlinear lattices: a model for a thermal rectifier. Phys. Rev. Lett. **88**(9), 094302 (2002)
21. B. Li, L. Wang, G. Casati, Thermal diode: rectification of heat flux. Phys. Rev. Lett. **93**(18), 184301 (2004)
22. C.W. Chang et al., Solid-state thermal rectifier. Science **314**(5802), 1121–1124 (2006)
23. B. Li, L. Wang, G. Casati, Negative differential thermal resistance and thermal transistor. Appl. Phys. Lett. **88**(14) (2006)
24. L. Wang, B. Li, Thermal logic gates: computation with phonons. Phys. Rev. Lett. **99**(17), 177208 (2007)

25. L. Wang, B. Li, Thermal memory: a storage of phononic information. Phys. Rev. Lett. **101** (26), 267203 (2008)
26. N. Yang, G. Zhang, B. Li, Carbon nanocone: a promising thermal rectifier. Appl. Phys. Lett. **93**(24), 243111 (2008)
27. N. Yang, G. Zhang, B. Li, Thermal rectification in asymmetric graphene ribbons. Appl. Phys. Lett. **95**(3), 033107 (2009)
28. E. Fermi, J.R. Pasta, S. Ulam, *Studies of Nonlinear Problems* (Los Alamos, Los Alamos Scientific Laboratory, 1955)
29. M. Porter et al., Fermi, Pasta, Ulam and the birth of experimental mathematics. Am. Sci. **97** (6) (2009)
30. D.B. Duncan et al., Solitons on lattices. Physica D **68**(1), 1–11 (1993)
31. Y.V. Kartashov, B.A. Malomed, L. Torner, Solitons in nonlinear lattices. Rev. Mod. Phys. **83** (1), 247 (2011)
32. P.G. Kevrekidis, Non-linear waves in lattices: past, present, future. IMA J. Appl. Math. **76**(3), 389–423 (2011)
33. S. Flach, A.V. Gorbach, Discrete breathers – advances in theory and applications. Phys. Rep. **467**(1–3), 1–116 (2008)
34. J.S. Russel, *Report on Waves*. Report of the 14th Meeting of the British Association for the Advancement of Science (1844), p. 311.
35. T. Dauxois, M. Peyrard, *Physics of solitons* (Cambridge University Press, Cambridge, 2006)
36. S. Aubry, Discrete breathers: localization and transfer of energy in discrete hamiltonian nonlinear systems. Physica D **216**(1), 1–30 (2006)
37. D.K. Campbell, S. Flach, Y.S. Kivshar, Localizing energy through nonlinearity and discreteness. Phys. Today **57**(1), 43–49 (2004)
38. S. Wiggins, *Introduction to Applied Nonlinear Systems and Chaos*, 2nd edn. (Springer, New York, NY, 2000)
39. R. Vijay, M.H. Devoret, I. Siddiqi, *Invited Review Article: The Josephson Bifurcation Amplifier*. Rev. Sci. Instrum. **80**(11) (2009)
40. S.H. Strogatz, *Nonlinear Dynamics and Chaos* (Perseus Publishing, Cambridge, MA, 1994)
41. M. Soljacic, J.D. Joannopoulos, Enhancement of nonlinear effects using photonic crystals. Nat. Mater. **3**(4), 211–219 (2004)
42. R.B. Karabalin et al., Signal amplification by sensitive control of bifurcation topology. Phys. Rev. Lett. **106**(9), 094102 (2011)
43. A.H. Nayfeh, D.T. Mook, *Nonlinear Oscillations* (Wiley, New York, 1979)
44. V.F. Nesterenko, Propagation of nonlinear compression pulses in granular media. J. Appl. Mech. Tech. Phys. [Zhurnal Prikladnoi Mekhaniki i Tehknicheskoi Fiziki], 1983. **24**(5 [vol.24, no.5]), pp. 733–743 [136–148]
45. A.N. Lazaridi, V.F. Nesterenko, Observation of a new type of solitary waves in one-dimensional granular medium. J. Appl. Mech. Tech. Phys. **26**(3), 405–408 (1985)
46. C. Coste, E. Falcon, S. Fauve, Solitary waves in a chain of beads under hertz contact. Phys. Rev. E **56**(5), 6104–6117 (1997)
47. C. Coste, B. Gilles, On the validity of Hertz contact law for granular material acoustics. Eur. Phys. J. B **7**(1), 155–168 (1999)
48. C. Daraio et al., Tunability of solitary wave properties in one-dimensional strongly nonlinear phononic crystals. Phys. Rev. E **73**(2), 026610 (2006)
49. N. Boechler, C. Daraio, *An Experimental Investigation of Acoustic Band Gaps and Localization in Granular Elastic Chains*, in *Proceedings of the Asme International Design Engineering Technical Conferences and Computers and Information in Engineering Conference*, 2010, pp. 271–276
50. N. Boechler et al., Tunable vibrational band gaps in one-dimensional diatomic granular crystals with three-particle unit cells. J. Appl. Phys. **109**(7), 074906 (2011)
51. N. Boechler et al., Discrete breathers in one-dimensional diatomic granular crystals. Phys. Rev. Lett. **104**(24), 244302 (2010)

52. G. Theocharis et al., Intrinsic energy localization through discrete gap breathers in one-dimensional diatomic granular crystals. Phys. Rev. E **82**(5), 056604 (2010)
53. K.R. Jayaprakash, Y. Starosvetsky, A.F. Vakakis, M. Peeters, G. Kerschen, Nonlinear normal modes and band gaps in granular chains with no pre-compression. Nonlinear Dynam. (2010). Available online: 10.1007/s11071-010-9809-0
54. V.F. Nesterenko et al., Anomalous wave reflection at the interface of two strongly nonlinear granular media. Phys. Rev. Lett. **95**(15), 158703 (2005)
55. C. Daraio et al., Energy trapping and shock disintegration in a composite granular medium. Phys. Rev. Lett. **96**(5), 058002 (2006).
56. E.B. Herbold et al., Tunable frequency band-gap and pulse propagation in a strongly nonlinear diatomic chain. Acta Mech. **205**, 85–103 (2009)
57. F. Fraternali, M.A. Porter, C. Daraio, Optimal design of composite granular protectors. Mech. Adv. Mater. Struct. **17**, 1–19 (2010)
58. D. Khatri, P. Rizzo, C. Daraio. *Highly Nonlinear Waves' Sensor Technology for Highway Infrastructures.* in *SPIE Smart Structures/NDE, 15th annual international symposium.* San Diego, CA, 2008
59. A. Spadoni, C. Daraio, Generation and control of sound bullets with a nonlinear acoustic lens. Proc. Natl. Acad. Sci. **107**, 7230 (2010)
60. N. Boechler, G. Theocharis, C. Daraio, Bifurcation-based acoustic switching and rectification. Nat. Mater. **10**(9), 665–8 (2011)
61. K.L. Johnson, *Contact Mechanics* (Cambridge University Press, Cambridge, 1985)
62. D. Sun, C. Daraio, S. Sen, The nonlinear repulsive force between two solids with axial symmetry. Phys. Rev. E **83**, 066605 (2011)
63. D. Ngo, D. Khatri, C. Daraio, Solitary waves in uniform chains of ellipsoidal particles. Phys. Rev. E **84**, 026610 (2011)
64. D. Khatri, D. Ngo, C. Daraio, Solitary waves in uniform chains of cylindrical particles. Granul. Matter (2011), in press
65. D. Ngo et al., *Highly nonlinear solitary waves in chains of hollow spherical particles* Granul. Matter (2012), in press
66. M. de Billy, Experimental study of sound propagation in a chain of spherical beads. J. Acoust. Soc. Am. **108**(4), 1486–1495 (2000)
67. A. Pankov, *Traveling Waves and Periodic Oscillations in Fermi-Pasta-Ulam Lattices* (Imperial College Press, London, 2005)
68. S. Flach, A. Gorbach, Discrete breathers in Fermi-Pasta-Ulam lattices. Chaos **15**(1), 015112 (2005)
69. Remoissenet, M., *Waves Called Solitons (Concepts and Experiments).* 3rd revised and enlarged edition ed (Springer, Berlin, 1999)
70. G.X. Huang, Z.P. Shi, Z.X. Xu, Asymmetric intrinsic localized modes in a homogeneous lattice with cubic and quartic anharmonicity. Phys. Rev. B **47**(21), 14561–14564 (1993)
71. V.J. Sanchez-Morcillo et al., *Second Harmonics, Instabilities and Hole Solitons in 1D Phononic Granular Chains.* in *Phononics 2011: First International conference on phononic crystals, metamaterials and optomechanics*, Santa Fe, New Mexico,USA, 2011
72. V. Tournat, V.E. Gusev, B. Castagnede, Self-demodulation of elastic waves in a one-dimensional granular chain. Phys. Rev. E **70**(5), 056603 (2004)
73. P. Rosenau, J.M. Hyman, Compactons – solitons with finite wavelength. Phys. Rev. Lett. **70** (5), 564–567 (1993)
74. R.S. MacKay, Solitary waves in a chain of beads under Hertz contact. Phys. Lett. A **251**(3), 191–192 (1999)
75. G. Friesecke, J.A.D. Wattis, Existence theorem for solitary waves on lattices. Commun. Math. Phys. **161**(2), 391–418 (1994)
76. J.Y. Ji, J.B. Hong, Existence criterion of solitary waves in a chain of grains. Phys. Lett. A **260** (1–2), 60–61 (1999)
77. S. Sen et al., Solitary waves in the granular chain. Phys. Rep. **462**(2), 21–66 (2008)

78. A. Chatterjee, Asymptotic solution for solitary waves in a chain of elastic spheres. Phys. Rev. E **59**(5), 5912–5919 (1999)
79. A. Rosas, K. Lindenberg, Pulse propagation in chains with nonlinear interactions. Phys. Rev. E **69**(1), 016615 (2004)
80. A. Rosas, K. Lindenberg, Pulse velocity in a granular chain. Phys. Rev. E **69**(3), 037601 (2004)
81. J.M. English, R.L. Pego, On the solitary wave pulse in a chain of beads. Proc. Am. Math. Soc. **133**(6), 1763–1768 (2005)
82. K. Ahnert, A. Pikovsky, Compactons and chaos in strongly nonlinear lattices. Phys. Rev. E **79**(2), 026209 (2009)
83. Y. Starosvetsky, A.F. Vakakis, Traveling waves and localized modes in one-dimensional homogeneous granular chains with no precompression. Phys. Rev. E **82**(2), 026603 (2010)
84. Y. Zhu, A. Shukla, M.H. Sadd, The effect of microstructural fabric on dynamic load transfer in two dimensional assemblies of elliptical particles. J. Mech. Phys. Solids **44**(8), 1283–1303 (1996)
85. C. Daraio, V.F. Nesterenko, Strongly nonlinear wave dynamics in a chain of polymer coated beads. Phys. Rev. E **73**(2), 026612 (2006)
86. C. Daraio et al., Strongly nonlinear waves in a chain of teflon beads. Phys. Rev. E **72**(1), 016603 (2005)
87. S. Job et al., How Hertzian solitary waves interact with boundaries in a 1D granular medium. Phys. Rev. Lett. **94**(17), 178002 (2005)
88. S. Job et al., Solitary wave trains in granular chains: experiments, theory and simulations. Granul. Matter **10**, 13–20 (2007)
89. F. Santibanez et al., Experimental evidence of solitary wave interaction in Hertzian chains. Phys. Rev. E **84**(2), 026604 (2011)
90. D. Ngo, C. Daraio, Nonlinear dynamics of chains of coated particles (2011), in preparation
91. A.C. Hladky-Hennion, M. de Billy, Experimental validation of band gaps and localization in a one-dimensional diatomic phononic crystal. J. Acoust. Soc. Am. **122**, 2594–2600 (2007)
92. A.C. Hladky-Hennion, G. Allan, M. de Billy, Localized modes in a one-dimensional diatomic chain of coupled spheres. J. Appl. Phys. **98**(5), 054909 (2005)
93. R.F. Wallis, Effect of free ends on the vibration frequencies of one-dimensional lattices. Phys. Rev. **105**(2), 540–545 (1957)
94. R. Livi, M. Spicci, R.S. MacKay, Breathers on a diatomic FPU chain. Nonlinearity **10**(6), 1421–1434 (1997)
95. P. Maniadis, A.V. Zolotaryuk, G.P. Tsironis, Existence and stability of discrete gap breathers in a diatomic beta Fermi-Pasta-Ulam chain. Phys. Rev. E **67**(4), 046612 (2003)
96. G.X. Huang, B.B. Hu, Asymmetric gap soliton modes in diatomic lattices with cubic and quartic nonlinearity. Phys. Rev. B **57**(10), 5746–5757 (1998)
97. M. Aoki, S. Takeno, A.J. Sievers, Stationary anharmonic gap modes in a one-dimensional diatomic lattice with quartic anharmonicity. J. Physical Soc. Japan **62**(12), 4295–4310 (1993)
98. G. Theocharis, N. Boechler, C. Daraio, *Control of Vibrational Energy in Nonlinear Granular Crystals*, in *Proceedings of Phononics*, Santa Fe, New Mexico, USA, 2011, pp. 170–171
99. M.A. Porter et al., Highly nonlinear solitary waves in heterogeneous periodic granular media. Physica D **238**, 666–676 (2009)
100. K.R. Jayaprakash, Y. Starosvetsky, A.F. Vakakis, New family of solitary waves in granular dimer chains with no precompression. Phys. Rev. E **83**(3 Pt 2), 036606 (2011)
101. Y. Man et al., Defect modes in one-dimensional granular crystals. Phys. Rev. E **85**, 037601 (2012)
102. G. Theocharis et al., Localized breathing modes in granular crystals with defects. Phys. Rev. E **80**, 066601 (2009)
103. S. Job et al., Wave localization in strongly nonlinear Hertzian chains with mass defect. Phys. Rev. E **80**, 025602(R) (2009)
104. P.C. Kevrekidis et al., Spontaneous symmetry breaking in photonic lattices: theory and experiment. Phys. Lett. A **340**(1–4), 275–280 (2005)

105. M. Albiez et al., Direct observation of tunneling and nonlinear self-trapping in a single bosonic Josephson junction. Phys. Rev. Lett. **95**(1), 010402 (2005)
106. G. Theocharis et al., Symmetry breaking in symmetric and asymmetric double-well potentials. Phys. Rev. E **74**(5), 056608 (2006)
107. S. Sen et al., Solitonlike pulses in perturbed and driven Hertzian chains and their possible applications in detecting buried impurities. Phys. Rev. E **57**(2), 2386–2397 (1998)
108. M. Manciu et al., Phys. A **274**, 607–618 (1999)
109. S. Sen et al., Int. J. Mod. Phys. B **19**(18), 2951–2973 (2005). Review, and references therein
110. E. Hascoet, H.J. Herrmann, Shocks in non-loaded bead chains with impurities. Eur. Phys. J. B **14**(1), 183–190 (2000)
111. Y. Starosvetsky, K.R. Jayaprakash, A.F. Vakakis, Scattering of solitary waves and excitation of transient breathers in granular media by light intruders. J. Appl. Mech. (2011), accepted for publication
112. M. Manciu, S. Sen, A.J. Hurd, Impulse propagation in dissipative and disordered chains with power-law repulsive potentials. Physica D **157**(3), 226–240 (2001)
113. A. Rosas, K. Lindenberg, Pulse dynamics in a chain of granules with friction. Phys. Rev. E **68** (4 Pt 1), 041304 (2003)
114. N.V. Brilliantov et al., Model for collisions in granular gases. Phys. Rev. E **53**(5), 5382–5392 (1996)
115. E.B. Herbold, V.F. Nesterenko, Shock wave structure in a strongly nonlinear lattice with viscous dissipation. Phys. Rev. E **75**(2), 021304 (2007)
116. A. Rosas et al., Observation of two-wave structure in strongly nonlinear dissipative granular chains. Phys. Rev. Lett. **98**(16), 164301 (2007)
117. R. Carretero-Gonzalez et al., Dissipative solitary waves in periodic granular media. Phys. Rev. Lett. **102**, 024102 (2009)
118. I.S. Pavlov, A.I. Potapov, G.A. Maugin, A 2D granular medium with rotating particles. Int. J. Solids Struct. **43**(20), 6194–6207 (2006)
119. V. Tournat et al., Elastic waves in phononic monolayer granular membranes. New J. Phys. **13**, 073042 (2011)
120. M.H. Sadd, J. Gao, A. Shukla, Numerical analysis of wave propagation through assemblies of elliptical particles. Comput. Geotech. **20**(3–4), 323–343 (1997)
121. A. Leonard, F. Fraternali, C. Daraio, Directional wave propagation in a highly nonlinear square packing of spheres. Exp. Mech. (2011), DOI: 10.1007/s11340-011-9544-6
122. J.D. Goddard, Nonlinear elasticity and pressure-dependent wave speed in granular media. Proc. R. Soc. Lond. A Math. Phys. Eng. Sci. **430**(1878), 105–131 (1990)
123. S. Sen, R.S. Sinkovits, Sound propagation in impure granular columns. Phys. Rev. E **54**(6), 6857–6865 (1996)
124. B. Velicky, C. Caroli, Pressure dependence of the sound velocity in a two-dimensional lattice of Hertz-Mindlin balls: mean-field description. Phys. Rev. E **65**(2), 021307 (2002)
125. B. Gilles, C. Coste, Low-frequency behavior of beads constrained on a lattice. Phys. Rev. Lett. **90**(17), 174302 (2003)
126. M. Nishida, K. Tanaka, T. Ishida, *DEM simulation of wave propagation in two-dimensional ordered array of particles*. Shock Waves, vol. 2, Proceedings, ed. by K.S.F. Hannemann. (2009), pp. 815–820
127. A. Merkel, V. Tournat, V. Gusev, Elastic waves in noncohesive frictionless granular crystals. Ultrasonics **50**(2), 133–138 (2010)
128. M. Nishida, Y. Tanaka, DEM simulations and experiments for projectile impacting two-dimensional particle packings including dissimilar material layers. Granul. Matter **12**(4), 357–368 (2010)
129. A. Leonard, C. Daraio, A. Awasthi et al., Effects of weak disorder on stress wave anisotropy in centered square nonlinear granular crystals. Phys. Rev. E **86**(3), 031305 (2012)
130. A. Leonard, C. Daraio, Varying stress wave anisotropy in centered square highly nonlinear granular crystals. Phys. Rev. Lett. **108**, 214301 (2012)
131. A. Shukla, C.Y. Zhu, M. Sadd, Angular-dependence of dynamic load-transfer due to explosive loading in granular aggregate chains. J. Strain Anal. Eng. Des. **23**(3), 121–127 (1988)

132. C. Daraio et al., Highly nonlinear pulse splitting and recombination in a two-dimensional granular network. Phys. Rev. E **82**(3), 036604 (2010)
133. D. Ngo, F. Fraternali, C. Daraio, Highly nonlinear solitary wave propagation in y-shaped granular crystals with variable branch angles. Phys. Rev. E **85**, 036602 (2012)
134. J. Yang, S. Dunatunga, C. Daraio, Amplitude-dependent attenuation of compressive waves in curved granular crystals constrained by elastic guides. Acta Mech. **223**, 549–562 (2012)
135. M. Eichenfield et al., Optomechanical crystals. Nature **462**(7269), 78–82 (2009)
136. S. Sadat-Saleh et al., Tailoring simultaneous photonic and phononic band gaps. J. Appl. Phys. **106**(7), 074912 (2009)

Chapter 8
Tunable Phononic Crystals and Metamaterials

O. Bou Matar, J.O. Vasseur, and Pierre A. Deymier

Abstract The objective of this chapter is to show how it would be possible to introduce a certain degree of tunability of the properties of phononic crystals. The main concepts underlying the conception of tunable phononic crystals are first introduced with simple models: the one-dimensional harmonic crystal with varying parameter and two coupled one-dimensional harmonic crystals. An overview of the literature on tunable phononic crystals is given. Three of the tuning methods proposed in the literature are described in some details. We also illustrate the new or enhanced functionalities open by the tuning of the phononic crystal properties. These applications include reconfigurable waveguides and tunable superlenses.

O. Bou Matar (✉)
International Associated Laboratory LEMAC: IEMN, UMR CNRS 8520, PRES Lille Nord de France, EC Lille, 59652 Villeneuve d'Ascq, France
e-mail: olivier.boumatar@iemn.univ-lille1.fr

J.O. Vasseur
Institut d'Electronique, de Micro-electronique et de Nanotechnologie (IEMN, UMR CNRS 8520), PRES Lille Nord de France, 59652 Villeneuve d'Ascq, France
e-mail: Jerome.Vasseur@univ-lille1.fr

P.A. Deymier
Department of Materials Science and Engineering, University of Arizona, Tucson, AZ 85721, USA
e-mail: deymier@email.arizona.edu

P.A. Deymier (ed.), *Acoustic Metamaterials and Phononic Crystals*,
Springer Series in Solid-State Sciences 173, DOI 10.1007/978-3-642-31232-8_8,
© Springer-Verlag Berlin Heidelberg 2013

8.1   Introduction to Tunability. One-Dimensional Tunable Harmonic Crystal

8.1.1   One-Dimensional Diatomic Harmonic Crystal With Varying Parameters

As an introduction to the concept of tunability of the phononic crystal properties, we first consider the 1-D diatomic harmonic crystal introduced in paragraph 2.2.2, but treat it as a system with variable properties. It has been demonstrated that the dispersion equation of diatomic harmonic crystal, constituted of an infinite chain of masses with alternatively a mass m_1 and a mass m_2 connected by harmonic springs with spring constant β, can be expressed as

$$\omega^2 = \frac{\omega_3^2}{2} \mp \sqrt{\frac{\omega_3^4}{4} - (\omega_1\omega_2)^2 \sin^2 ka} \qquad (8.1)$$

where ω_1, ω_2, and ω_3 are characteristic frequencies given by: $\omega_1 = \sqrt{\frac{2\beta}{m_1}}, \omega_2 = \sqrt{\frac{2\beta}{m_2}}$, and $\omega_3 = \sqrt{2\beta\left(\frac{1}{m_1} + \frac{1}{m_2}\right)}$ if one choose $m_1 > m_2$. Two kinds of parameters appear in the dispersion equation (8.1): geometrical parameters, i.e., the separation distance between the masses at rest a, and property parameters, i.e., the spring constant β and masses m_1 and m_2. This opens the possibility to tune the 1-D harmonic crystal band structure by changing its geometry, or by varying part or all of the properties of its constituents (spring and/or masses). First, we consider a 1-D harmonic crystal in which the lattice parameter is tunable from a to a', with $a' = 1.5a$. The original, corresponding to the lattice parameter a, and modified band structures are displayed in Fig. 8.1a by solid and dashed lines, respectively. In this case, the phase velocity of the acoustic branch at low frequency and the slope of the negative optical branch are drastically modified. Nevertheless, the width of the gap appearing between ω_1 and ω_2 does not change.

Consider now a 1-D harmonic crystal with a value of the mass m_1 doubled, and with all the other parameters kept constant. The corresponding band structures are displayed in Fig. 8.1b. In this case, the phase velocity of the acoustic branch at low frequency and the slope of the negative optical branch can once again be tuned, but also the gap width. In the example presented in Fig. 8.1b, the gap width is increased by 20 %.

The tuning of the physical properties of the constituents of the phononic crystal can be made by the application of an external stimulus, such as the temperature, an electrical or magnetic field, etc. Some time, these external stimuli interact with the acoustical vibrations as coupled modes. A simple, but instructive, 1-D model of such coupling, a system of two coupled 1-D harmonic crystals, is now presented.

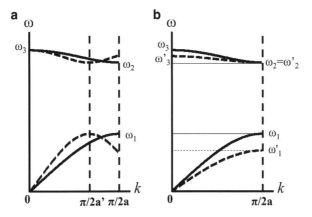

Fig. 8.1 Schematic representation of the modifications of the band structure of a 1-D diatomic harmonic crystal in the irreducible Brillouin zone induced by changing (**a**) the geometrical parameter a, and (**b**) the property parameter m_1. The *solid lines* represent the original band structure, and the *dashed lines* the modified band structure

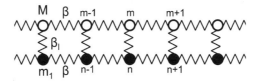

Fig. 8.2 Schematic illustration of the system of two coupled 1-D harmonic crystals. The atoms of one of the 1-D harmonic crystal have a mass m_1, and the atoms of the other one have a mass M. The force constant of the springs of each 1-D harmonic crystal is β. The force constant of the coupling springs is β_I. The periodicity of the crystal is a

8.1.2 Two Coupled One-Dimensional Mono-Atomic Harmonic Crystals

The system of two coupled one-dimensional harmonic crystals is illustrated in Fig. 8.2.

In absence of external forces, the equations describing the motion of atom "n" of the first 1-D harmonic crystal, and "m" of the second 1-D harmonic crystal are given by

$$m_1 \ddot{u}_n = \beta(u_{n+1} - 2u_n + u_{n-1}) + \beta_I(u_m - u_n),$$
$$M \ddot{u}_m = \beta(u_{m+1} - 2u_m + u_{m-1}) - \beta_I(u_m - u_n),$$
(8.2)

Here, β_I is the coupling spring constant. We seek solutions in the form of propagating waves with different amplitudes for atoms of each 1-D harmonic crystal as their masses are different:

$$u_n = Ae^{i\omega t}e^{ikna},$$
$$u_m = Be^{i\omega t}e^{ikma}, \tag{8.3}$$

where k is a wavenumber and ω is an angular frequency. Inserting solutions of the form given by (8.3) into (8.2) leads, after some algebraic manipulations and using the relation $2i\sin\theta = e^{i\theta} - e^{-i\theta}$, to the following set of linear equations in A and B:

$$\left(m_1\omega^2 + \beta\left(2i\sin\frac{ka}{2} \right)^2 - \beta_I \right)A + \beta_I e^{ik(m-n)a}B = 0,$$

$$\beta_I e^{ik(n-m)a}A + \left(M\omega^2 + \beta\left(2i\sin\frac{ka}{2} \right)^2 - \beta_I \right)B = 0. \tag{8.4}$$

This is an Eigen value problem in ω^2 which admits non-trivial solutions when the determinant of the matrix composed of the linear coefficient in (8.4) is equal to zero:

$$\begin{vmatrix} m_1\omega^2 + \beta\left(2i\sin\frac{ka}{2} \right)^2 - \beta_I & \beta_I e^{ik(m-n)a} \\ \beta_I e^{ik(n-m)a} & M\omega^2 + \beta\left(2i\sin\frac{ka}{2} \right)^2 - \beta_I \end{vmatrix} = 0 \tag{8.5}$$

Setting $\alpha = \omega^2$, (8.5) takes the form of the following quadratic equation:

$$\alpha^2 + \left(\beta\left(2i\sin\frac{ka}{2} \right)^2 - \beta_I \right)\left(\frac{1}{m_1} + \frac{1}{M} \right)\alpha + \frac{\beta^2}{m_1 M}\left(2i\sin\frac{ka}{2} \right)^4 = 0 \tag{8.6}$$

which admits two solutions:

$$\omega^2 = \left(2\beta\sin^2\frac{ka}{2} + \frac{\beta_I}{2} \right)\left(\frac{1}{m_1} + \frac{1}{M} \right)$$
$$\pm \sqrt{\left(2\beta\sin^2\frac{ka}{2} + \frac{\beta_I}{2} \right)^2\left(\frac{1}{m_1} + \frac{1}{M} \right)^2 - \frac{16\beta^2}{m_1 M}\sin^4\frac{ka}{2}} = 0 \tag{8.7}$$

These two solutions are periodic in wave number k with a period of $\frac{\pi}{a}$. These solutions are represented graphically in the band structure (solid lines) of Fig. 8.3a over the interval $k \in \left[0, \frac{\pi}{a}\right]$ for a ratio $\frac{\beta_I}{\beta} = 0.1$. For comparison, the band structure in the case of uncoupled modes, i.e., $\beta_I = 0$, is also displayed in Fig. 8.3 as dashed lines. It appears that the effect of coupling arises mainly in the region where the dispersion curves of the two uncoupled modes intersect, i.e., at low frequencies in the considered case. This is a universal property of modes coupling. For weak coupling, as displayed in Fig. 8.3, the upper limits for angular frequency of the two

Fig. 8.3 (a) Band structure of the system of two coupled 1-D mono-atomic harmonic crystals. (b) Zoom of the band structure around the low-frequency range. The *dashed lines* represent the uncoupled case ($\beta_I = 0$)

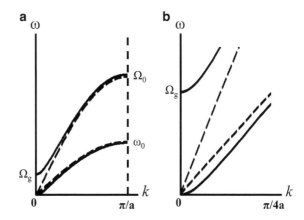

modes, at the edge of the Brillouin zone, are very close to those obtained in the case of uncoupled modes, i.e., $\omega_0 = 2\sqrt{\frac{\beta}{m_1}}$ and $\Omega_0 = 2\sqrt{\frac{\beta}{M}}$. But, at $k = 0$ a gap appears in the interval of frequency $[0, \Omega_g]$, where $\Omega_g = \sqrt{\beta_I \left(\frac{1}{m_1} + \frac{1}{M}\right)}$, for the mode with the highest phase velocity, corresponding to the chain of atoms with masses M. This behavior is very similar to those observed in magnetoacoustic waves, where a magnetoacoustic gap appears in the spin waves branch near a phase transition [1].

If we enlarge on the band structures of the coupled modes at low frequencies, as shown in Fig. 8.3b, the influence of the coupling on the mode with the lowest phase velocity is enlightened: its low-frequency phase velocity, corresponding to the slop of band structure curve near $k = 0$, can be dramatically reduced. Then, if we consider the chain of masses M as an external stimulus acting on a 1-D mono-atomic harmonic crystal, with masses m_1, it clearly appears that the external stimulus can be used to tune the acoustic properties of the harmonic crystal.

8.2 Literature Review

Phononic crystal may have applications in numerous technological fields. Nevertheless, for enhanced functionality it appears necessary to introduce a certain degree of frequency tunability of phononic crystal properties. As shown in the 1-D harmonic crystal example, tunability could be achieved by changing the geometry of the phononic crystal or by varying the elastic characteristics of the constitutive materials through application of external stimuli.

In 2-D, contrary to the harmonic crystal, the geometry of a phononic crystal can be changed not only by modifying the filling fraction, but also by the rotation of square inclusions [2]. In 2-D periodic arrays of rotating square solid rods, variations up to 60 % of the relative band width of the first gap have been predicted for filling fractions less than 0.5, and even more for higher filling fractions [3]. The large tuning capability of the properties of these rotating rods phononic crystals opens the opportunity to develop devices with enhanced or new functionality. An example of refraction control in such 2-D phononic crystal will be presented at the end of this chapter [4]. Isotropic materials are usually chosen as constituents of phononic crystals where a change in geometry is utilized to tune the band structure. The isotropy of the materials imposes the choice of inclusions that do not have an axial symmetry to modify the geometry of the phononic crystal when the rods are rotated. But, when anisotropic inclusions are considered, even cylindrical inclusions can be used to obtain similar results [5].

Other authors [6] exploit the change of the structure, i.e., the lattice and the form of the inclusions, of a phononic crystal made of holes in an elastomeric matrix, due to an external stress to alter the band structure. Periodic elastomeric structures can reversely undergo large strain deformations and dramatic transformations in their periodic pattern with only a small applied stress. A 2-D periodically patterned SU-8, the material of choice for microelectromechanical system fabrication, has been used as the matrix of a phononic crystal where the internal stress arises due to swelling of a solvent. This system shows a significant change in the phononic band structure, specifically in the opening of a new band gap in the GHz range [7]. Micrometric 3-D elastomeric network/air structures have been realized by interference lithography [8].

Now, if the used elastomer is a dielectric elastomer, an electric field can be used to deform the structure. For example, one can modify the size of dielectric elasto-mer cylindrical inclusions by applying an electric field [9]. A square lattice arrangement of dielectric elastomer tube in air has been shown to open the possi-bility to change the refraction from positive to negative with the increase in the applied electric field [10]. The application to the conception of a tunable narrow pass band filter based on a 1-D phononic crystal with a dielectric elastomer layer has also been presented [11].

However, the main part of these approaches requires physical contact with the phononic crystal. Another proposed solution requires using active materials as constituents of the composite material. In this case, the geometry of the phononic crystal is fixed and only the constituent properties, i.e., density and elastic constants, are varied. Then one can expect that the elastic contrast, and subse-quently the crystal properties, e.g., the bang gaps frequencies and widths, the negative refraction behaviors, could be controlled by an external stimulus.

Following this way, some authors [12] have studied how the piezoelectric effect can influence the elastic properties of the system and therefore can change the dispersion curves and in particular the gaps. For an arrangement of piezoelec-tric cylinders embedded in a polymer matrix, they show that the effect is signifi-cant for large filling fractions but negligibly small for small ones [13]. Moreover, a strong influence of the polarization direction on the width and starting frequency of the first band gap of a phononic crystal consisting of rectangular piezoelectric ceramics placed periodically in an epoxy matrix has been reported [14]. As regions of polarizations, alternatively oriented toward the top (up) or bottom (down), in plates or films, can be obtained in ceramics at the millimeter scales or with domain engineering of ferroelectric films at the micrometer scale. This high polarization direction sensitivity can open new opportunity for the design of integrated tunable phononic devices. Following this way, the concept of a switchable phononic crystal filter using polarization-patterned piezoelectric solids has been proposed [15].

Several studies have reported changes in the band structures of magnetoelectroelastic phononic crystals when the coupling between magnetic, electric, and elastic phenomena is taken into account [16]. Nevertheless, noticeable changes can be obtained only by modifying the geometry in the considered piezoelectric/piezomagnetic layered composites. In fact, crystals presenting true piezomagnetism, i.e., a linear dependence of stress, or strain, on a magnetic field, are quite rare, and piezomagnetic behavior is often observed in magnetostrictive media around an equilibrium state imposed by an external static magnetic field. The band structure of a two-dimensional phononic crystal constituted of a square array of Terfenol-D square rods embedded in an epoxy resin matrix can be controlled by application of an external magnetic field [17, 18]. Indeed, the elastic properties of magnetoacoustic material are very sensitive to its magnetic state and on the applied magnetic field. For instance, in giant magnetostrictive material, such as Terfenol-D, this dependence can lead to more than 50 % variation of some of the elastic constants, even at ultrasonic frequencies [19], without any contact by a magnetic field. Magnetoacoustic phononic crystal properties will be presented in some detail in Sect. 8.3.3 and their application to the design of a reconfigurable waveguide in Sect. 8.4.1. The use of an external magnetic field to tune the properties of colloidal phononic crystals with paramagnetic particles integrated in the crystal has also been proposed [20].

Other authors have considered the effect of temperature on the elastic moduli of the constituents of the phononic crystal, as for example in the case of holes containing air in a Quartz background [21]. The tuning of negative refraction of a sonic crystal constituted of steel rods in air background, induced by the variations of the air density and sound speed with temperature changes in the range $-40°$ to $100\,°C$, has also been studied [22]. Generally, the elastic moduli changes as a function of temperature are quite small, with the exception of those close to a phase transition. Using a phase transition around $35\,°C$ in $Ba_{0.7}Sr_{0.3}TiO_3$, a tunable ferroelectric phononic crystal, with an epoxy matrix, has been designed and realized by a dice and fill technique [23]. More details on ferroelectric phononic crystal will be given in Sect. 8.3.2. A fluid/solid phase transition with temperature can also be used as shown in [24], where an anodic aluminum oxide containing periodically arranged cylindrical nanopores infiltrated by PVDF polymer has been studied. Here, the solid/fluid transition of the PVDF, when the temperature changes from $25\,°C$ to $180\,°C$, is shown to induce an on/off switch of some of the passing band in the band structure of the realized hypersound phononic crystal.

Some authors have proposed the use of electrorheological materials in conjunction with the application of an external electric field [25]. As in electrorheological material this is the shear modulus that can be controlled by an electric field, only the mode with transverse polarization can be tuned. In the same spirit, one can imagine to use magnetorheological material with the application of a magnetic field [26].

The last envisaged method is to incorporate non-linear media as constituents of a phononic crystal. Indeed, in non-linear materials the application of an external static stress induces a variation of the effective elastic constants [27, 28]. As for most of methods implying variations of the geometry, the use of non-linear materials requires physical contact with the phononic crystal. More details on this topic have been presented in Chap. 7.

8.3 Two-Dimensional Tunable Phononic Crystals

8.3.1 Tunable Phononic Crystals Created by Rotating Square Rods

The first method of tuning 2-D phononic crystal that we will consider in more details is the physical rotation of square inclusions periodically positioned in air, first proposed by Goffaux and Vigneron [2]. Here, we only consider the low solid filling fraction, i.e., the case of isolated solid rods in air, as shown in Fig. 8.4. Indeed, this configuration corresponds to the only one with potential application for tunable phononic crystals. For filling fractions higher than 0.5, and above a critical value of the rotation angle, $\theta_c = \cos^{-1}(\sqrt{f})$, the rods can no more be rotated due to contacts between neighboring rods.

Because, in this system of square solid rods in air, the high-density contrast between solid and air authorizes the use of the condition of elastic rigidity to the solid rods, all the calculations presented in this part have been made with the PWE method applied to fluid inclusions in fluid. The structure factor for the square lattice of square rods with rotation angle θ can be written as [29]

$$I\left(\vec{G}\right) = f\sin\left(\tilde{G}_x \frac{d}{2}\right)\sin\left(\tilde{G}_y \frac{d}{2}\right), \tag{8.8}$$

with

$$\begin{aligned}
\tilde{G}_x &= G_x\cos\theta + G_y\sin\theta, \\
\tilde{G}_y &= -G_x\sin\theta + G_y\cos\theta,
\end{aligned} \tag{8.9}$$

where d is the edge length of the square rods, and G_x and G_y are the components of the reciprocal vectors.

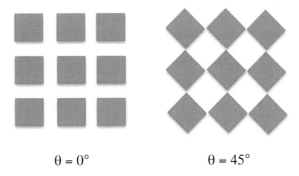

$\theta = 0°$ $\theta = 45°$

Fig. 8.4 Transverse cross-section of two-dimensional phononic crystals constituted of square section hard rods in air arranged in a square lattice. The figure represents an array having the same orientation as the empty lattice (*left part*), and the same array rotated by an angle $\theta = 45°$ (*right part*)

The elastic parameters chosen for all the calculations are $\rho_{air} = 1\,\text{kg.m}^{-3}$, $\rho_{rod} = 1,500\,\text{kg.m}^{-3}$, $c_{air} = 340\,\text{m.s}^{-1}$, and $c_{rod} = 2,000\,\text{m.s}^{-1}$. We investigate the acoustic band structures as a function of the rotation angle and filling fraction. To ensure a good convergence, 625 plane waves have been used in all the acoustic band structure calculations.

Two examples of band structures, obtained for a square-lattice two-dimensional phononic crystal consisting of square solid rods in air with a filling fraction $f = 0.5$ and edge length $d = 0.5$ mm, are displayed in Fig. 8.5 for a rotating angle (a) $\theta = 0$ and (b) $\theta = 45°$. As seen in Fig. 8.5a, no absolute acoustic band gap exists for $\theta = 0$ in the first eight bands. In fact, as shown in Fig. 8.6, no absolute band gap appears between the first two bands for any filling fraction for this orientation. Rotating the square rods from an angle $\theta = 0$ to $45°$, a large absolute phononic band gap appears between the first and second bands.

Figure 8.6 shows the numerical results of the normalized width of the lowest band gap, between the first and second bands, as a function of the rotation angle θ of the rods for five different filling fractions $f = 0.30, 0.35, 0.40, 0.45$, and 0.50. The normalized width of the band gap is taken as the gap width $\Delta\omega$ divided by the mid gap frequency ω_g. It can be seen that for a given filling fraction the normalized gap width increases with the rotation angle, and, for a given angle, it increases with the filling fraction. Moreover, for a given filling fraction, the absolute acoustic band gap appears only above a certain angle. This gap-opening angle gradually decreases as the filling fraction increases.

The widening of the gap can be explained by the change of the geometry [2]. Indeed, at $0°$ the space left between the rods is sufficiently large to allow for propagation of the waves in the structure with little wave interferences. In this case, no gap appears in the band structure. On the other hand, when the rods are rotated this space is reduced, and the destructive wave interferences are increased. This finally leads to first the appearance of a gap and then to its widening.

In the present method, a physical contact with the phononic crystal is needed in order to tune its properties. This limitation can be overcome, as shown in the

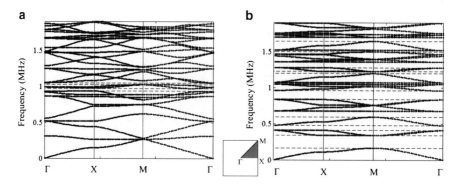

Fig. 8.5 Acoustic band structure for a square-lattice two-dimensional phononic crystal consisting of square solid rods rotated with an angle (**a**) $\theta = 0°$ and (**b**) $\theta = 45°$, in air. The filling fraction $f = 0.5$ and edge length $d = 0.5$ mm

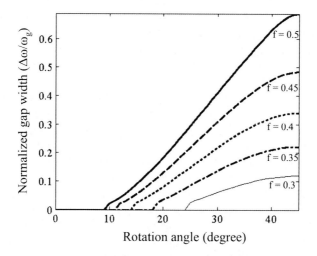

Fig. 8.6 Evolution of the normalized lowest absolute band gap width as a function of the rotation angle θ for five different filling fractions $f = 0.30, 0.35, 0.40, 0.45, 0.50$, respectively

Literature review section 8.2, by the introduction of an active material as one, or even both, of the constituents of the phononic crystal.

8.3.2 Tunable Ferroelectric Phononic Crystals

Ferroelectric ceramics are one kind of such active material. Perovskite ferroelectrics, such as $Ba_{0.7}Sr_{0.3}TiO_3$, undergo phase transformation around Curie temperature T_C, accompanied by huge variations of the material properties, as the acoustic velocities.

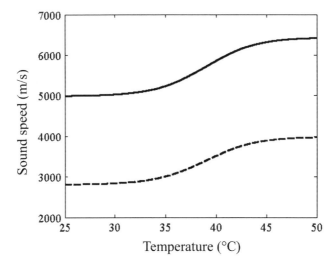

Fig. 8.7 Temperature dependence of the longitudinal (*solid line*) and transverse (*dashed line*) sound velocities in $Ba_{0.7}Sr_{0.3}TiO_3$

For $Ba_{0.7}Sr_{0.3}TiO_3$, the Curie temperature is around 35 °C [23, 30], a suitable value for real device operations. Figure 8.7 displays the variation of the longitudinal c_l and transverse c_t acoustic velocities as a function of temperature. These curves corresponding to fit from the experimental data presented in [23] show increases of about 20 % and 30 % for c_l and c_t, respectively, in a temperature range going from 35 to 45 °C and overlapping the Curie temperature of the material.

The two-dimensional phononic crystal constituted of $Ba_{0.7}Sr_{0.3}TiO_3$ square rods in an epoxy matrix, proposed and realized by a dice-and-fill technique by Jim et al. [23], is considered. The filling fraction is $f = 0.57$ and the period is $a = 267$ μm. Epoxy has been chosen as the matrix constituent due to its large contrast in acoustic properties (both density and acoustic velocities) with $Ba_{0.7}Sr_{0.3}TiO_3$. This is generally convenient for the emergence of a large absolute band gap. Moreover, epoxy shows an infinitesimal velocity change in the considered temperature range (less than 0.03 %).

The band structure of the *XY* modes of propagation of this ferroelectric/epoxy phononic crystal has been calculated by the PWE method with 625 plane waves (see Chap. 10), considering isotropic inclusions in an isotropic matrix. In this configuration the out-of-plane *Z* modes are decoupled from the in plane *XY* modes. Results are displayed in Fig. 8.8 for two different temperatures: (a) 35 °C and (b) 45 °C. Two remarks should be made. First, the width of the first absolute band gap, appearing between the first and second bands, increases from 3 MHz to 4.4 MHz when the temperature is increased by 10 °C around the Curie temperature of the inclusions. This corresponds to a relative increase of about 30 %. Second, the lowest band edge at 5 MHz almost does not depend on the temperature.

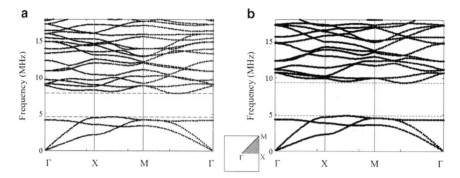

Fig. 8.8 Band structure of the *XY* modes of propagation of a square-lattice 2D phononic crystal consisting of $Ba_{0.7}Sr_{0.3}TiO_3$ square rods embedded in an epoxy matrix at a temperature of (**a**) 35 °C and (**b**) 45 °C. The filling ratio $f = 0.57$ and period $a = 267$ μm

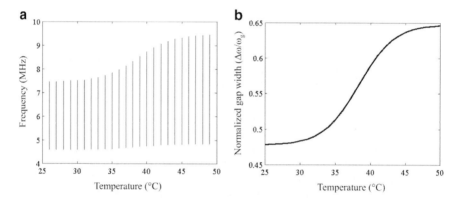

Fig. 8.9 Evolution of (**a**) the absolute elastic band gaps, and (**b**) the normalized absolute elastic band gaps width as a function of the temperature

The evolution of the first absolute elastic band gap of the square lattice of $Ba_{0.7}Sr_{0.3}TiO_3$ square rods embedded in an epoxy matrix as a function of temperature is summarized in Fig. 8.9a. This figure confirms the fact that only the upper band edge follows the temperature variation of the acoustic properties of the inclusions.

Plotting, now, the evolution of the normalized band gap width as a function of the temperature (see Fig. 8.9b), it becomes clear that this evolution follows the variation of the acoustic velocities of the $Ba_{0.7}Sr_{0.3}TiO_3$ inclusions. These variations of about 30 % are of the same order of magnitude as the ones obtained by a mechanical means (rotation of square rods), but now with a solid state tuning scheme. It has to be noted that, in the case of the considered ferroelectric material, the acoustical velocity variations are induced by a structural phase transition, a tetragonal-to-cubic transition.

8.3.3 Tunable Magnetoacoustic Phononic Crystals

The possibility of controling and tuning of the band structures of phononic crystals offered by the introduction of an active magnetoacoustic material and the application of an external magnetic field are now presented. Three means to obtain large elastic property variations in magnetoacoustic materials can be envisaged: giant magnetostriction, spin reorientation transition [1], and ferromagnetic resonance effects. Here, only the first two will be described. The magnetoacoustic coupling is taken into account through the consideration of an equivalent piezomagnetic material model with elastic C_{ijkl}, piezomagnetic q_{lij} and magnetic permeability μ_{ij} tensors varying as a function of the amplitude and orientation of the applied magnetic field [17, 18]. Considering a uniformly oriented magnetization, the equivalent piezomagnetic material formulation leading to equations similar to the ones classically used for piezoelectric materials is

$$
\rho_0 \frac{\partial^2 u_i}{\partial t^2} = \frac{\partial \sigma_{ij}}{\partial x_j},
$$
$$
\frac{\partial b_i}{\partial x_i} = 0,
$$
(8.10)

with

$$
\sigma_{ij} = C_{ijkl}(H)\frac{\partial u_k}{\partial x_l} + q_{lij}(H)\frac{\partial \varphi_m}{\partial x_l},
$$
(8.11)

$$
b_i = q_{ikl}(H)\frac{\partial u_k}{\partial x_l} - \mu_{il}(H)\frac{\partial \varphi_m}{\partial x_l},
$$
(8.12)

where ρ_0 is the mass density, u_i and b_i are the ith component of the particle displacement and magnetic induction, x_i denotes the Eulerian coordinates, σ_{ij} are the stress tensor components, and φ_m is the magnetic potential. This formulation enables the direct use of PWE and FE methods developed for the calculation of phononic crystal characteristics in piezoelectric media [31].

According to the Bloch-Floquet theorem, the displacement vector and the magnetic potential can be expanded in infinite Fourier series

$$
u_i(r,t) = \sum_G u^i_{k+G} e^{j(\omega t - k.r - G.r)},
$$
$$
\varphi_m(r,t) = \sum_G \varphi_{k+G} e^{j(\omega t - k.r - G.r)},
$$
(8.13)

where $r = (x, y, z)$ is the vector position, ω is the circular frequency, G are the reciprocal lattice vectors, and k is the wave vector. Moreover, due to the periodicity the material constants $\rho(r)$, $C_{ijkl}(r)$, $q_{lij}(r)$, and $\mu_{ij}(r)$ are also expanded as Fourier series

$$\alpha(r) = \sum_G \alpha_G e^{-jG.r}. \tag{8.14}$$

Inserting these expansions, (8.13)–(8.14) in (8.10)–(8.12), using orthogonality property of Fourier series components and collecting terms yields the following generalized eigenvalue equation [31]

$$\omega^2 \tilde{R}\tilde{U} = \Gamma_i \tilde{A}_{il} \Gamma_l \tilde{U}, \tag{8.15}$$

where $\tilde{U}$ is a vector gathering the Fourier amplitudes of the generalized displacement $u = (u_1, u_2, u_3, \varphi_m)$, $\tilde{R}$, $\tilde{A}_{il}$ are $4N$ x $4N$ matrix involving only material constants, and Γ_i are diagonal matrices involving the wave vector and vectors of the reciprocal lattice. The detailed expressions of all these matrices are given in [31]. By solving (8.15) for ω as a function of the wave vector k in the first Brillouin zone of the considered lattice, the band structures can be calculated.

Results of contactless tunability of the absolute band gaps are presented for a two-dimensional phononic crystal constituted of Terfenol-D square rod embedded in an epoxy matrix.

The evolution of the effective elastic coefficients, piezomagnetic constants, and magnetic permeability for a Terfenol-D rod as a function of the amplitude of an external magnetic field applied along the rod axis is displayed in Fig. 8.10a. The Terfenol-D parameters used in all the calculations correspond to the ones of commercially available data [18]. As the external magnetic field is applied in the Z direction, parallel to the rod axis, only two elastic coefficients C_{44} and C_{55} and two piezomagnetic constants q_{24} and q_{15} display strong variations as a function of the magnetic field. The order of magnitude of the predicted transverse elastic coefficient variations is in good agreement with the one measured in Terfenol-D [19]. The variations of the diagonal terms of the effective magnetic permeability tensor are also displayed in Fig. 8.10a. The slowness polar diagrams calculated for elastic waves propagating in the XY plane, perpendicular to the rod axis, using the effective piezomagnetic material properties are shown in Fig. 8.10b for three increasing values of the amplitude of the external magnetic field: 1 kOe (dashed line), 10 kOe (dotted line), and 20 kOe (solid line). In this configuration, only the out of plane transverse wave, propagating with the velocity and with displacement directed along the applied static magnetic field, is coupled to this magnetic field.

We study now the influence of the introduction of a magnetoelastic medium on the properties of phononic crystals. The calculations have been made for a square lattice of Terfenol-D square rods of section $d = 1$ mm embedded in an epoxy matrix. The matrix is constituted of epoxy resin, considered as isotropic and with the following parameters: $\rho_0 = 1,142\,\text{kg/m}^3$, $C_{11} = 7.54\,\text{GPa}$, and $C_{44} = 1.48\,\text{GPa}$. 441 plane waves have been used in the PWE calculation of the band structure.

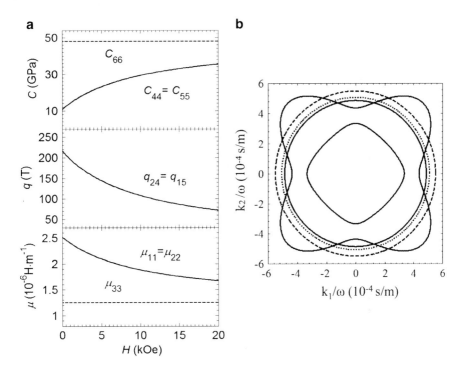

Fig. 8.10 (a) Evolution of the effective elastic moduli, piezomagnetic constants, and magnetic permeabilities of a Terfenol-D rod as a function of the static external magnetic field applied along the rod axis (Z). The effective elastic and piezomagnetic constants are expressed in Voigt notation. (b) Slowness polar diagram for propagation in a Terfenol-D rod with a static external magnetic field, applied along the rod axis (Z), of 1 kOe (*dashed line*), 10 kOe (*dotted line*), and 20 kOe (*solid line*)

Comparison with FEM results has confirmed the rather good convergence of the Fourier series.

We study the evolution of the band structure, as a function of the amplitude and the orientation of the external magnetic field, induced by the variations of the effective parameters of the Terfenol-D rods. The band structures displayed in Fig. 8.11 give a typical example of the magnetic field influence when applied along the rod axis. With a filling factor $f = (d/a)^2 = 0.35$ and an applied field $H_{ext} = 3$ kOe, the phononic crystal possesses an absolute band gap in the 0–1 MHz frequency range, as shown in Fig. 8.11b. When the external field is increased to 10 kOe, a second absolute band gap, ranging from approximately 0.76 to 0.8 MHz, appears (Fig. 8.11a).

Moreover, the frequency range of the first absolute band gap is slightly increased. More precisely, the application of a magnetic field with an amplitude higher than 6 kOe increases the bandwidth of the band gap, and opens a second one in the 0–1 MHz frequency range, as shown in Fig. 8.12a. So elastic waves are evanescent waves in this phononic crystal at 0.8 MHz when the field becomes higher than 6 kOe. When the filling factor is increased to 0.5, as shown in

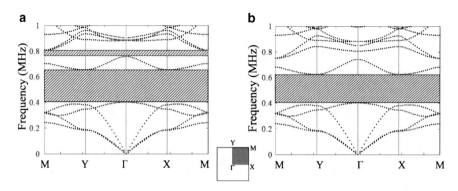

Fig. 8.11 Band structure of a square lattice of Terfenol-D square rods with a filling factor $f = (d/a)^2 = 0.35$, embedded in an epoxy matrix for two applied static magnetic fields along the rod axis Z: (**a**) $H_{ext} = 10\,\text{kOe}$ and (**b**) $H_{ext} = 3\,\text{kOe}$

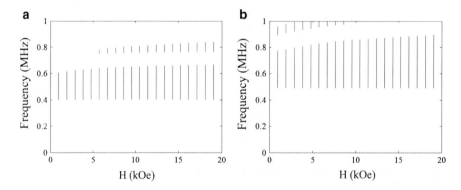

Fig. 8.12 Evolution of the absolute elastic band gaps of a square lattice of Terfenol D square rods embedded in an epoxy matrix as a function of the amplitude of the applied static magnetic field along the rod axis Z, for a filling factor (**a**) $f = 0.35$ and (**b**) $f = 0.5$

Fig. 8.12b, the process is inverted: the second absolute band gap disappears when the amplitude of the external magnetic field is increased. In both cases, the phononic crystal behaves as a switch controlled without any contact by an external applied magnetic field. Nevertheless, in this case, where the magnetic field is applied along the rod axis, the band gap width variation remains lower than 25 %. Moreover, a careful look at the band structures of Fig. 8.11 has shown that only modes polarized along Z are coupled to the external field [18].

Considering now a magnetic field applied in a direction perpendicular to the Terfenol-D rod, the evolution of the parameters of the effective piezomagnetic material as a function of the amplitude of the external magnetic field are presented in Fig. 8.13a. Contrary to the previous case, all the Christoffel tensor components are now field dependent. As shown in Fig. 8.13b, displaying the slowness polar diagrams for an external field of 4 kOe (dashed line), 10 kOe (dotted line), and

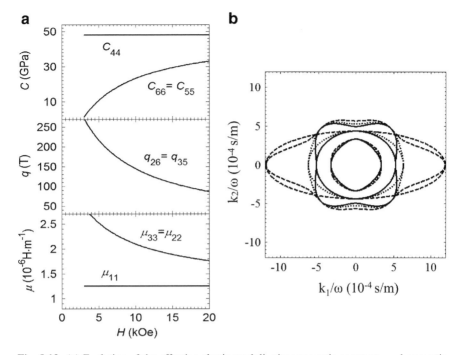

Fig. 8.13 (a) Evolution of the effective elastic moduli, piezomagnetic constants, and magnetic permeabilities of a Terfenol-D rod as a function of the static external magnetic field applied along the X axis. The effective elastic and piezomagnetic constants are expressed in Voigt notation. (b) Slowness polar diagram for propagation in a Terfenol-D rod with a static external magnetic field, applied along the X axis, of 4 kOe (*dashed line*), 10 kOe (*dotted line*), and 20 kOe (*solid line*)

20 kOe (solid line), this leads to the fact that all the modes are now affected by the magnetic field, even if the influence on the quasi-longitudinal mode is still low. Moreover, the induced velocity variations of the transverse waves, both in plane and out of plane, become very large for propagation in the X direction, due to the presence of a magnetic spin reorientation transition (SRT). It is well known that the magnetoelastic coupling can become significant near a SRT if the magnetic mode of frequency ω_0 that interacts with the sound is a soft mode leading to a coupling coefficient $\varsigma = 1$ at the transition. At the SRT, the equilibrium orientation of the magnetization suddenly changes. For the considered Terfenol-D sample, the SRT corresponds to $H_{ext} = 2.56$ kOe. Close to the SRT, e.g., for $H_{ext} = 4$ kOe, for a propagation along the external magnetic field direction, the phase velocity of the transverse waves tends to zero.

On the other hand, when the propagation is perpendicular to the external field direction, only the in plane transverse wave velocity shows a slight variation as a function of the magnetic field amplitude. The difference between these two cases arises from the existence of a dynamic dipole field created by the magnetoelastic wave propagation. So the long-range dipole interaction can considerably weaken the magnetoelastic coupling at the SRT for waves propagating in an arbitrary direction.

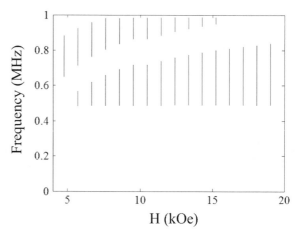

When the magnetic field is applied perpendicularly to the Terfenol-D rod axis, the absolute band gap evolution displayed in Fig. 8.14 shows more important variations than in the previously considered case where the magnetic field was along the rod axis. This can be directly linked to the SRT described in the preceding part. The calculations have been made for decreasing external magnetic field amplitude down to 4 kOe, not too close to the SRT. Indeed, below and close to the SRT, the used assumption of uniformly oriented magnetization becomes doubtful.

8.4 Applications of Tunable Phononic Crystals

We have seen how a sufficient, e.g., at least 10 %, level of tunability can be introduced in the properties of 2-D phononic crystals. This tuning capability opens the opportunity to design and create phononic crystal devices with new or enhanced functionalities.

As mentioned in Chap. 1, some authors have demonstrated that the removal of inclusions along some pathway in the phononic crystal produces acoustic waveguides [32, 33]. Acoustic waves that would not propagate otherwise in a phononic crystal can be guided with minimal loss along such waveguides. Low-loss transmission can be achieved in linear waveguides as well as guides with sharp bends. That opens possibilities for the design of devices allowing the filtering or the demultiplexing of acoustic waves at the scale of the wavelength [34]. More specifically it has been shown numerically and experimentally that such structures manufactured at the micrometer scale behave as high Q micromechanical resonators with high resonance frequencies and can be integrated in devices for wireless communications and sensing applications [35]. Moreover some dispersion curves in the band structure of a phononic crystal may present a negative curvature i.e., the Poynting vector and the wave vector, associated to energy flux and phase velocity are opposite in sign. This property may lead to the negative refraction of acoustic waves for frequencies falling in the frequency domain of the band with negative curvature. Negative refraction allows for the focusing of acoustic waves with a resolution lower than the diffraction limit [36] as well as for the autocollimation of an acoustic beam [37].

Two examples of how tuning capability could improve the potentiality of such waveguiding and negative refraction flat lens devices are now presented.

All the calculations presented in this section have been performed using the Finite Element (FE) method to simulate the propagation of acoustic waves in the designed phononic crystal devices.

8.4.1 Tunable and Reconfigurable Waveguides

The first application considered is the use of a 2-D magnetoacoustic tunable phononic crystal to design a completely reconfigurable waveguiding device. The building block of this system is a 7 by 10 array of 0.5 mm radius Terfenol-D cylindrical rods embedded in an epoxy matrix, as shown in Fig. 8.15. A filling factor of 0.6 is chosen in order to obtain a large band gap around 1 MHz when an external static magnetic field of 20 kOe is applied along the rod axis, see Fig. 8.16a. The 2-D phononic crystal is sandwiched between two homogeneous parts composed of epoxy. To simulate infinite media, Perfectly Matched Layers (PMLs) are implemented on the left and right sides of the calculation domain. Moreover, periodic boundary conditions are used on the upper and lower sides.

As shown in Sect. 8.3.3, the band structure of the realized magnetoacoustic phononic crystal can be tuned by changing the amplitude (or the direction) of the applied external magnetic field. In the present configuration, when the amplitude of the magnetic field applied along the rod axis, i.e., the Z axis, is reduced to 1 kOe, then a transmission band appears in the gap near 1 MHz, as displayed in Fig. 8.16b. In fact, the two modes, which constitute this transmission band, correspond to out of plane transversely polarized modes.

We first consider a straight waveguide created by applying locally a static magnetic field of 1 kOe (in place of 20 kOe) on one row of cylinders along the propagation direction (X axis), as shown in Fig. 8.15. From an experimental point of view, the external magnetic field can be applied locally on each cylinder using a magnetic writing head. The length of the obtained waveguide is 10 periods and its width is one period. A plane wave, containing components in the three directions, with a frequency of 970 kHz is launched from the left side of the 2-D phononic crystal. The obtained three components of the particle displacement u_1, u_2, and u_3 are displayed in Fig. 8.16c. This figure clearly demonstrates that only the out of plane transversely polarized mode is transmitted through the waveguide.

Classically, waveguides are created in phononic crystal by removing one row of inclusions. In order to understand the difference between the two kinds of waveguide, the band structures for the waveguide modes along the ΓX direction are calculated with the FE method by defining a supercell of five periods in the

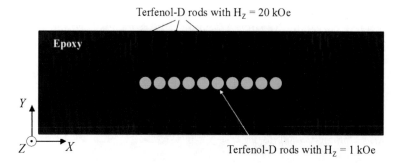

Fig. 8.15 Structure of a linear waveguide created, in a square-lattice 2D phononic crystal, by applying an external magnetic field of 1 kOe on one row of cylindrical Terfenol-D inclusions along the X direction. The phononic crystal is constituted of cylindrical Terfenol-D rods of 0.5 mm radius embedded in an epoxy matrix with a filling factor $f = 0.6$. The applied static magnetic field is 20 kOe along the Z axis

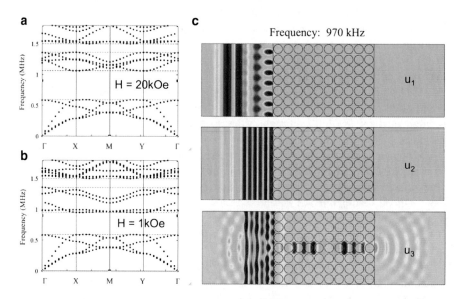

Fig. 8.16 Band structure of a square-lattice 2D phononic crystal constituted of cylindrical Terfenol-D rods of 0.5 mm radius embedded in an epoxy matrix with a filling factor $f = 0.6$. The applied static magnetic field is (**a**) 20 kOe and (**b**) 1 kOe along the Z axis. (**c**) Three components of the particle displacement of a plane wave with a frequency of 970 kHz impinging on a square-lattice 2D phononic crystal containing a linear waveguide. Only the out of plane transversely polarized mode is transmitted through the waveguide

Y direction, and reported in Fig. 8.17. In the case of a waveguide created by applying locally a static magnetic field of 1 kOe, only two flat modes appear in the band gap of the phononic crystal, leading to the emergence of a narrow passing band. As it will be shown latter on, such flat modes can be used for multiplexing or demultiplexing applications [34]. For a waveguide realized by removing one row of

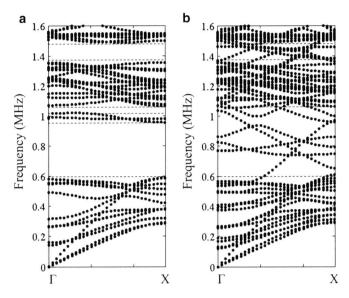

Fig. 8.17 (a) Band structure of a square-lattice 2D phononic crystal, containing a linear waveguide obtained by applying an external magnetic field of 1 kOe on one row of cylindrical Terfenol-D inclusions along the X direction, performed by considering a supercell of 5 periods along the Y direction. The phononic crystal is constituted of cylindrical Terfenol-D rods of 0.5 mm radius embedded in an epoxy matrix with a filling factor $f = 0.6$. The applied static magnetic field is 20 kOe along the Z axis. (b) Band structure of a square-lattice 2D phononic crystal containing a linear waveguide obtained by removing one row of cylindrical Terfenol-D inclusions along the X direction. The calculation is performed by considering a supercell of 5 periods along the Y direction

rods, see Fig. 8.17b, the number of modes appearing inside the band gap is considerably higher than in the previous case. Generally, to decrease this number of modes, the width of the waveguide needs to be reduced.

The signal transmitted along the waveguide is recorded at its end and integrated along its width. The transmission is then calculated by normalizing this signal with respect to the case where a homogeneous epoxy medium is considered. The calculated transmission is displayed as a function of frequency in Fig. 8.18a. We can observe full transmission of out of plane elastic waves for certain frequencies within the phononic crystal stop band. Zooming on this passing band, as shown in Fig. 8.18b, we can see oscillations of the transmission coefficient as a function of frequency typical of phononic waveguide, induced by the roughness, with a periodicity a, of the guide wall.

Now, if we apply the localized 1 kOe static magnetic field on a succession of Terfenol-D rods forming a complex path, we can for example design a bent waveguide, as shown in Fig. 8.19a. Calculating, as in the case of a straight waveguide, the particle displacement induced by an impinging plane wave at a frequency of 970 kHz, we can see that the wave follows the guide even in the sharp corner (90°).

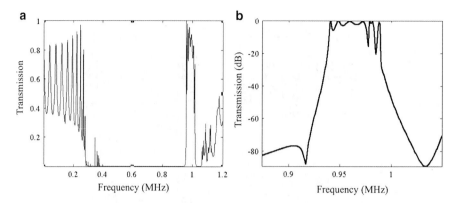

Fig. 8.18 (a) Transmission through the waveguide of Fig. 8.17a for an out of plane transversely polarized incident waves. (**b**) Zoom of the transmission in dB around the passing band introduced by the linear guide for out of plane transversely polarized modes

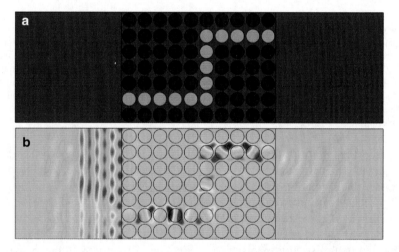

Fig. 8.19 (a) Structure of a square-lattice 2D phononic crystal, containing a bended waveguide obtained by applying an external magnetic field of 1 kOe (blue rods) along the Z direction on cylindrical Terfenol-D inclusions. The phononic crystal is constituted of cylindrical Terfenol-D rods of 0.5 mm radius embedded in an epoxy matrix with a filling factor $f = 0.6$. The applied static magnetic field is 20 kOe (red rods) along the Z axis. (**b**) Out of plane component u_3 of the particle displacement of a plane wave with a frequency of 970 kHz impinging on the waveguide

With the same structure of Terfenol-D rods embedded in an epoxy matrix, we can design a Y-shaped waveguide, as shown in Fig. 8.20a. The left part of the waveguide contains cylinders of Terfenol-D with two different applied magnetic fields along the Z direction: 1 kOe (blue rods) and 2 kOe (green rods). In the right

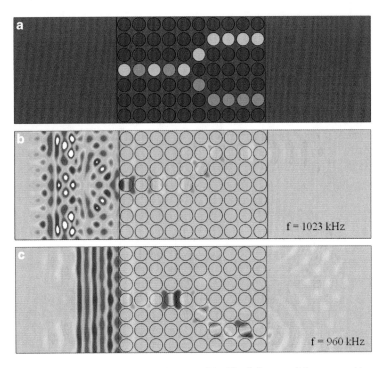

Fig. 8.20 (a) Schematic of the Y-shaped waveguide. The left part of the waveguide contains cylinders of Terfenol-D with two different applied magnetic fields along the Z direction: 1 kOe (*blue rods*) and 2 kOe (*green rods*). The phononic crystal is constituted of cylindrical Terfenol-D rods of 0.5 mm radius embedded in an epoxy matrix with a filling factor $f = 0.6$. The applied static magnetic field is 20 kOe along the Z axis (*red rods*). Each branch of the Y contains one type of cylinder. Representation of the out of plane component of the displacement field for the Y-shaped waveguide at two frequencies of (**b**) 1,023 kHz, and (**c**) 960 kHz

part, each branch of the Y contains one type of cylinder. Applying a 2 kOe magnetic field moves the passing band of the 2D-phononic crystal to higher frequencies. As the passing band created in the band gap is sufficiently narrow we can find frequencies moving from the passing band to the band gap, or inversely, when the amplitude of the applied magnetic field is changed. The plot of the out of plane component of the displacement field for the Y-shaped waveguide is represented in Fig. 8.20 for two different frequencies: 1,023 kHz (b) and 960 kHz (c). These spectra show that the superposed waves supported by the mixed waveguide are separated and directed toward the two branches of the Y junction. This system can be used as a demultiplexer or a multiplexer if used in the reversed direction.

In conclusion, we have seen that an array of Terfenol-D arranged in a square lattice and embedded in an epoxy matrix can be used as a reconfigurable device for guiding, multiplexing, or demultiplexing acoustic waves.

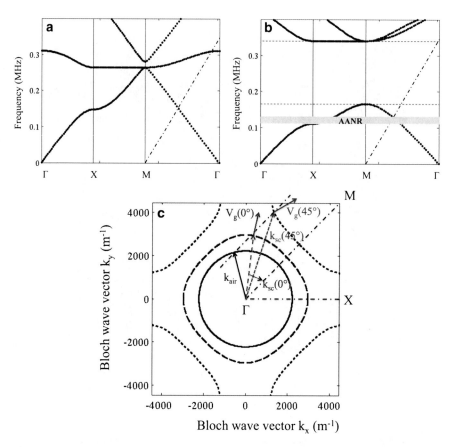

Fig. 8.21 Acoustic band structure for a square-lattice two-dimensional phononic crystal consisted of square solid rods rotated with an angle (**a**) $\theta = 0°$ and (**b**) $\theta = 45°$, in air. The filling fraction $f = 0.5$ and edge length $d = 0.5$ mm. (**c**) EFSs k space of air (*solid*) and the phononic crystal with both $\theta = 0°$ (*dashed*) and $\theta = 45°$ (*dotted*) at 120 kHz. V_g is the group velocity in the phononic crystal

8.4.2 Tunable Negative Refraction Lenses

Due to emerging applications of negative refraction, such as the realization of flat lenses and lenses with resolution beyond the diffraction limit, i.e., superlenses, it becomes highly desirable to obtain some degrees of tunability in wave refraction.

We consider a square-lattice 2-D phononic crystal constituted of square solid rods that can be rotated in air, as in Sect. 8.3.1. The filling fraction is $f = 0.5$ and the square rods edge length is $d = 0.5$ mm. In such phononic crystal, the negative refraction is realized without employing a negative index or a backward wave effect, i.e., the phase and group velocities are not opposites, but is due to a negative phononic "effective mass" effect [4, 38]. For the frequency 120 kHz, the equi-

Fig. 8.22 Pressure field
generated at 120 kHz by a
point source placed at a
distance of a from the left of
the phononic crystal slab. The
phononic crystal is
constituted of square solid
rods in air with a filling
fraction $f = 0.5$ and edge
length $= 0.5$ mm. The angle
of rotation of the rods is
(**a**) $\theta = 0°$ and (**b**) $\theta = 45°$

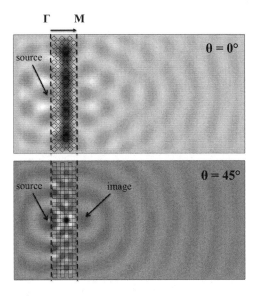

frequency surfaces (EFS) obtained by the PWE method, for the two different angles
$\theta = 0°$ and $45°$, are displayed in Fig. 8.21c. As explained in Chap. 4, the anisotropy
of EFS determines the refraction of acoustic waves at the interface between the air
and the phononic crystal. As shown in Fig. 8.21c, in the case of a phononic crystal
with surface normal oriented along the $\Gamma - M$ direction, when the rotation angle is
$\theta = 0°$, the shape of the EFS is convex in the vicinity of the point Γ with an outward-
pointing group velocity, $V_g(0°)$, leading to positive refraction. When $\theta = 45°$, the
shape of the EFS becomes square-like and centered at the point M. As shown in
Fig. 8.21c the group velocity, $V_g(45°)$ is inward-pointing to the point M. Therefore,
the refraction is negative for this angle of rotation. So, by rotating the rods from
$\theta = 0°$ to $45°$, the refraction can be changed from positive to negative.

The acoustic band structures, calculated by the PWE method, are displayed in
Fig. 8.21a for $\theta = 0°$ and Fig. 8.21b for $\theta = 45°$. As shown, for the larger angle of
rotation the first acoustic band is compressed, corresponding to an increase of the
anisotropy, and is generally well suited to the apparition of an all-angle negative
refraction (AANR) region [38]. Here, the required conditions for AANR effect
are obtained when the solid rods are rotated with an angle of $45°$, in a frequency
range highlighted in Fig. 8.21b. From Fig. 8.21a, we can notice that the AANR
region is absent when the angle $\theta = 0°$, although the negative refraction region is
larger. A tunable acoustic superlens can be designed through such a tunable
AANR effect.

In this perspective, we consider a 6-layer square-lattice 2-D phononic slab as
shown in Fig. 8.22. The surface perpendicular to the phononic crystal slab is
oriented along the $\Gamma - M$ direction. The pressure fields generated at 120 kHz by
a point source placed at a distance equals to the lattice parameter from the left of the
phononic crystal slab calculated by an FE method for the two angles of rotation of

the rods, $\theta = 0°$ and $\theta = 45°$, are displayed in Fig. 8.22. It appears that when the rods are rotated by an angle of 45°, a perfect image of the source is obtained, due to the AANR effect. Now, when the angle $\theta = 0°$ the image completely disappears. Therefore, by rotating the solid rods composing the phononic crystal slab, the superlens can be switched on/off.

All the described behaviors of refraction control of an acoustic wave in a square rod phononic crystal have been verified experimentally by Feng et al. [4]. Moreover, similar results have been predicted by Yang et al. [10] for a tunable phononic crystal made of dielectric elastomer cylindrical actuators.

8.5 Summary

We have shown that tunability of phononic crystal properties could be achieved by changing their geometry or by varying the elastic characteristics of their constitutive materials through application of external stimuli. Variation of the relative band gap width of more than 50 % could be attained by the physical rotation of square inclusions periodically positioned in air or by using an active material, such as a ferroelectric or a magnetoacoustic material, as one of the constituent of the phononic crystal. The introduction of an active material constituent opens the possibility of easy controllability of the properties of a phononic crystal without any physical contact. More specifically one can achieve additional functionalities such as the switching of transmission in a defined frequency range, the control of refraction properties, and the reconfiguration of waveguide and multiplexer.

References

1. E.A. Turov, V.G. Shavrov, Broken symmetry and magnetoacoustic effects in ferro- and antiferromagnetics. Sov. Phys. Usp. **26**, 593–611 (1983)
2. C. Goffaux, J.P. Vigneron, Theoretical study of a tunable phononic band gap system. Phys. Rev. B **64**, 075118 (2001)
3. X. Li, F. Wu, H. Hu, S. Zhong, Y. Liu, Large acoustic band gaps created by rotating square rods in two-dimensional periodic composites. J. Phys. D Appl. Phys. **36**, L15–L17 (2003)
4. L. Feng, X.-P. Liu, M.-H. Lu, Y.-B. Chen, Y.-W. Mao, J. Zi, Y.-Y. Zhu, S.-N. Zhu, N.-B. Ming, Refraction control of acoustic waves in a square rod-constructed tunable sonic crystal. Phys. Rev. B **73**, 193101 (2006)
5. S.-C.S. Lin, T.J. Huang, Tunable phononic crystals with anisotropic inclusions. Phys. Rev. B **83**, 174303 (2011)
6. K. Bertoldi, M.C. Boyce, Mechanically triggered transformations of phononic band gaps in periodic elastomeric structures. Phys. Rev. B **77**, 052105 (2008)
7. J.-H. Jang, C.Y. Koh, K. Bertoldi, M.C. Boyce, E.L. Thomas, Combining pattern instability and shape-memory hysteresis for phononic switching. Nano Lett. **9**(5), 2113–2119 (2009)
8. J.-H. Jang, C.K. Ullal, T. Gorishnyy, V.V. Tsukruk, E.L. Thomas, Mechanically tunable three-dimensional elastomeric network/air structures via interference lithography. Nano Lett. **6**(4), 740–743 (2006)

9. W.-P. Yang, L.-W. Chen, The tunable acoustic band gaps of two-dimensional phononic crystals with a dielectric elastomer cylindrical actuator. Smart Mater. Struct. **17**, 015011 (2008)

10. W.-P. Yang, L.-Y. Wu, L.-W. Chen, Refractive and focusing behaviours of tunable sonic crystals with dielectric elastomer cylindrical actuators. J. Phys. D Appl. Phys. **41**, 135408 (2008)

11. L.-Y. Wu, M.-L. Wu, L.-W. Chen, The narrow pass band filter of tunable 1D phononic crystals with a dielectric elastomer layer. Smart Mater. Struct. **18**, 015011 (2009)

12. Z. Hou, F. Wu, Y. Liu, Phononic crystals containing piezoelectric material. Solid State Commun. **130**(11), 745–749 (2004)

13. Y.-Z. Wang, F.-M. Li, Y.-S. Wang, K. Kishimoto, W.-H. Huang, Tuning of band gaps for a two-dimensional piezoelectric phononic crystal with a rectangular lattice. Acta Mech. Sin. **25**, 65–71 (2009)

14. X.-Y. Zou, Q. Chen, B. Liang, J.-C. Cheng, Control of the elastic wave bandgaps in two-dimensional piezoelectric periodic structures. Smart Mater. Struct. **17**, 015008 (2008)

15. C.J. Rupp, M.L. Dunn, K. Maute, Switchable phononic wave filtering, guiding, harvesting, and actuating in polarization-patterned piezoelectric solids. Appl. Phys. Lett. **96**, 111902 (2010)

16. Y.-Z. Wang, F.-M. Li, W.-H. Huang, X. Jiang, Y.-S. Wang, K. Kishimoto, Wave band gaps in two-dimensional piezoelectric/piezomagnetic phononic crystals. Int. J. Solids Struct. **45** (14–15), 4203–4210 (2008)

17. O. Bou Matar, J.F. Robillard, J. Vasseur, A.-C. Hladky-Hennion, P.A. Deymier, P. Pernod, V. Preobrazhensky, Band gap tunability of magneto-elastic phononic crystal. J. Appl. Phys. **111**, 054901 (2012)

18. J.-F. Robillard, O. Bou Matar, J.O. Vasseur, P.A. Deymier, M. Stippinger, A.-C. Hladky-Hennion, Y. Pennec, B. Djafari-Rouhani, Tunable magnetoelastic phononic crystals. Appl. Phys. Lett. **95**, 124104 (2009)

19. J.R. Cullen, S. Rinaldi, G.V. Blessing, Elastic versus magnetoelastic anisotropy in rare earth-iron alloys. J. Appl. Phys. **49**(3), 1960–1965 (1978)

20. J. Baumgartl, M. Zvyagolskaya, C. Bechinger, Tailoring of phononic band structures in colloidal crystals. Phys. Rev. Lett. **99**, 205503 (2007)

21. Z.-G. Huang, T.-T. Wu, Temperature effect on the bandgaps of surface and bulk acoustic waves in two-dimensional phononic crystals. IEEE Trans. Ultrason. Ferroelectr. Freq. Control **52**(3), 365–370 (2005)

22. L.-Y. Wu, W.-P. Yang, L.-W. Chen, The termal effects on the negative refraction of sonic crystals. Phys. Lett. A **372**, 2701–2705 (2008)

23. K.L. Jim, C.W. Leung, S.T. Lau, S.H. Choy, H.L.W. Chan, Thermal tuning of phononic bandstructure in ferroelectric ceramic/epoxy phononic crystal. Appl. Phys. Lett. **94**, 193501 (2009)

24. A. Sato, Y. Pennec, N. Shingne, T. Thurn-Albrecht, W. Knoll, M. Steinhart, B. Djafari-Rouhani, G. Fytas, Tuning and switching the hypersonic phononic properties of elastic impedance contrast nanocomposites. ACS Nano **4**(6), 3471–3481 (2010)

25. J.-Y. Yeh, Control analysis of the tunable phononic crystal with electrorheological material. Physica B **400**, 137–144 (2007)

26. B. Wu, R. Wei, C. He, H. Zhao, Research on two-dimensional phononic crystal with magnetorheological material. IEEE Int. Ultrason. Symp. Proc. 1484–1486 (2008)

27. A. Evgrafov, C.J. Rupp, M.L. Dunn, K. Maute, Optimal synthesis of tunable elastic wave-guides. Comput. Methods Appl. Mech. Eng. **198**, 292–301 (2008)

28. M. Gei, A.B.. Movchan, D. Bigoni. Band-gap shift and defect-induced annihilation in prestressed elastic structures. J. Appl. Phys. **105**, 063507 (2009)

29. F. Wu, Z. Liu, Y. Liu, Acoustic band gaps created by rotating square rods in a 2D lattice. Phys. Rev. B **66**, 046628 (2002)

30. Y. Cheng, X.J. Liu, D.J. Wu, Temperature effects on the band gaps of Lamb waves in a one-dimensional phononic-crystal plate. J. Acoust. Soc. Am. **129**(3), 1157–1160 (2011)

31. M. Wilm, S. Ballandras, V. Laude, T. Pastureaud, A full 3D plane wave-expansion model for 1-3 piezoelectric composite structures. J. Acoust. Soc. Am. **112**(3), 943–952 (2002)
32. A. Khelif, B. Djafari-Rouhani, J.O. Vasseur, P.A. Deymier, P. Lambin, L. Dobrzynski, Transmittivity through straight and stublike waveguides in a two-dimensional phononic crystal. Phys. Rev. B **65**(17), 174308 (2002)
33. T. Miyashita, C. Inoue. Numerical investigations of transmission and waveguide properties of sonic crystals by finite-difference time-domain method. Jpn. J. Appl. Phys., Part 1, 40(Part 1, No. 5B), 3488–3492 (2001)
34. Y. Pennec, B. Djafari-Rouhani, J.O. Vasseur, A. Khelif, P.A. Deymier. Tunable filtering and demultiplexing in phononic crystals with hollow cylinders. Phys. Rev. E **69**(4), 046608 (2004)
35. S. Mohammadi, A.A. Eftekhar, W.D. Hunt, A. Adibi. High-Q micromechanical resonators in a two-dimensional phononic crystal slab. Appl. Phys. Lett. **94**(5), 051906 (2009)
36. A. Sukhovich, B. Merheb, K. Muralidharan, J.O. Vasseur, Y. Pennec, P.A. Deymier, J.H. Page, Experimental and theoretical evidence for subwavelength imaging in phononic crystals. Phys. Rev. Lett. **102**(15), 154301 (2009)
37. I. Perez-Arjona, V.J. Sanchez-Morcillo, J. Redondo, V. Espinosa, K. Staliunas, Theoretical prediction of the nondiffractive propagation of sonic waves through periodic acoustic media. Phys. Rev. B **75**(1), 014304 (2007)
38. C. Luo, S.G. Johnson, J.D. Joannopoulos, J.B. Pendry, All-angle negative refraction without negative effective index. Phys. Rev. B **65**, 201104 (2002)

Chapter 9
Nanoscale Phononic Crystals and Structures

N. Swinteck, Pierre A. Deymier, K. Muralidharan, and R. Erdmann

Abstract The objective of this chapter is to explore advances in the development of phononic crystals and phononic structures at the nanoscale. The downscaling of phononic structures to nanometric dimensions requires an atomic treatment of the constitutive materials. At the nanoscale, the propagation of phonons may not be completely ballistic (wave-like) and nonlinear phenomena such as phonon–phonon scattering occur. We apply second-order perturbation theory to a one-dimensional anharmonic crystal to shed light on phonon self-interaction and three-phonon scattering processes. We emphasize the competition between dispersion effects induced by the structure, anharmonicity of the atomic bonds, and boundary scattering. These phenomena are illustrated by several examples of atomistic models of nanoscale phononic structures simulated using the method of molecular dynamics (MD). Special attention is also paid to size effects.

9.1 Introduction

Nanofabrication techniques can be used to structure matter in a way that affects the propagation of phononic excitations such as high frequency (short wavelength) thermal phonons. Modulating the thermal properties of materials by creating a nanoscale composite structure is an approach that has been extensively studied in the case of superlattices [1–3]. These stacks of nanoscale layers have been shown experimentally and theoretically [4, 5] to impact thermal transport due to scattering effects of phonons.

N. Swinteck (✉) • P.A. Deymier • K. Muralidharan • R. Erdmann
Department of Materials Science and Engineering, University of Arizona, Tucson,
AZ 85721, USA
e-mail: swinteck@email.arizona.edu; deymier@email.arizona.edu; Krishna@email.arizona.edu;
fluid.thought@gmail.com

P.A. Deymier (ed.), *Acoustic Metamaterials and Phononic Crystals*,
Springer Series in Solid-State Sciences 173, DOI 10.1007/978-3-642-31232-8_9,
© Springer-Verlag Berlin Heidelberg 2013

While superlattices are actually one-dimensional phononic structures, only a few studies have investigated 2D and 3D nanophononic structures. Most studies on 2D and 3D phononic crystals (PCs) have focused on macroscopic elastic systems. However this large body of knowledge suggests a possibility of designing dispersive properties by downscaling PCs to nanodimensions to affect the propagation characteristics of phonons with frequencies exceeding the THz range [6]. Recently, Gillet et al. [7] have reported simulations of atomic-level phononic structures made of three-dimensional lattices of Ge quantum dots in a Si matrix. They have shown a decrease of the thermal conductivity by several orders of magnitude due to the periodic structure of the system. Davis and Hussein [8] have considered three-dimensional nanoscale phononic crystals formed from silicon and cubic voids of vacuum. The voids are arranged on a simple cubic lattice with a lattice constant an order of magnitude larger than that of the bulk crystalline silicon primitive cell. This study showed that dispersion at the phononic crystal unit cell level plays a noticeable role in determining the thermal conductivity and that boundary scattering can also be a dominant factor. Control of high-frequency thermal phonons via structural periodicity requires preserving elastic Bragg scattering and is a significant challenge because of the possible loss of phonon coherence due to inherent inelastic scattering resulting from the anharmonicity of interatomic bonds. Band-structure effects will be highest at low temperatures where there is less anharmonic scattering [5] but one has to operate at often undesirably low temperatures [9]. For applications at ambient temperature and phononic crystal dimensions that can be fabricated with relative ease, the transition between Bragg- and inelastic-dominated scattering depends on the characteristic length of the phononic crystal and the Debye temperature of the constitutive material. This latter quantity relates directly to the phonon coherence length. Two-dimensional materials such as graphene or boron nitride (BN) sheets are therefore particularly suited for such applications due to their high phonon coherence length. Atomistic computational methods have been employed to shed light on the transport behavior of thermal phonons in models of graphene antidot super-lattice structures composed of periodic arrays of holes [10]. The phonon lifetime and thermal conductivity as a function of the crystal filling fraction and temperature were calculated in this study. These calculations indicated coherent phononic effects even at room temperature.

The first section of this chapter focuses on the relationship between wave interactions and dispersion in one-dimensional anharmonic crystals. This is done using second-order pertubation theory as well as numerial simulations of molecular dynamics (MD) models of nanoscale phononic systems. Details on the perturbation theory approach are given for pedagogical reasons. Subsequent sections show that coherent phononic effects due to period arrays of scatterers and/or asymmetric scatterers are achievable in nanostructured two-dimensional high-Debye temperature materials such as graphene and BN sheets. Attention is also paid to the competition between phonon–phonon scattering and boundary scattering.

9.2 Anharmonic One-Dimensional Atomic Structures

9.2.1 Perturbation Theory of the Mono-Atomic Anharmonic Crystal

In a harmonic crystal, the vibrational modes do not interact. Anharmonic lattice dynamics methods have been applied to introduce phonon interactions in three-dimensional crystals as perturbations to the harmonic solution [11–13]. Anharmonic forces lead to mode-dependent frequency shifts and introduce finite phonon life-time (i.e., line-width). In this section, we consider the anharmonic one-dimensional monoatomic crystal as a simple model to shed light on the effect of nonlinear interatomic forces on the vibrational modes that this medium can support. Amplitude-dependent self-interaction of a wave in a monoatomic and diatomic chain of masses and springs with nonlinear cubic forces has been studied [14]. It was shown that the dispersion curves undergo frequency shifts dependent on the amplitude of the wave. The interaction between two different waves in a nonlinear monoatomic chain results in the formation of different dispersion branches that are amplitude and frequency dependent [15]. Here, we employ second-order perturbation theory based on multiple time scale analysis [16, 17] and provide a detailed derivation of the anharmonic modes.

A schematic illustration of the 1D monoatomic crystal is shown in Fig. 9.1a. The potential energy function detailing the interaction between neighboring masses in the 1D crystal is shown in Fig. 9.1b. The parameter (ε) characterizes the strength of nonlinearity in the springs connecting the masses. As ε increases in magnitude a region of instability emerges in the potential energy function.

The equation of motion for the quadratically nonlinear monoatomic chain is represented by (9.1):

$$m\frac{d^2 u_n(t)}{dt^2} = \beta(u_{n+1} - 2u_n + u_{n-1}) + \varepsilon\left[(u_{n+1} - u_n)^2 - (u_n - u_{n-1})^2\right], \quad (9.1)$$

where m is mass, $u_n(t)$ is the displacement from equilibrium of the nth mass, β is linear stiffness, and ε is a small parameter characterizing quadratic nonlinearity. The time variable (t) is replaced by a collection of variables $\tau = (\tau_0, \tau_1, \tau_2)$ whereby: $\tau_0 = t, \tau_1 = \varepsilon t, \tau_2 = \varepsilon^2 t$. Under this condition, (9.1) becomes

$$\frac{d^2 u_n(\tau_0, \tau_1, \tau_2)}{dt^2} = \omega_n^2(u_{n+1} - 2u_n + u_{n-1}) + \frac{\varepsilon}{m}\left[(u_{n+1} - u_n)^2 - (u_n - u_{n-1})^2\right], \quad (9.2)$$

where $\omega_n = \sqrt{\frac{\beta}{m}}$

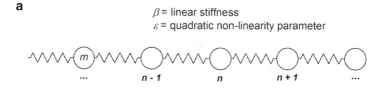

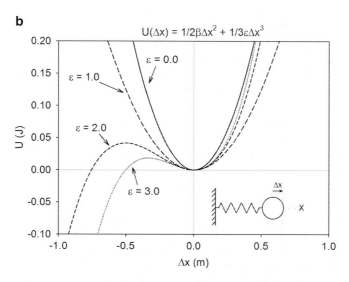

Fig. 9.1 (**a**) Schematic representation of 1D crystal with linear stiffness β and quadratic nonlinearity parameter ε. (**b**) The potential energy function describing the 1D crystal

The dependent variable in (9.2), $u_n(\tau)$, is expressed as an asymptotic expansion at multiple time scales:

$$u_n(\tau) = u_n^{(0)}(\tau) + \varepsilon u_n^{(1)}(\tau) + \varepsilon^2 u_n^{(2)}(\tau) + \text{higher order terms} \qquad (9.3)$$

With this (9.2) is decomposed into equations for each order of expansion of ε, namely, the following set of equations:

$$O(\varepsilon^0) : \frac{\partial^2 u_n^{(0)}}{\partial \tau_0{}^2} = \omega_n{}^2 \left(u_{n+1}^{(0)} - 2u_n^{(0)} + u_{n-1}^{(0)} \right)$$

$$O(\varepsilon^1) : \frac{\partial^2 u_n^{(1)}}{\partial \tau_0{}^2} + 2\frac{\partial^2 u_n^{(0)}}{\partial \tau_0 \partial \tau_1} = \omega_n{}^2 \left(u_{n+1}^{(1)} - 2u_n^{(1)} + u_{n-1}^{(1)} \right)$$
$$+ \frac{1}{m} \left[u_{n+1}^{(0)} u_{n+1}^{(0)} - 2u_{n+1}^{(0)} u_n^{(0)} + 2u_{n-1}^{(0)} u_n^{(0)} - u_{n-1}^{(0)} u_{n-1}^{(0)} \right]$$

$$O(\varepsilon^2): \quad \frac{\partial^2 u_n^{(2)}}{\partial \tau_0^2} + 2\frac{\partial^2 u_n^{(1)}}{\partial \tau_0 \partial \tau_1} + 2\frac{\partial^2 u_n^{(0)}}{\partial \tau_0 \partial \tau_2} + \frac{\partial^2 u_n^{(0)}}{\partial \tau_1^2} = \omega_n^2 \left(u_{n+1}^{(2)} - 2u_n^{(2)} + u_{n-1}^{(2)} \right)$$

$$+ \frac{2}{m} \left[u_{n+1}^{(1)} u_{n+1}^{(0)} - u_{n+1}^{(1)} u_n^{(0)} - u_{n+1}^{(0)} u_n^{(1)} + u_{n-1}^{(1)} u_n^{(0)} + u_{n-1}^{(0)} u_n^{(1)} - u_{n-1}^{(1)} u_{n-1}^{(0)} \right]$$

9.2.1.1 Self-Interaction

We first address the self-interaction of a vibrational mode, that is, the effect of the lattice deformation on itself. To solve the ε^0-equation, a general solution of the following form is proposed:

$$u_{n,G}^{(0)}(\tau_0, \tau_1, \tau_2) = A_0(\tau_1, \tau_2)\, e^{ikna} e^{-i\omega_0 \tau_0} + \bar{A}_0(\tau_1, \tau_2)\, e^{-ikna} e^{i\omega_0 \tau_0}, \qquad (9.4)$$

where

$$A_0(\tau_1, \tau_2) = \alpha(\tau_1, \tau_2)\, e^{-i\varphi(\tau_1,\tau_2)}$$

$$\bar{A}_0(\tau_1, \tau_2) = \alpha(\tau_1, \tau_2)\, e^{i\varphi(\tau_1,\tau_2)}$$

$A_0(\tau_1,\tau_2)$ is a complex quantity that permits slow time evolution of amplitude and phase and $\alpha(\tau_1, \tau_2)$ and $\varphi(\tau_1, \tau_2)$ are real-valued functions. Inserting (9.4) into the ε^0-order equation yields the well-known dispersion relationship for the harmonic system (9.5):

$$\omega_0^2 = \omega_n^2 \left(2 - e^{ika} - e^{-ika} \right) = \frac{\beta}{m}[2 - 2\cos(ka)] \qquad (9.5)$$

Equation (9.4) is now utilized in the ε^1-order equation to resolve the general solution for $u_n^{(1)}$. The ε^1-order equation is written as follows:

$$\frac{\partial^2 u_n^{(1)}}{\partial \tau_0^2} + \omega_n^2 \left(2u_n^{(1)} - u_{n+1}^{(1)} - u_{n-1}^{(1)} \right) = 2i\omega_0 \left[\frac{\partial A_0}{\partial \tau_1} e^{ikna} e^{-i\omega_0 \tau_0} - \frac{\partial \bar{A}_0}{\partial \tau_1} e^{-ikna} e^{i\omega_0 \tau_0} \right]$$

$$+ \frac{1}{m}\left[\left(e^{i2ka} - 2e^{ika} + 2e^{-ika} - e^{-i2ka} \right) \left(A_0 A_0 e^{i2kna} e^{-i2\omega_0 \tau_0} - \bar{A}_0 \bar{A}_0 e^{-i2kna} e^{i2\omega_0 \tau_0} \right) \right]$$

It is assumed that the solution to the homogeneous equation of the ε^1-order equation takes similar form to the general solution of the ε^0-order equation. Under this assumption, terms on the RHS of the ε^1-order equation with functional form $e^{i\omega_0 \tau_0}$ or $e^{-i\omega_0 \tau_0}$ contribute to secular behavior. These terms are eliminated by setting them equal to zero. Accordingly, A_0 and $\bar{A}_0$ are considered to be independent functions of τ_1. This modifies the form of the general solution to the ε^0-equation:

$$u_{n,G}^{(0)}(\tau_0, \tau_2) = A_0(\tau_2)\, e^{ikna} e^{-i\omega_0\tau_0} + \bar{A}_0(\tau_2)\, e^{-ikna} e^{i\omega_0\tau_0}, \tag{9.6}$$

where

$$A_0(\tau_2) = \alpha(\tau_2) e^{-i\varphi(\tau_2)}$$

$$\bar{A}_0(\tau_2) = \alpha(\tau_2)\, e^{i\varphi(\tau_2)}$$

The homogeneous solution to the ε^1-order equation takes the following form:

$$u_{n,H}^{(1)}(\tau_0, \tau_2) = B_0(\tau_2)e^{ikna} e^{-i\omega_0\tau_0} + \bar{B}_0(\tau_2)e^{-ikna} e^{i\omega_0\tau_0} \tag{9.7}$$

The particular solution to the ε^1-order equation is of the form:

$$u_{n,P}^{(1)}(\tau_0, \tau_2) = C_0(\tau_2)e^{i2kna} e^{-i2\omega_0\tau_0} + \bar{C}_0(\tau_2)e^{-i2kna} e^{i2\omega_0\tau_0} \tag{9.8}$$

Inserting (9.8) into the ε^1-order equation and relating like terms reveals relationships for the exponential pre-factors $C_0(\tau_2)$ and $\bar{C}_0(\tau_2)$. Equation (9.8) becomes

$$u_{n,P}^{(1)}(\tau_0, \tau_2) = \frac{2i(\sin(2ka) - 2\sin(ka))}{\beta((2-2\cos) - 4(2-2\cos(ka)))}\left[A_0^{\,2}e^{i2kna} e^{-i2\omega_0\tau_0} - \bar{A}_0^{\,2}e^{-i2kna} e^{i2\omega_0\tau_0}\right].$$

The general solution to the ε^1-order equation is a sum of the homogeneous $(u_{n,H}^{(1)})$ and particular solutions $(u_{n,P}^{(1)})$:

$$\begin{aligned}
u_{n,G}^{(1)}(\tau_0, \tau_2) = {} & B_0 e^{ikna} e^{-i\omega_0\tau_0} + \bar{B}_0\, e^{-ikna} e^{i\omega_0\tau_0} \\
& + \frac{2i(\sin(2ka) - 2\sin(ka))}{\beta((2-2\cos(2ka)) - 4(2-2\cos(ka)))} \\
& \times \left[A_0 A_0 e^{i2kna} e^{-i2\omega_0\tau_0} - \bar{A}_0\bar{A}_0 e^{-i2kna} e^{i2\omega_0\tau_0}\right]
\end{aligned}$$

The values for B_0 and $\bar{B}_0$ are found from initial conditions. With the general solutions to the ε^0-equation and the ε^1-equation, the ε^2-order equation is developed. Inserting $u_{n,G}^{(0)}$ and $u_{n,G}^{(1)}$ into the ε^2-order equation, utilizing the expressions for $A_0(\tau_2)$ and $\bar{A}_0(\tau_2)$, and noting that $u_n^{(0)}$ and $u_n^{(1)}$ are independent functions of τ_1, the ε^2-order equation is written as

$$\frac{\partial^2 u_n^{(2)}}{\partial \tau_0^2} + \omega_n^{\,2}\left(2u_n^{(2)} - u_{n+1}^{(2)} - u_{n-1}^{(2)}\right) =$$

$$e^{ikna}e^{-i\omega_0\tau_0}\left(2\omega_0\alpha\frac{\partial\varphi}{\partial\tau_2}e^{-i\varphi} + 2i\omega_0 e^{-i\varphi}\frac{\partial\alpha}{\partial\tau_2}\right)$$

$$+ e^{-ikna}e^{i\omega_0\tau_0}\left(2\omega_0\alpha\frac{\partial\varphi}{\partial\tau_2}e^{i\varphi} - 2i\omega_0 e^{i\varphi}\frac{\partial\alpha}{\partial\tau_2}\right)$$

$$+ \frac{2}{m}\{\left[\left(e^{i2ka} - 2e^{ika} + 2e^{-ika} - e^{-i2ka}\right)\left(A_0 B_0 e^{i2kna}e^{-i2\omega_0\tau_0} - \bar{A}_0\bar{B}_0 e^{-i2kna}e^{i2\omega_0\tau_0}\right)\right]$$

$$+ \left[\left(e^{i3ka} - e^{i2ka} - e^{ika} + e^{-ika} + e^{-i2ka} - e^{-i3ka}\right)\right.$$

$$\times\left(A_0 C_0 e^{i3kna}e^{-i3\omega_0\tau_0} - \bar{A}_0\bar{C}_0 e^{-i3kna}e^{i3\omega_0\tau_0}\right)\right]$$

$$+ \left[\left(e^{i2ka} - 2e^{ika} + 2e^{-ika} - e^{-i2ka}\right)\left(A_0\bar{C}_0 e^{-ikna}e^{i\omega_0\tau_0} - \bar{A}_0 C_0 e^{ikna}e^{-i\omega_0\tau_0}\right)\right]\}$$

The homogeneous solution to the ε^2-order equation is similar in form to the general solution of the ε^0-equation and the homogeneous solution of the ε^1-equation. Accordingly, terms on the RHS of the ε^2-order equation with functional form $e^{i\omega_0\tau_0}$ or $e^{-i\omega_0\tau_0}$ contribute to secular behavior and must be eliminated. Setting exponential pre-factors equal to zero yields the following relationships for $\alpha(\tau_2)$ and $\varphi(\tau_2)$:

$$\alpha(\tau_2) = \alpha_0 \tag{9.9}$$

$$\varphi(\tau_2) = -\frac{\alpha^2}{\omega_0\beta m}\cdot\frac{4(\sin(2ka) - 2\sin(ka))^2}{(2 - 2\cos(2ka)) - 4(2 - 2\cos(ka))}\tau_2 + \varphi_0, \tag{9.10}$$

where α_0 and φ_0 are constants determined from initial plane wave conditions. The general solution to the ε^0-equation [(9.6)] is considered again with (9.9) and (9.10) utilized in expressions for A_0 and $\bar{A}_0$. Here, the constant φ_0 can be set equal to zero without loss of generality.

$$u_{n,G}^{(0)}(\tau_0, \tau_2) = \alpha_0 e^{i\left(kna - \left(\omega_0 - \varepsilon^2\frac{\alpha^2}{\omega_0\beta m}\cdot\frac{4(\sin(2ka) - 2\sin(ka))^2}{(2 - 2\cos(2ka)) - 4(2 - 2\cos(ka))}\right)\tau_0\right)}$$

$$+ \alpha_0 e^{-i\left(kna - \left(\omega_0 - \varepsilon^2\frac{\alpha^2}{\omega_0\beta m}\cdot\frac{4(\sin(2ka) - 2\sin(ka))^2}{(2 - 2\cos(2ka)) - 4(2 - 2\cos(ka))}\right)\tau_0\right)}$$

This result shows that the 0th order term in the asymptotic expansion of u_n shows the harmonic dispersion curve to be shifted by a quantity that has quadratic dependence on the strength of the nonlinearity parameter ε.

9.2.1.2   Three-Wave Interactions

Here we consider the interaction between three waves with different wave vectors and frequencies. The analysis begins with the equation of motion [(9.1)] from the single-wave dispersion analysis. The displacement of the nth mass is represented by a superposition of wave modes each with a unique, time and wave vector-dependent amplitude factor [(9.11)]:

$$u_n(t) = \sum_k A(k,t)e^{ikna} \qquad (9.11)$$

Here we use a discrete summation over the wave numbers instead of an integral over a continuum of wave vectors. This is done to help the reader to conceptualize the interactions between specific phonons and to facilitate the comparison with the MD models presented subsequently. Indeed, MD simulations are limited to finite size systems for which the phonon modes do not form a continuum but a discrete set of possible wave vectors. Inserting (9.11) into the equation of motion for the 1D monoatomic crystal yields a modified equation of motion [(9.12)].

$$= -4\beta \sum_k A(k,t)e^{ikna}\sin^2\left(\frac{ka}{2}\right) + \varepsilon\left[\sum_{k'}\sum_{k''}A(k',t)A(k'',t)e^{i(k'+k'')na}f(k',k'')\right],$$
$$(9.12)$$

where $f(k',k'') = -8i\sin\left(\frac{k'a}{2}\right)\sin\left(\frac{k''a}{2}\right)\sin\left(\frac{(k'+k'')a}{2}\right)$. Equation (9.12) is multiplied by e^{-ik^*na} and a summation over all n masses is imposed. This procedure selects the mode k^* as reference wave vector. With $\omega_n^2 = \frac{4\beta}{m}$, (9.12) becomes

$$\frac{d^2A(k^*,t)}{dt^2} + \omega_n^2\sin^2\left(\frac{k^*a}{2}\right)A(k^*,t) = \frac{\varepsilon}{m}\sum_{k'}\sum_{k''}A(k',t)A(k'',t)f(k',k'')\delta_{k'+k'',k^*}$$
$$(9.13)$$

$\delta_{k'+k'',k^*}$ imposes the wave vector conservation rule $k^* = k' + k'' + mG$ where m is an integer and G is a reciprocal lattice vector of the periodic structure. We do not label G in the delta function for the sake of simplicity of the notation. For $m = 0$, one has the so-called normal three phonon scattering process. The case of $m \neq 0$ corresponds to umklapp processes where $k' + k''$ is located outside the first Brillouin zone. In (9.13), the variable τ is introduced, where $\tau = \omega_n t$. Single time variables (τ) are replaced by a collection of variables $\tau = (\tau_0, \tau_1, \tau_2)$ whereby: $\tau_0 = \tau$, $\tau_1 = \varepsilon\tau$, $\tau_2 = \varepsilon^2\tau$. Additionally, $A(k^*, \tau)$ is replaced by an asymptotic expansion whereby:

$$A(k^*,\tau) = A_0(k^*,\tau) + \varepsilon A_1(k^*,\tau) + \varepsilon^2 A_2(k^*,\tau)$$
$$A(k^*,\tau_0,\tau_1,\tau_2) = A_0(k^*,\tau_0,\tau_1,\tau_2) + \varepsilon A_1(k^*,\tau_0,\tau_1,\tau_2) + \varepsilon^2 A_2(k^*,\tau_0,\tau_1,\tau_2)$$

With these considerations, (9.13) is separated into expressions at order ε^0, ε^1, and ε^2:

$O(\varepsilon^0)$:

$$\frac{\partial^2 A_0(k^*,\tau)}{\partial\tau_0^2} + \sin^2\left(\frac{k^*a}{2}\right)A_0(k^*,\tau) = 0$$

$O(\varepsilon^1)$:

$$\frac{\partial^2 A_1(k^*,\tau)}{\partial \tau_0^2} + 2\frac{\partial^2 A_0(k^*,\tau)}{\partial \tau_1 \partial \tau_0} + \sin^2\left(\frac{k^* a}{2}\right) A_1(k^*,\tau)$$

$$= \frac{1}{m\omega_n^2}\sum_{k'}\sum_{k''} f(k',k'')\delta_{k'+k'',k^*}[A_0(k',\tau)A_0(k'',\tau)]$$

$O(\varepsilon^2)$:

$$\frac{\partial^2 A_2(k^*,\tau)}{\partial \tau_0^2} + 2\frac{\partial^2 A_1(k^*,\tau)}{\partial \tau_1 \partial \tau_0} + 2\frac{\partial^2 A_0(k^*,\tau)}{\partial \tau_2 \partial \tau_0} + \frac{\partial^2 A_0(k^*,\tau)}{\partial \tau_1^2} + \sin^2\left(\frac{k^* a}{2}\right) A_2(k^*,\tau)$$

$$= \frac{1}{m\omega_n^2}\sum_{k'}\sum_{k''} f(k',k'')\delta_{k'+k'',k^*}[A_0(k',\tau)A_1(k'',\tau) + A_1(k',\tau)A_0(k'',\tau)]$$

To solve the ε^0-equation, a general solution of the following form is proposed:

$$A_0(k^*,\tau_0,\tau_1,\tau_2) = a_0(k^*,\tau_1,\tau_2)e^{i\omega_0^* \tau_0} + \bar{a}_0(k^*,\tau_1,\tau_2)e^{-i\omega_0^* \tau_0} \qquad (9.14)$$

Inserting (9.14) into the ε^0-equation offers the expected relationship between ω_0^* and k^*: $\omega_0^{*2} = \sin^2\left(\frac{k^* a}{2}\right)$. Inserting (9.14) into the ε^1-equation offers an expression to solve for $A_1(k^*,\tau)$. After rearranging and utilizing the following definitions

$$A_0(k',\tau_0,\tau_1,\tau_2) = a_0(k',\tau_1,\tau_2)e^{i\omega'_0 \tau_0} + \bar{a}_0(k',\tau_1,\tau_2)e^{-i\omega'_0 \tau_0}$$

$$A_0(k'',\tau_0,\tau_1,\tau_2) = a_0(k'',\tau_1,\tau_2)e^{i\omega''_0 \tau_0} + \bar{a}_0(k'',\tau_1,\tau_2)e^{-i\omega''_0 \tau_0}$$

the ε^1-equation becomes

$$\frac{\partial^2 A_1(k^*,\tau)}{\partial \tau_0^2} + \omega_0^{*2} A_1(k^*,\tau) = -2i\omega_0^*\left[\frac{\partial a_0^*}{\partial \tau_1}e^{i\omega_0^* \tau_0} - \frac{\partial \bar{a}_0^*}{\partial \tau_1}e^{-i\omega_0^* \tau_0}\right]$$

$$+ \frac{1}{m\omega_n^2}\sum_{k'}\sum_{k''} f(k',k'')\delta_{k'+k'',k^*}[a'_0 a''_0 e^{i(\omega'_0+\omega''_0)\tau_0} + a'_0 \bar{a}''_0 e^{i(\omega'_0-\omega''_0)\tau_0}$$

$$+ \bar{a}'_0 a''_0 e^{-i(\omega'_0-\omega''_0)\tau_0} + \bar{a}'_0 \bar{a}''_0 e^{-i(\omega'_0+\omega''_0)\tau_0}]$$

where terms like $a_0^*, a'_0, a'' \ldots$ etc. are compact representations for $a_0(k^*,\tau_1,\tau_2)$, $a_0(k',\tau_1,\tau_2), a_0(k'',\tau_1,\tau_2) \ldots$ etc. A homogeneous solution to the ε^1-equation is proposed:

$$A_{1,H}(k^*,\tau_0,\tau_2) = a_1(k^*,\tau_2)e^{i\omega_0^* \tau_0} + \bar{a}_1(k^*,\tau_2)e^{-i\omega_0^* \tau_0} = a_1^* e^{i\omega_0^* \tau_0} + \bar{a}_1^* e^{-i\omega_0^* \tau_0} \quad (9.15)$$

The forcing terms on the right hand side (RHS) of the ε^1-equation with functional form $e^{i\omega_0^* \tau_0}$ or $e^{-i\omega_0^* \tau_0}$ contribute to secular behavior. These terms must be eliminated such that the final representation of $A(k^*,\tau)$ is well behaved (e.g. contains no terms that temporally grow without bound). These terms are set to zero by making a_0 and $\bar{a}_0$

functions of k^* and τ_2 only. With this stipulation, an appropriate form of the particular solution to the ε^1-equation is:

$$A_{1,P}(k^*,\tau) = \frac{1}{m\omega_n^2} \sum_{k'} \sum_{k''} f(k',k'')\delta_{k'+k'',k^*}$$

$$\times \left[b_1 e^{i(\omega'_0+\omega''_0)\tau_0} + \bar{b}_1 e^{-i(\omega'_0+\omega''_0)\tau_0} + c_1 e^{i(\omega'_0-\omega''_0)\tau_0} + \bar{c}_1 e^{-i(\omega'_0-\omega''_0)\tau_0} \right]$$

(9.16)

The exponential pre-factors $b_1, \bar{b}_1, c_1, \bar{c}_1$ have dependency on $k', k'', \tau_2, \omega_0^*, \omega'_0, \omega''_0$. Substituting (9.16) into the ε^1-equation and relating like terms reveals the exponential pre-factors: $b_1, \bar{b}_1, c_1, \bar{c}_1$

$$b_1 = \frac{a_0(k',\tau_2)a_0(k'',\tau_2)}{\omega_0^{*2} - (\omega'_0+\omega''_0)^2}; \quad \bar{b}_1 = \frac{\bar{a}_0(k',\tau_2)\bar{a}_0(k'',\tau_2)}{\omega_0^{*2} - (\omega'_0+\omega''_0)^2}$$

$$c_1 = \frac{a_0(k',\tau_2)\bar{a}_0(k'',\tau_2)}{\omega_0^{*2} - (\omega'_0-\omega''_0)^2}; \quad \bar{c}_1 = \frac{\bar{a}_0(k',\tau_2)a_0(k'',\tau_2)}{\omega_0^{*2} - (\omega'_0-\omega''_0)^2}$$

In the long wavelength limit, angular frequency has nearly linear dependence on wave vector. In considering the stipulated wave vector relationship inside the double summation in (9.16), $(k'+k''=k^*)$, it is conceivable that $\omega_0(k') + \omega_0(k'') = \omega_0(k^*)$ or $\omega_0(k') - \omega_0(k'') = \omega_0(k^*)$. In this instance, the denominator terms in the expressions for $b_1, \bar{b}_1, c_1, \bar{c}_1$ will go to zero. To avoid this complication, following the procedure stipulated by Khoo et al. [17], a small imaginary part φ is introduced in the denominator. At the final result of the calculation a limit will be taken as $\varphi \to 0$. The general solution to the ε^1-equation is a sum of the homogeneous and particular solutions:

$$A_1(k^*,\tau_0,\tau_2) = a_1^* e^{i\omega_0^*\tau_0} + \bar{a}_1^* e^{-i\omega_0^*\tau_0} + \frac{1}{m\omega_n^2} \sum_{k'} \sum_{k''} f(k',k'')\delta_{k'+k'',k^*}$$

$$\times \left[\frac{a'_0 a''_0}{g_1^*} e^{i(\omega'_0+\omega''_0)\tau_0} + \frac{\bar{a}'_0 \bar{a}''_0}{g_1^*} e^{-i(\omega'_0+\omega''_0)\tau_0} + \frac{a'_0 \bar{a}''_0}{g_2^*} e^{i(\omega'_0-\omega''_0)\tau_0} \right.$$

$$\left. + \frac{\bar{a}'_0 a''_0}{g_2^*} e^{-i(\omega'_0-\omega''_0)\tau_0} \right],$$

(9.17)

where $g_1^* = \omega_0^{*2} - (\omega'_0+\omega''_0)^2 + i\varphi$; $g_2^* = \omega_0^{*2} - (\omega'_0-\omega''_0)^2 + i\varphi$

The ε^2-equation is reduced to the following expressions because $A_0(k^*,\tau)$ and $A_1(k^*,\tau)$ are independent of τ_1:

$$\frac{\partial^2 A_2(k^*, \tau_0, \tau_1, \tau_2)}{\partial \tau_0^2} + \omega_0^{*2} A_2(k^*, \tau_0, \tau_1, \tau_2)$$

$$= -2i\omega_0^* \frac{\partial a_0(k^*, \tau_2)}{\partial \tau_2} e^{i\omega_0^* \tau_0} + 2i\omega_0^* \frac{\partial \bar{a}_0(k^*, \tau_2)}{\partial \tau_2} e^{-i\omega_0^* \tau_0}$$

$$+ \frac{1}{m\omega_n^2} \sum_{k'} \sum_{k''} f(k', k'') \delta_{k'+k'',k^*} [A_0(k', \tau_0, \tau_2) A_1(k'', \tau_0, \tau_2)$$

$$+ A_1(k', \tau_0, \tau_2) A_0(k'', \tau_0, \tau_2)]$$

As before, the solution to the homogeneous equation of the ε^2-equation is of the form:

$$A_{2,H}(k^*, \tau_0, \tau_2) = a_2(k^*, \tau_2) e^{i\omega_0^* \tau_0} + \bar{a}_2(k^*, \tau_2) e^{-i\omega_0^* \tau_0}$$

Terms on the RHS of the ε^2-equation with functional form $e^{i\omega_0^* \tau_0}$ or $e^{-i\omega_0^* \tau_0}$ contribute to secular behavior. Using equations (9.14) and (9.17) to develop the RHS of the ε^2-equation gives (9.18):

$$\frac{\partial^2 A_2(k^*, \tau_0, \tau_1, \tau_2)}{\partial \tau_0^2} + \omega_0^{*2} A_2(k^*, \tau_0, \tau_1, \tau_2)$$

$$= -2i\omega_0^* \frac{\partial a_0(k^*, \tau_2)}{\partial \tau_2} e^{i\omega_0^* \tau_0} + 2i\omega_0^* \frac{\partial \bar{a}_0(k^*, \tau_2)}{\partial \tau_2} e^{-i\omega_0^* \tau_0}$$

$$+ \frac{1}{m\omega_n^2} \sum_{k'} \sum_{k''} f(k', k'') \delta_{k'+k'',k^*} \left(a'_0 a''_1 e^{i(\omega'_0 + \omega''_0)\tau_0} + a'_0 \bar{a}''_1 e^{i(\omega'_0 - \omega''_0)\tau_0} \right.$$

$$\left. + \bar{a}'_0 a''_1 e^{-i(\omega'_0 - \omega''_0)\tau_0} + \bar{a}'_0 \bar{a}''_1 e^{-i(\omega'_0 + \omega''_0)\tau_0} \right)$$

$$+ \frac{1}{m\omega_n^2} \sum_{k'} \sum_{k''} f(k', k'') \delta_{k'+k'',k^*} \left(a''_0 a'_1 e^{i(\omega''_0 + \omega'_0)\tau_0} + a''_0 \bar{a}'_1 e^{i(\omega''_0 - \omega'_0)\tau_0} \right.$$

$$\left. + \bar{a}''_0 a'_1 e^{-i(\omega''_0 - \omega'_0)\tau_0} + \bar{a}''_0 \bar{a}'_1 e^{-i(\omega''_0 + \omega'_0)\tau_0} \right)$$

$$+ \frac{1}{m\omega_n^2} \sum_{k'} \sum_{k''} f(k', k'') \delta_{k'+k'',k^*} \left\{ \left[\frac{1}{m\omega_n^2} \sum_{k_1} \sum_{k_2} f(k_1, k_2) \delta_{k_1+k_2,k''} \right. \right.$$

$$\left[\frac{a'_0 a_0^{(1)} a_0^{(2)}}{g''_1} e^{i(\omega_0^{(1)} + \omega_0^{(2)} + \omega'_0)\tau_0} + \frac{a'_0 \bar{a}_0^{(1)} \bar{a}_0^{(2)}}{g''_1} e^{-i(\omega_0^{(1)} + \omega_0^{(2)} - \omega'_0)\tau_0} \right.$$

$$+ \frac{a'_0 a_0^{(1)} \bar{a}_0^{(2)}}{g''_2} e^{i(\omega_0^{(1)} - \omega_0^{(2)} + \omega'_0)\tau_0} + \frac{a'_0 \bar{a}_0^{(1)} a_0^{(2)}}{g''_2} e^{-i(\omega_0^{(1)} - \omega_0^{(2)} - \omega'_0)\tau_0}$$

$$\frac{\bar{a}'_0 a_0^{(1)} a_0^{(2)}}{g''_1} e^{i(\omega_0^{(1)} + \omega_0^{(2)} - \omega'_0)\tau_0} + \frac{\bar{a}'_0 \bar{a}_0^{(1)} \bar{a}_0^{(2)}}{g''_1} e^{-i(\omega_0^{(1)} + \omega_0^{(2)} + \omega'_0)\tau_0}$$

$$\left. \left. + \frac{\bar{a}'_0 a_0^{(1)} \bar{a}_0^{(2)}}{g''_2} e^{i(\omega_0^{(1)} - \omega_0^{(2)} - \omega'_0)\tau_0} + \frac{\bar{a}'_0 \bar{a}_0^{(1)} a_0^{(2)}}{g''_2} e^{-i(\omega_0^{(1)} - \omega_0^{(2)} + \omega'_0)\tau_0} \right] \right]$$

$$+\left[\frac{1}{m\omega_n^2}\sum_{k_1}\sum_{k_2}f(k_1,k_2)\delta_{k_1+k_2,k'}\right.$$

$$\times\left[\frac{a_0''^{(1)}a_0^{(2)}}{g'_1}e^{i(\omega_0^{(1)}+\omega_0^{(2)}+\omega''_0)\tau_0}+\frac{a_0''\bar{a}_0^{(1)}\bar{a}_0^{(2)}}{g'_1}e^{-i(\omega_0^{(1)}+\omega_0^{(2)}-\omega''_0)\tau_0}\right.$$

$$+\frac{a_0''a_0^{(1)}\bar{a}_0^{(2)}}{g'_2}e^{i(\omega_0^{(1)}-\omega_0^{(2)}+\omega''_0)\tau_0}+\frac{a_0''\bar{a}_0^{(1)}a_0^{(2)}}{g'_2}e^{-i(\omega_0^{(1)}-\omega_0^{(2)}-\omega''_0)\tau_0}$$

$$+\frac{\bar{a}_0''a_0^{(1)}a_0^{(2)}}{g'_1}e^{i(\omega_0^{(1)}+\omega_0^{(2)}-\omega''_0)\tau_0}+\frac{\bar{a}_0''\bar{a}_0^{(1)}\bar{a}_0^{(2)}}{g'_1}e^{-i(\omega_0^{(1)}+\omega_0^{(2)}+\omega''_0)\tau_0}$$

$$\left.\left.+\frac{\bar{a}_0''a_0^{(1)}\bar{a}_0^{(2)}}{g'_2}e^{i(\omega_0^{(1)}-\omega_0^{(2)}-\omega''_0)\tau_0}+\frac{\bar{a}_0''\bar{a}_0^{(1)}a_0^{(2)}}{g'_2}e^{-i(\omega_0^{(1)}-\omega_0^{(2)}+\omega''_0)\tau_0}\right]\right]\right\}$$

$$(9.18)$$

There is notable similarity between the terms on the RHS of the ε^1-equation that was solved to yield (9.17) and the third and fourth terms on the RHS of (9.18). These terms are treated with the same procedure as that used for the ε^1-equation. Accordingly, they will not contribute to secular terms.

The objective is to identify terms in the ε^2-equation with $e^{i\omega_0^*\tau_0}$ or $e^{-i\omega_0^*\tau_0}$ dependency. This will be done by systematically evaluating all wave vector pairs $\{k_1,k_2\}$ that satisfy the wave vector constraints stipulated by (9.18). Specifically,

$$\delta_{k'+k'',k^*}\delta_{k_1+k_2,k''}\rightarrow k'+k_1+k_2=k^*$$
$$\delta_{k'+k'',k^*}\delta_{k_1+k_2,k'}\rightarrow k''+k_1+k_2=k^*.$$

If a certain pair of wave vectors satisfies the above-mentioned wave vector constraints, then an analysis will be carried through to see if these wave vectors give rise to terms with $e^{i\omega_0^*\tau_0}$ or $e^{-i\omega_0^*\tau_0}$ dependence. As before, terms with $e^{i\omega_0^*\tau_0}$ or $e^{-i\omega_0^*\tau_0}$ dependence will be removed.

In (9.18), inside the summation over k',k'', there are two summations over k_1,k_2. For the first summation over k_1,k_2, two conditions must be met: (1) $k'+k''=k^*$ and (2) $k_1+k_2=k''$.

The only possible combinations for k_1,k_2 that give wave vector relationships that are compatible with $\delta_{k'+k'',k^*}$ are shown as Condition A and Condition B:

$$\text{Condition A}: k_1=-k', k_2=k^*\text{ and }-k'+k^*=k''$$

$$\text{Condition B}: k_1=k^*, k_2=-k\prime, \text{ and } k^*-k'=k''$$

Now that wave vector constraints are satisfied, an analysis is carried out to see if any terms with $e^{i\omega_0^*\tau_0}$ or $e^{-i\omega_0^*\tau_0}$ dependence arise in the first summation over k_1,k_2.

The following frequency relationships are present in the first summation over k_1, k_2 in (9.18):

(i). $\omega_0^{(1)} + \omega_0^{(2)} + \omega_0'$

(ii). $\omega_0^{(1)} + \omega_0^{(2)} - \omega_0'$

(iii). $\omega_0^{(1)} - \omega_0^{(2)} + \omega_0'$

(iv). $\omega_0^{(1)} - \omega_0^{(2)} - \omega_0'$

Applying Condition A to these frequency relationships show two relationships that offer terms with $e^{i\omega_0^* \tau_0}$ or $e^{-i\omega_0^* \tau_0}$ dependence:

$$\text{Condition A}: k_1 = -k' \rightarrow \omega_0^{(1)} = \omega_0' \text{ and } k_2 = k^* \rightarrow \omega_0^{(2)} = \omega_0^*$$

Applying Condition A to frequency relationships leads to:

(i). $\omega_0^{(1)} + \omega_0^{(2)} + \omega_0' \rightarrow \omega_0' + \omega_0^* + \omega_0' = \omega_0^* + 2\omega_0'$

(ii). $\omega_0^{(1)} + \omega_0^{(2)} - \omega_0' \rightarrow \omega_0' + \omega_0^* - \omega_0' = \omega_0^*$

(iii). $\omega_0^{(1)} - \omega_0^{(2)} + \omega_0' \rightarrow \omega_0' - \omega_0^* + \omega_0' = -\omega_0^* + 2\omega_0'$

(iv). $\omega_0^{(1)} - \omega_0^{(2)} - \omega_0' \rightarrow \omega_0' - \omega_0^* - \omega_0' = -\omega_0^*$

As a result, with Condition A, the following terms in the first summation over k_1, k_2 contribute to secular terms:

$$\frac{a_0' \bar{a}_0^{(1)} \bar{a}_0^{(2)}}{g''_1} e^{-i(\omega_0^{(1)} + \omega_0^{(2)} - \omega_0')\tau_0} = \frac{a_0' \bar{a}_0 \bar{a}_0^*}{g''_1} e^{-i(\omega_0' + \omega_0^* - \omega_0')\tau_0} = \frac{a_0' \bar{a}_0 \bar{a}_0^*}{g''_1} e^{-i(\omega_0^*)\tau_0}$$

$$\frac{\bar{a}_0' a_0^{(1)} a_0^{(2)}}{g''_1} e^{i(\omega_0^{(1)} + \omega_0^{(2)} - \omega_0')\tau_0} = \frac{\bar{a}_0' a_0 a_0^*}{g''_1} e^{i(\omega_0' + \omega_0^* - \omega_0')\tau_0} = \frac{\bar{a}_0' a_0 a_0^*}{g''_1} e^{i(\omega_0^*)\tau_0}$$

$$\frac{a_0' \bar{a}_0^{(1)} a_0^{(2)}}{g''_2} e^{-i(\omega_0^{(1)} - \omega_0^{(2)} - \omega_0')\tau_0} = \frac{a_0' \bar{a}_0 a_0^*}{g''_2} e^{-i(\omega_0' - \omega_0^* - \omega_0')\tau_0} = \frac{a_0' \bar{a}_0^{(1)} a_0^{(2)}}{g''_2} e^{i(\omega_0^*)\tau_0}$$

$$\frac{\bar{a}_0' a_0^{(1)} \bar{a}_0^{(2)}}{g''_2} e^{i(\omega_0^{(1)} - \omega_0^{(2)} - \omega_0')\tau_0} = \frac{\bar{a}_0' a_0 \bar{a}_0^*}{g''_2} e^{i(\omega_0' - \omega_0^* - \omega_0')\tau_0} = \frac{\bar{a}_0' a_0^{(1)} \bar{a}_0^{(2)}}{g''_2} e^{-i(\omega_0^*)\tau_0}$$

Applying Condition B to these frequency relationships show two different relationships that offer terms with $e^{i\omega_0^* \tau_0}$ or $e^{-i\omega_0^* \tau_0}$ dependence:

$$\text{Condition B}: k_1 = k^* \rightarrow \omega_0^{(1)} = \omega_0^*, k_2 = -k' \rightarrow \omega_0^{(2)} = \omega_0'$$

Applying Condition B to frequency relationships leads to:

(i). $\omega_0^{(1)} + \omega_0^{(2)} + \omega_0' \rightarrow \omega_0^* + \omega_0' + \omega_0' = \omega_0^* + 2\omega_0'$

(ii). $\omega_0^{(1)} + \omega_0^{(2)} - \omega_0' \rightarrow \omega_0^* + \omega_0' - \omega_0' = \omega_0^*$

(iii). $\omega_0^{(1)} - \omega_0^{(2)} + \omega_0' \rightarrow \omega_0^* - \omega_0' + \omega_0' = \omega_0^*$

(iv). $\omega_0^{(1)} - \omega_0^{(2)} - \omega_0' \rightarrow \omega_0^* - \omega_0' - \omega_0' = \omega_0^* - 2\omega_0'$

As a result, with Condition B, the following terms in the first summation over k_1, k_2 contribute to secular terms:

$$\frac{a'_0 \bar{a}_0^{(1)} \bar{a}_0^{(2)}}{g''_1} e^{-i(\omega_0^{(1)} + \omega_0^{(2)} - \omega'_0)\tau_0} = \frac{a'_0 \bar{a}_0 \bar{a}_0}{g''_1} e^{-i(\omega_0^* + \omega'_0 - \omega'_0)\tau_0} = \frac{a'_0 \bar{a}_0^* \bar{a}_0}{g''_1} e^{-i(\omega_0^*)\tau_0}$$

$$\frac{\bar{a}'_0 a_0^{(1)} a_0^{(2)}}{g''_1} e^{i(\omega_0^{(1)} + \omega_0^{(2)} - \omega'_0)\tau_0} = \frac{\bar{a}'_0 a_0 a_0}{g''_1} e^{i(\omega_0^* + \omega'_0 - \omega'_0)\tau_0} = \frac{\bar{a}'_0 a_0^* a_0}{g''_1} e^{i(\omega_0^*)\tau_0}$$

$$\frac{a'_0 a_0^{(1)} \bar{a}_0^{(2)}}{g''_2} e^{i(\omega_0^{(1)} - \omega_0^{(2)} + \omega'_0)\tau_0} = \frac{a'_0 a_0 \bar{a}_0}{g''_2} e^{i(\omega_0^* - \omega'_0 + \omega'_0)\tau_0} = \frac{a'_0 a_0^* \bar{a}_0}{g''_2} e^{i(\omega_0^*)\tau_0}$$

$$\frac{\bar{a}'_0 \bar{a}_0^{(1)} a_0^{(2)}}{g''_2} e^{-i(\omega_0^{(1)} - \omega_0^{(2)} + \omega'_0)\tau_0} = \frac{\bar{a}'_0 \bar{a}_0 a_0}{g''_2} e^{-i(\omega_0^* - \omega'_0 + \omega'_0)\tau_0} = \frac{\bar{a}'_0 \bar{a}_0^* a_0}{g''_2} e^{-i(\omega_0^*)\tau_0}$$

For the second summation over k_1, k_2, two conditions must be met:
(1) $k' + k'' = k^*$
(2) $k_1 + k_2 = k'$

The only possible combinations for k_1, k_2 that give wave vector relationships that are compatible with $\delta_{k'+k'',k^*}$ are shown as Condition C and Condition D:

$$\text{Condition C}: k_1 = -k'', k_2 = k*, \text{ and } -k'' + k* = k'$$

$$\text{Condition D}: k_1 = k^*, k_2 = -k'', \text{ and } k^* - k'' = k$$

Now that wave vector constraints are satisfied, an analysis is carried out to see if any terms with $e^{i\omega_0^*\tau_0}$ or $e^{-i\omega_0^*\tau_0}$ dependency arise in the second summation over k_1, k_2. The following frequency relationships are present in the second summation over k_1, k_2 in (9.18):
(v). $\omega_0^{(1)} + \omega_0^{(2)} + \omega_0''$
(vi). $\omega_0^{(1)} + \omega_0^{(2)} - \omega_0''$
(vii). $\omega_0^{(1)} - \omega_0^{(2)} + \omega_0''$
(viii). $\omega_0^{(1)} - \omega_0^{(2)} - \omega_0''$

Applying Condition C to these frequency relationships show two relationships that offer terms with $e^{i\omega_0^*\tau_0}$ or $e^{-i\omega_0^*\tau_0}$ dependence:

$$\text{Condition C}: k_1 = -k'' \rightarrow \omega_0^{(1)} = \omega_0'' \quad \text{and} \quad k_2 = k^* \rightarrow \omega_0^{(2)} = \omega_0^*$$

Applying Condition C to frequency relationships leads to
(v). $\omega_0^{(1)} + \omega_0^{(2)} + \omega_0'' \rightarrow \omega_0'' + \omega_0^* + \omega_0'' = \omega_0^* + 2\omega_0''$
(vi). $\omega_0^{(1)} + \omega_0^{(2)} - \omega_0'' \rightarrow \omega_0'' + \omega_0^* - \omega_0'' = \omega_0^*$
(vii). $\omega_0^{(1)} - \omega_0^{(2)} + \omega_0'' \rightarrow \omega_0'' - \omega_0^* + \omega_0'' = -\omega_0^* + 2\omega_0''$
(viii). $\omega_0^{(1)} - \omega_0^{(2)} - \omega_0'' \rightarrow \omega_0'' - \omega_0^* - \omega_0'' = -\omega_0^*$

As a result, with Condition C, the following terms in the second summation over k_1, k_2 contribute to secular terms:

$$\frac{a_0'' \bar{a}_0^{(1)} \bar{a}_0^{(2)}}{g'_1} e^{-i(\omega_0^{(1)} + \omega_0^{(2)} - \omega''_0)\tau_0} = \frac{a_0'' \bar{a}_0'' \bar{a}_0^*}{g'_1} e^{-i(\omega''_0 + \omega_0^* - \omega''_0)\tau_0} = \frac{a_0'' \bar{a}_0'' \bar{a}_0^*}{g'_1} e^{-i(\omega_0^*)\tau_0}$$

$$\frac{\bar{a}_0'' a_0^{(1)} a_0^{(2)}}{g'_1} e^{i(\omega_0^{(1)} + \omega_0^{(2)} - \omega''_0)\tau_0} = \frac{\bar{a}_0'' a_0'' a_0^*}{g'_1} e^{i(\omega''_0 + \omega_0^* - \omega''_0)\tau_0} = \frac{\bar{a}_0'' a_0^* a_0''}{g'_1} e^{i(\omega_0^*)\tau_0}$$

$$\frac{a_0'' \bar{a}_0^{(1)} a_0^{(2)}}{g'_2} e^{-i(\omega_0^{(1)} - \omega_0^{(2)} - \omega''_0)\tau_0} = \frac{a_0'' \bar{a}_0'' a_0^*}{g'_2} e^{-i(\omega''_0 - \omega_0^* - \omega''_0)\tau_0} = \frac{a_0'' \bar{a}_0'' a_0^*}{g'_2} e^{i(\omega_0^*)\tau_0}$$

$$\frac{\bar{a}_0'' a_0^{(1)} \bar{a}_0^{(2)}}{g'_2} e^{i(\omega_0^{(1)} - \omega_0^{(2)} - \omega''_0)\tau_0} = \frac{\bar{a}_0'' a_0'' \bar{a}_0^*}{g'_2} e^{i(\omega''_0 - \omega_0^* - \omega''_0)\tau_0} = \frac{\bar{a}_0'' a_0'' \bar{a}_0^*}{g'_2} e^{-i(\omega_0^*)\tau_0}$$

Applying Condition D to these frequency relationships show two different relationships that offer terms with $e^{i\omega_0^* \tau_0}$ or $e^{-i\omega_0^* \tau_0}$ dependence:

$$\text{Condition D}: k_1 = k^* \rightarrow \omega_0^{(1)} = \omega_0^* \text{ and } k_2 = -k'' \rightarrow \omega_0^{(2)} = \omega''_0$$

Applying Condition D to frequency relationships leads to

(v). $\omega_0^{(1)} + \omega_0^{(2)} + \omega''_0 \rightarrow \omega_0^* + \omega''_0 + \omega''_0 = \omega_0^* + 2\omega''_0$

(vi). $\omega_0^{(1)} + \omega_0^{(2)} - \omega''_0 \rightarrow \omega_0^* + \omega''_0 - \omega''_0 = \omega_0^*$

(vii). $\omega_0^{(1)} - \omega_0^{(2)} + \omega''_0 \rightarrow \omega_0^* - \omega''_0 + \omega''_0 = \omega_0^*$

(viii). $\omega_0^{(1)} - \omega_0^{(2)} - \omega''_0 \rightarrow \omega_0^* - \omega''_0 - \omega''_0 = \omega_0^* - 2\omega''_0$

As a result, with Condition D, the following terms in the second summation over k_1, k_2 contribute to secular terms:

$$\frac{a_0'' \bar{a}_0^{(1)} \bar{a}_0^{(2)}}{g'_1} e^{-i(\omega_0^{(1)} + \omega_0^{(2)} - \omega''_0)\tau_0} = \frac{a_0'' \bar{a}_0^* \bar{a}_0''}{g'_1} e^{-i(\omega_0^* + \omega''_0 - \omega''_0)\tau_0} = \frac{a_0'' \bar{a}_0^* \bar{a}_0''}{g'_1} e^{-i(\omega_0^*)\tau_0}$$

$$\frac{\bar{a}_0'' a_0^{(1)} a_0^{(2)}}{g'_1} e^{i(\omega_0^{(1)} + \omega_0^{(2)} - \omega''_0)\tau_0} = \frac{\bar{a}_0'' a_0^* a_0''}{g'_1} e^{i(\omega_0^* + \omega''_0 - \omega''_0)\tau_0} = \frac{\bar{a}_0'' a_0^* a_0''}{g'_1} e^{i(\omega_0^*)\tau_0}$$

$$\frac{a_0'' a_0^{(1)} \bar{a}_0^{(2)}}{g'_2} e^{i(\omega_0^{(1)} - \omega_0^{(2)} + \omega''_0)\tau_0} = \frac{a_0'' a_0^* \bar{a}_0''}{g'_2} e^{i(\omega_0^* - \omega''_0 + \omega''_0)\tau_0} = \frac{a_0'' a_0^* \bar{a}_0''}{g'_2} e^{i(\omega_0^*)\tau_0}$$

$$\frac{\bar{a}_0'' \bar{a}_0^{(1)} a_0^{(2)}}{g'_2} e^{-i(\omega_0^{(1)} - \omega_0^{(2)} + \omega''_0)\tau_0} = \frac{\bar{a}_0'' \bar{a}_0^* a_0''}{g'_2} e^{-i(\omega_0^* - \omega''_0 + \omega''_0)\tau_0} = \frac{\bar{a}_0'' \bar{a}_0^* a_0''}{g'_2} e^{-i(\omega_0^*)\tau_0}$$

In assuming that terms $[a_0(k'), \bar{a}_0(k'), a_0(-k'), \bar{a}_0(-k'), \ldots, \text{etc.}]$ in (9.18) behave as follows:

$$a_0(k', \tau_2) = a_0(k', 0)e^{i\beta(k')\tau_2}$$

$$\bar{a}_0(k', \tau_2) = \bar{a}_0(k', 0)e^{-i\beta(k')\tau_2}$$

$$a_0(-k', \tau_2) = a_0(-k', 0)e^{i\beta(-k')\tau_2}$$

$$\bar{a}_0(-k', \tau_2) = \bar{a}_0(-k', 0)e^{-i\beta(-k')\tau_2}$$

$$\vdots$$

etc.

Additionally,

$$a_0(k', 0) = a_0(-k', 0)$$

$$\bar{a}_0(k', 0) = \bar{a}_0(-k', 0)$$

$$\beta(k') = \beta(-k')$$

Equation (9.18) can be rewritten in the form of (9.19):

$$\frac{\partial^2 A_2(k^*, \tau_0, \tau_1, \tau_2)}{\partial \tau_0^2} + \omega_0^{*2} A_2(k^*, \tau_0, \tau_1, \tau_2)$$

$$= \left\{ -2i\omega_0^* \frac{\partial a_0(k^*, \tau_2)}{\partial \tau_2} + a_0(k^*, 0)e^{i\beta(k^*)\tau_2} \left(\frac{1}{m\omega_n^2}\right)^2 \sum_{k'} \sum_{k''} f(k', k'')\delta_{k'+k'', k^*} \right.$$

$$\times \left[2f(-k', k^*)\delta_{-k'+k^*, k''} a_0(k', 0)\bar{a}_0(k', 0)\left[\frac{1}{g''_1} + \frac{1}{g''_2}\right] \right.$$

$$\left. \left. + 2f(-k'', k^*)\delta_{-k''+k^*, k'} a_0(k'', 0)\bar{a}_0(k'', 0)\left[\frac{1}{g'_1} + \frac{1}{g'_2}\right] \right] \right\} e^{i\omega_0^*\tau_0}$$

$$+ \left\{ 2i\omega_0^* \frac{\partial \bar{a}_0(k^*, \tau_2)}{\partial \tau_2} + \bar{a}_0(k^*, 0)e^{-i\beta(k^*)\tau_2} \left(\frac{1}{m\omega_n^2}\right)^2 \right.$$

$$\times \sum_{k'} \sum_{k''} f(k', k'')\delta_{k'+k'', k^*} \left[2f(-k', k^*)\delta_{-k'+k^*, k''} a_0(k', 0)\bar{a}_0(k', 0) \right.$$

$$\left. \left. \times \left[\frac{1}{g''_1} + \frac{1}{g''_2}\right] + 2f(-k'', k^*)\delta_{-k''+k^*, k'} a_0(k'', 0)\bar{a}_0(k'', 0)\left[\frac{1}{g'_1} + \frac{1}{g'_2}\right] \right] \right\} e^{-i\omega_0^*\tau_0}$$

$$+ \text{ other terms which will not give } e^{i\omega_0^*\tau_0} \text{ or } e^{-i\omega_0^*\tau_0} \text{ dependence} \qquad (9.19)$$

The terms in front of $e^{i\omega_0^*\tau_0}$ and $e^{-i\omega_0^*\tau_0}$ are set to zero. This is shown by equations (9.20) and (9.21):

$$a_0(k^*,0)e^{i\beta(k^*)\tau_2}\left(\frac{1}{m\omega_n^2}\right)^2\sum_{k'}\sum_{k''}$$

$$f(k',k'')\delta_{k'+k'',k^*}\left[2f(-k',k^*)\delta_{-k'+k^*,k''}a_0(k',0)\bar{a}_0(k',0)\right.$$

$$\times\left[\frac{1}{g''_1}+\frac{1}{g''_2}\right]+2f(-k'',k^*)\delta_{-k''+k^*,k'}a_0(k'',0)\bar{a}_0(k'',0)\left[\frac{1}{g'_1}+\frac{1}{g'_2}\right]\right]$$

$$=-2\omega_0^2\beta(k^*)a_0(k^*,0)e^{i\beta(k^*)\tau_2}$$

$$(9.20)$$

$$\bar{a}_0(k^*,0)e^{-i\beta(k^*)\tau_2}\left(\frac{1}{m\omega_n^2}\right)^2\sum_{k'}\sum_{k''}$$

$$f(k',k'')\delta_{k'+k'',k^*}\left[2f(-k',k^*)\delta_{-k'+k^*,k''}a_0(k',0)\bar{a}_0(k',0)\right.$$

$$\times\left[\frac{1}{g''_1}+\frac{1}{g''_2}\right]+2f(-k'',k^*)\delta_{-k''+k^*,k'}a_0(k'',0)\bar{a}_0(k'',0)\left[\frac{1}{g'_1}+\frac{1}{g'_2}\right]\right]$$

$$=-2\omega_0^2\beta(k^*)\bar{a}_0(k^*,0)e^{-i\beta(k^*)\tau_2}$$

$$(9.21)$$

From (9.20) and (9.21), the same expression for β^* results [(9.22)]:

$$\beta(k^*)=-\frac{1}{2\omega_0^*}\left(\frac{1}{m\omega_n^2}\right)^2\sum_{k'}\sum_{k''}$$

$$f(k',k'')\delta_{k'+k'',k^*}\left[2f(-k',k^*)\delta_{-k'+k^*,k''}a_0(k',0)\bar{a}_0(k',0)\right.$$

$$\times\left[\frac{1}{g''_1}+\frac{1}{g''_2}\right]+2f(-k'',k^*)\delta_{-k''+k^*,k'}a_0(k'',0)\bar{a}_0(k'',0)\left[\frac{1}{g'_1}+\frac{1}{g'_2}\right]\right]$$

$$(9.22)$$

Recall that φ appears in the terms containing g''_1, g''_2, g'_1, g'_2 The limit of (9.22) is taken as $\varphi \to 0$. The following definition is utilized [17]:

$$\lim_{\theta\to 0}\frac{1}{(x\pm i\theta)}=\left(\frac{1}{x}\right)_{pp}\mp i\pi\delta(x)$$

where pp denotes principle part. The real and imaginary parts of (9.22) are shown as (9.23) and (9.24), respectively.

$$\mathrm{Re}(\beta^*) = \Delta_{k^*} = \frac{-64}{\omega_0^*}\left(\frac{1}{m\omega_n^2}\right)^2 \sum_{k'}\sum_{k''} \sin^2\left(\frac{k'a}{2}\right)\sin^2\left(\frac{k''a}{2}\right)\sin^2\left(\frac{k^*a}{2}\right)$$

$$\left\{a_0'\bar{a}'_0\left[\left(\frac{1}{\omega_0''^2 - (\omega_0^* + \omega_0')^2}\right)_{pp} + \left(\frac{1}{\omega_0''^2 - (\omega_0^* - \omega_0')^2}\right)_{pp}\right]\right.$$

$$\left. + a_0''\bar{a}''_0\left[\left(\frac{1}{\omega_0'^2 - (\omega_0^* + \omega_0'')^2}\right)_{pp} + \left(\frac{1}{\omega_0'^2 - (\omega_0^* - \omega_0'')^2}\right)_{pp}\right]\right\}$$

$$(9.23)$$

$$\mathrm{Im}(\beta^*) = \Gamma_{k^*} = \frac{32\pi}{\omega_0^*}\left(\frac{1}{m\omega_n^2}\right)^2 \sum_{k'}\sum_{k''} \sin^2\left(\frac{k'a}{2}\right)\sin^2\left(\frac{k''a}{2}\right)\sin^2\left(\frac{k^*a}{2}\right)$$

$$\times \left\{\delta(\omega_0^* + \omega_0' + \omega_0'')\left[\frac{a_0'\bar{a}'_0}{\omega_0''} + \frac{a_0''\bar{a}''_0}{\omega_0'}\right] - \delta(\omega_0^* + \omega_0' - \omega_0'')\right.$$

$$\times \left[\frac{a_0'\bar{a}'_0}{\omega_0''} - \frac{a_0''\bar{a}''_0}{\omega_0'}\right] + \delta(\omega_0^* - \omega_0' + \omega_0'')\left[\frac{a_0'\bar{a}'_0}{\omega_0''} - \frac{a_0''\bar{a}''_0}{\omega_0'}\right]$$

$$\left. - \delta(\omega_0^* - \omega_0' - \omega_0'')\left[\frac{a_0'\bar{a}'_0}{\omega_0''} + \frac{a_0''\bar{a}''_0}{\omega_0'}\right]\right\}$$

$$(9.24)$$

In the above expressions for the real and imaginary parts of β^*

$$a_0'\bar{a}'_0 = a_0(k',0)\bar{a}_0(k',0)$$

$$a_0''\bar{a}'' = a_0(k'',0)\bar{a}_0(k'',0)$$

From here, the general solution to the ε^0-equation [(9.14)] is considered with the new found results for $a_0(k^*,\tau_2)$ and $\bar{a}_0(k^*,\tau_2)$:

$$a_0(k^*,\tau_2) = a_0(k^*,0)e^{i\beta(k^*)\tau_2}$$

$$\bar{a}_0(k^*,\tau_2) = \bar{a}_0(k^*,0)e^{-i\beta(k^*)\tau_2}$$

Equation (9.14) is written as follows:

$$A_0(k^*,\tau_0,\tau_2) = a_0(k^*,\tau_2)e^{i\omega_0^*\tau_0} + \bar{a}_0(k^*,\tau_2)e^{-i\omega_0^*\tau_0}$$

Utilizing the new found results for $a_0(k^*,\tau_2)$ and $\bar{a}_0(k^*,\tau_2)$, one arrives at the following expression:

$$A_0(k^*,\tau_0,\tau_2) = a_0(k^*,0)e^{i\left(\omega_0^*\tau_0 + \beta(k^*)\tau_2\right)} + \bar{a}_0(k^*,0)e^{-i\left(\omega_0^*\tau_0 + \beta(k^*)\tau_2\right)}$$

Writing the above expression strictly in terms of τ_0, where $\tau_2 = \varepsilon^2 \tau_0$ gives the following representation for $A_0(k^*, \tau_0)$

$$A_0(k^*, \tau_0) = a_0(k^*, 0)e^{i\left(\omega_0^* \tau_0 + \varepsilon^2 \beta(k^*) \tau_0\right)} + \bar{a}_0(k^*, 0)e^{-i\left(\omega_0^* \tau_0 + \varepsilon^2 \beta(k^*) \tau_0\right)}$$

β^* is expressed in terms of its real and imaginary parts to yield the final representation for $A_0(k^*, \tau_0)$:

$$\beta(k^*) = \Delta_{k^*} + i\Gamma_{k^*}$$

$$A_0(k^*, \tau_0) = a_0(k^*, 0)e^{i\left(\omega_0^* \tau_0 + \varepsilon^2 (\Delta_{k^*} + i\Gamma_{k^*}) \tau_0\right)} + \bar{a}_0(k^*, 0)e^{-i\left(\omega_0^* \tau_0 + \varepsilon^2 (\Delta_{k^*} + i\Gamma_{k^*}) \tau_0\right)}$$

$$A_0(k^*, \tau_0) = a_0(k^*, 0)e^{i\left(\left(\omega_0^* + \varepsilon^2 \Delta_{k^*}\right) \tau_0\right)} e^{-\varepsilon^2 \Gamma_{k^*} \tau_0}$$

$$+ \bar{a}_0(k^*, 0)e^{-i\left(\left(\omega_0^* + \varepsilon^2 \Delta_{k^*}\right) \tau_0\right)} e^{\varepsilon^2 \Gamma_{k^*} \tau_0} \tag{9.25}$$

Three-wave interaction leads therefore to an additional frequency shift proportional to the square of the strength of the nonlinearity. Moreover, three-wave interaction leads to a damping of each wave, that is, a finite lifetime. This result is the classical mechanics equivalent of that reported within the framework of quantum mechanics [11–13, 17].

9.2.2   Molecular Dynamics Simulation and Spectral Energy Density Approach

In this section we shed additional light on the three phonon scattering processes in one-dimensional anharmonic crystals using the numerical method of MD. MD is a simulation technique for computing the thermodynamic as well as kinetic properties of a classical many-body system [18]. Classical MD methods consist of solving numerically Newton's equations of motion of a collection of N interacting particles or atoms. The most critical component of an MD simulation is the interatomic potential from which interatomic forces may be derived. The equation of motion of each individual atom is solved numerically in time to obtain the trajectories of the system, namely, the time evolution of the positions and momenta of every particle. In some systems the computational task of solving the equations of motion scales at best linearly with the number of particles, N, and more generally as N^2. Periodic boundary conditions (PBC) are often used to reduce the computational problem size. PBC consist of repeating periodically in all directions of space a "small" simulation cell. One allows interaction between the N atoms within the simulation cell and also between atoms inside the simulation cell and atoms in the periodically repeated "image" cells. Interactions are cut-off to less than half the minimum characteristic length of the simulation cell to avoid spurious effects such as interaction of an atom with its own image. This method effectively reduces the effects that may be associated with surfaces associated with a finite size system. However, while trying to mimic the behavior of an infinite system, the simulated system still possesses the characteristics of a finite system. For instance, the finiteness of an MD system with PBC leads to a discretization of the phonon modes and a suppression of the modes with wavelength longer than the simulation cell length. This is easily seen by considering a 1D monoatomic system composed of N atoms interacting via a nearest neighbor harmonic (or anharmonic) potential. In this case, imposing PBC leads to atom N interacting with atom 1 thus forming a ring. Modes with wavelengths exceeding the length $L = Na$, where a is the interatomic spacing, are not compatible with the constraint of the ring geometry and cannot be supported by that structure. The finite number of modes will also impact the number of three phonon interactions that may take place in a finite simulation cell. The discrete phonon modes may not allow the requirement of frequency conservation. These points will be illustrated with numerical simulations of the 1D anharmonic monoatomic crystal.

For the present discussion, the equation of motion [(9.1)] for a toy system is integrated by MD techniques with PBC using the velocity Verlet algorithm under the microcanonical ensemble (constant energy)[18]. This scheme ensures that energy is conserved within 0.5%. Harmonic MD simulations of the 1D monoatomic crystal utilize $\beta = 1.0$ N/m and $\varepsilon = 0.0$ N/m^2 whereas anharmonic simulations utilize $\beta = 1.0$ N/m and $\varepsilon = [0.9-3.7]$ N/m^2. The 1D crystal consists of a chain of 1.0 kg masses spaced periodically 1.0 meter apart. These parameters can be easily scaled down to represent an atomic system. To initiate a simulation, every mass in the MD simulation cell is randomly displaced from its equilibrium position. The maximum value in which a mass can be displaced is constrained such that instabilities do not emerge in the potential energy function. MD simulations are run for 2^{21} time steps with a timestep of 0.01 s. For post-processing spectral energy density (SED) calculations, velocity data is collected for each mass in the simulation cell over the entire simulation time.

The SED method is a technique for predicting phonon dispersion relations and lifetimes from the atomic velocities of the particles in a crystal generated by classical MD [19]. The SED method offers a comprehensive description of phonon properties because individual phonon modes can be isolated for analysis and is computationally affordable for the systems that will be examined in this section. Formally, the expression for SED is written as follows:

$$\Phi(\vec{k}, \omega) = \frac{1}{4\pi\tau_0 N} \sum_{\alpha} \sum_{b}^{B} m_b \left| \int_0^{\tau_0} \sum_{n_{x,y,z}}^{N} v_\alpha \left(\frac{n_{x,y,z}}{b} ; t \right) \times e^{(i\vec{k}\cdot\vec{r_0} - i\omega t)} dt \right|^2$$

where τ_0 represents the length of time over which velocity data is collected from a given MD simulation, N is the total number of unit cells represented in the MD simulation, and $v_\alpha \left(\frac{n_{x,y,z}}{b} ; t \right)$ represents the velocity of atom b (of mass m_b in unit cell $n_{x,y,z}$) in the α-direction. For a specified wave vector $(\vec{k})$, the spectrum relating

SED to frequency is found by adding the square of the absolute value of the Fourier transform of the discrete temporal signal $f(t) = \sum_{n_{x,y,z}}^{N_T} v_\alpha \left(\frac{n_{x,y,z}}{b} ; t \right) \times e^{(i\vec{k}\cdot\vec{r_0})}$ for every $[\alpha, b]$ pair. A SED value represents the average kinetic energy per unit cell as a function of wave vector and frequency. A peak in the spectrum relating SED to frequency signifies a vibrational eigenmode for wave vector $(\vec{k})$. The shape of the frequency spread for eigenmode $(\vec{k})$ is represented with the Lorentzian function:

$$\Phi(\vec{k}, \omega) = \frac{I}{1 + [(\omega - \omega_c)/\gamma]^2}$$

where I is the peak magnitude, ω_c is the frequency at the center of the peak, and γ is the half-width at half-maximum. The lifetime for phonon mode $(\vec{k})$ is defined as $\tau = 1/2\gamma$[19]. Nondegenerate wave vector modes are dependent on the size of the MD simulation cell and are written as follows: $k_i = 2\pi n_i/aN_i$, where a is the lattice constant, N_i is the total number of unit cells in the i-direction, and n_i is an integer ranging from $-N_i+1$ to N_i. The robust nature of the SED method is used to quantify specific phonon modes in several configurations of the 1D anharmonic crystal in the following subsections.

9.2.3   One-Dimensional Anharmonic Monoatomic Crystal

To begin with, the band structure generated by the SED method is shown for the 1D harmonic monoatomic crystal (Fig. 9.2). Figure 9.2 shows contours of constant SED over the wave vector-frequency plane.

There are 101 discrete, nondegenerate wave vectors resolved between the center of the irreducible Brillouin Zone and the zone edge at $k = \pi/a$. In the band structure, there is a nearly linear region that accounts for the propagative characteristics of long wavelength excitations in the 1D harmonic crystal. At larger wave vector values, a departure from the linear behavior is apparent and the phase velocity of propagative phonon modes is markedly different from the group velocity. This is similar to the expected dispersion behavior of the infinite monoatomic harmonic crystal. At the edge of the irreducible Brillouin zone, a SED-frequency plot is reported. A peak in the spectrum shows this vibrational mode contributing significantly to the average kinetic energy per unit cell. A Lorentzian function is fit to this peak and shows a finite value for half-width at half-maximum (γ) because the fast Fourier transform scheme used in the SED calculation involves a signal sampled over a finite time window. This value for half-width at half-maximum is subsequently used as a lower bound for the error on lifetime estimated with the SED method. This error amounts to one interval in the discrete frequency scale. The band structure of the harmonic system is highlighted in the long wavelength regime; Fig. 9.3 zooms in on a region of the dispersion curve near $k = \pi/10a$.

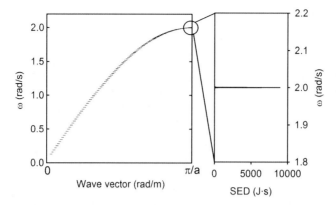

Fig. 9.2 (*Left*) Band Structure of 1D harmonic monoatomic crystal. (*Right*) SED-frequency plot showing wave vector mode $k = \pi/a$

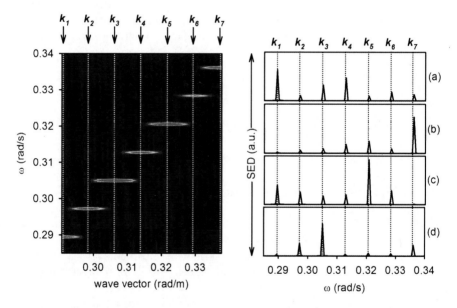

Fig. 9.3 (*Left*) Band Structure (constant SED contours) for 1D harmonic monoatomic crystal near $k = \pi/10a$. (*Right*) SED-frequency plots for four MD simulations differing in their initial random configurations

In Fig. 9.3 on the right hand side, four SED-frequency plots are shown (plots a–d). Each plot represents a different MD simulation of the 1D harmonic monoatomic crystal. Each MD simulation begins with a random starting configuration for atomic displacements in the 1D crystal. It is observable from these four plots that for a given wave vector, the SED takes on different values. This is due to the fact that for a harmonic crystal, energy contained within a particular mode cannot be passed to

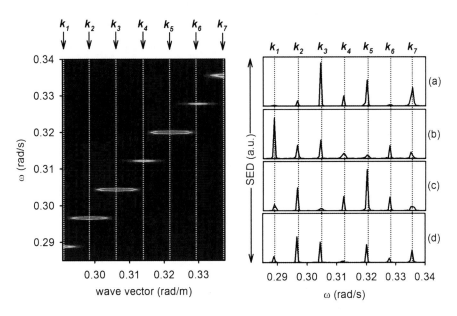

Fig. 9.4 (*Left*) Band Structure (constant SED contours) for 1D anharmonic monoatomic crystal near $k = \pi/10a$. (*Right*) SED-frequency plots for four MD simulations differing in their initial random configurations

other modes of vibration. This highlights the sensitivity of the vibrational modes of the harmonic crystal on the initial configuration. Consequently, to obtain a nonbiased band structure, multiple MD simulations must be run such that an average can be taken of the different SED values for each discrete, nondegenerate wave vector mode. An average of plots (a–d) is shown on the left hand side of Fig. 9.3 with the color of the contours signifying SED intensity. A Lorentzian function is fit to each of the peaks in the left hand figure and shows the same value for half-width at half-maximum as that calculated in Fig. 9.2. For comparison, the band structure of the 1D anharmonic monoatomic crystal near $k = \pi/10a$ is shown in Fig. 9.4. Here the parameter characterizing the degree of anharmonicity in the 1D crystal is $\varepsilon = 3.0$ (see Fig. 9.1b).

Similar to Fig. 9.3, the four plots on the right hand side of Fig. 9.4 represent SED-frequency plots generated from four different MD simulations. The SED intensity for a given mode varies from simulation to simulation, which indicates that energy does not easily exchange between modes of vibration in the 1D anharmonic crystal. In contrast, though, there are some peaks in the SED-frequency spectra that show slightly larger values for half-width at half-maximum. However, it is critical that averages be taken for SED data extracted from several MD simulations such that an accurate quantification of phonon lifetime can be realized. The contour map on the left hand side of Fig. 9.4 represents an average over plots (a–d). Lorentzian functions are fit to the peaks in this figure. The half-width at half-maximum for all peaks is found to be comparable to the harmonic case. With a

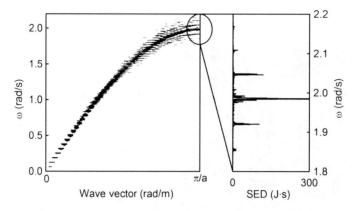

Fig. 9.5 (*Left*) Band Structure for 1D anharmonic monoatomic crystal. (*Right*) SED-frequency plot showing wave vector mode $k = \pi/a$. $\varepsilon = 3.0$ and initial random displacement does not exceed 10 % of a

random initial displacement of the masses of at most 10 % of the lattice spacing "a," the total energy of the anharmonic system is only 1.3 % higher than that of the harmonic system. Under this condition, the system can be considered to be weakly anharmonic and second order perturbation theory is applicable. In other words, the system studied here belongs to the category of weak coupling and is not expected to behave like the Fermi-Pasta-Ulam model where strong nonlinearity leads to persistent recurring vibrational modes[20]. Considering the final expression for $A_0(k^*, \tau_0)$ in Sect. 9.2.1 (9.25), which represents the 0th order term in the asymptotic expansion of $A(k^*, \tau)$ describing three-wave interactions, Γ_{k^*} (9.24) corresponds to a decay constant for mode k^*. Half-width at half-maximum calculations of peaks in SED-frequency spectra embody Γ_{k^*}. In the long wavelength regime, Γ_{k^*} is small because of squared sinusoidal terms inside the double summation over k' and k''. Accordingly, one should not expect large values for half-width at half-maximum in the long wavelength limit. The complete band structure for the 1D anharmonic monoatomic crystal is shown in Fig. 9.5. The band structure is generated from SED averages taken from four MD simulations.

In Fig. 9.5, it seems that each nondegenerate wave vector is associated with multiple eigenfrequencies due to the fact that multiple peaks appear in the SED. At the edge of the irreducible Brillouin zone, an intense central peak is seen along with multiple, less intense symmetrical satellite peaks. These satellite peaks emerge when the anharmonicity of the system is adequately sampled (i.e., large amplitudes of vibration). Equation (9.17) of Sect. 9.2.1 is utilized to explain the appearance of these satellite peaks. This equation represents the 1st order term in the asymptotic expansion of $A(k^*, \tau)$ describing three-wave interactions. Inside the double summation over (k', k'') in (9.17), conservation of wave vectors is imposed: $\delta_{k'+k'',k^*} \rightarrow k' + k'' = k^*$. If the mode of interest is $k^* = \pi/a$, then conservation of wave vector can be satisfied by adding nondegenerate wave vector pairs that yield k^*. With N = 400, nondegenerate wave vectors are limited to the following: $k_i = \frac{n_i}{400} \cdot \frac{2\pi}{a}$. If only wave

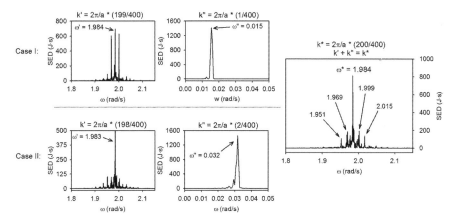

Fig. 9.6 (*Top, left*) SED-frequency plots for wave vector modes k and k corresponding to Case I. (*Bottom, left*) SED-frequency plots for wave vector modes k and k corresponding to Case II. (*Right*) SED-frequency plot corresponding to $k^* = \pi/a$. For Cases I and II, wave vectors k and k satisfy wave vector conservation for mode k^*. The frequencies of modes k and k add (or subtract) to yield near-resonance peaks near $\omega^*(k^*)$. Notice the frequency scale difference between short and long wavelength modes

vectors contained between the center of the irreducible Brillouin zone and the zone edge are considered, then n_i ranges from 0 to 200. As a first example, to satisfy wave vector conservation, consider two wave vectors: (1) the first nondegenerate wave vector before the zone edge at $(k = \pi/a)$ and (2) the first nondegenerate wave vector after the center of the irreducible Brillouin zone at $(k = 0)$. This pair of wave vectors is shown as Case I and satisfies wave vector conservation: (Case I) $k' = \frac{199}{400} \cdot \frac{2\pi}{a}$, $k'' = \frac{1}{400} \cdot \frac{2\pi}{a}$, $k^* = \frac{200}{400} \cdot \frac{2\pi}{a}$.

As a second example, consider (1) the second nondegenerate wave vector before the zone edge at $(k = \pi/a)$ and (2) the second nondegenerate wave vector after the center of the irreducible Brillouin zone at $(k = 0)$. This pair of wave vectors is defined as Case II and satisfies wave vector conservation: (Case II) $k' = \frac{198}{400} \cdot \frac{2\pi}{a}$, $k'' = \frac{2}{400} \cdot \frac{2\pi}{a}$, $k^* = \frac{200}{400} \cdot \frac{2\pi}{a}$. We note that both cases do not conserve frequency. In both cases, since the dispersion relationship for the 1D anharmonic monoatomic crystal is not strictly linear, the frequency of mode k' plus (or minus) the frequency of mode k'' will not exactly equal the frequency of mode k^*. Instead, the addition (or subtraction) of the frequencies associated with modes k' and k'' will be slightly greater than (or less than) the frequency of mode k^*. This forces the denominator of the pre-exponential factors in (9.17) to become small, thereby contributing to a large value of $A_1(k^*, \tau_0, \tau_2)$. The presence of nonzero $A_1(k^*, \tau_0, \tau_2)$ indicates that discrete, near-resonance modes are initiated for short wavelength phonons (k') interacting with long wavelength phonons (k''). On the left hand side of Fig. 9.6, we show nondegenerate wave vector modes k' and k'' corresponding to Case I (top) and Case II (bottom). On the right hand side, Fig. 9.6 shows the modes at $k^* = \pi/a$.

In this image the satellite peaks coincide with discrete, near-resonance modes. The central peak frequencies of modes k' and k'' add (or subtract) to yield satellite peaks to the central peak for $k^* = \pi/a$. The primary satellite peaks at 1.999 and 1.969 rad/s come from Case I. The secondary satellite peaks at 2.015 and 1.951 rad/s come from Case II. Tertiary, quaternary, and other higher order satellite peaks exist and are revealed if the scale on the right hand SED plot were adjusted. The magnitude of the satellite peaks depends upon the "distance" from the central peak at $k = \pi/a$ in accordance with their near resonant character. This distance depends upon the size of the MD simulation. For an MD simulation with $N = 100$ atoms, there are 51 discrete, nondegenerate wave vector modes available between the center of the irreducible Brillouin zone and the zone edge. For $N = 1,000$ atoms, there are 501 available modes. The resolution in wave vector-space is finer for larger MD systems as is the resolution in frequency-space. Higher frequency resolution results in smaller spacing between satellite peaks. This is shown in Fig. 9.7. As the number of atoms (N) increases, the satellite peaks congregate around the central peak and increase in relative amplitude. In the limit of an infinite system all satellite peaks merge into the central peak.

For a phonon mode to decay, wave vector and frequency conservation rules must be satisfied. For short wavelength phonon modes, these constraints are pathologically difficult to satisfy because the monoatomic dispersion curve is not linear. The central frequency peaks in Fig. 9.7 represent the resonance mode of wave vector $k = \pi/a$. The satellite peaks in Fig. 9.7 represent frequency-nonconserving near-resonance modes spawned from nonlinear wave interactions between short wavelength phonons and long wavelength phonons. The lifetime of phonon mode $k = \pi/a$ comes from fitting a Lorentzian function to the central peak. As Fig. 9.7 shows, the half-width at half-maximum for phonon mode $k = \pi/a$ is rather insensitive to the number of atoms in the MD simulation cell. It is found that the half-width at half-maximum for $k = \pi/a$ is the same order of magnitude as the error estimate found from the harmonic case in Fig. 9.2. As a result, lifetime of high-frequency phonon modes in the anharmonic monoatomic crystal is inherently long because wave vector and frequency conservation constraints cannot be satisfied.

In comparing the anharmonic band structure with the harmonic band structure at $(k = \pi/a)$, there is an obvious shift in frequency of the central peak. The perturbation analysis of the single-wave dispersion has shown that the anharmonic dispersion curve is frequency-shifted (with respect to the harmonic dispersion curve) by a quantity that has quadratic dependence on the strength of the nonlinearity parameter ε. Fig. 9.8 shows a plot mapping the frequency shift relative to the harmonic system for several values of ε for a MD simulation cell consisting of $N = 200$ atoms. In Fig. 9.8, three different curves are rendered. Each curve represents a different magnitude for the initial random displacement imposed upon the masses in the 1D crystal in terms of percentage lattice spacing. The magnitude of the initial displacement controls the amplitude of the phonon modes. For triangles, the maximum value a mass can be displaced is 10 % of the lattice spacing. For squares

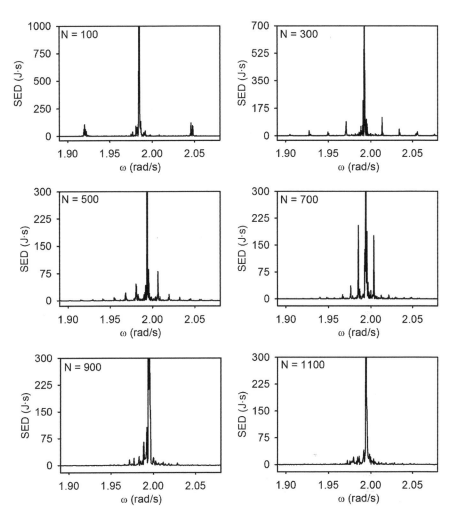

Fig. 9.7 SED-frequency plots for 1D anharmonic monoatomic crystal at $k = \pi/a$ for MD systems of varying sizes. The parameter characterizing the degree of anharmonicity in the 1D crystal is $\varepsilon = 3.0$

and circles, displacement values are 5 % and 2 %, respectively. Quadratic dependence is observed for values of ε ranging from 0.0 to 3.7. Beyond $\varepsilon = 3.7$, the potential energy function becomes completely unstable.

Analysis of the weakly anharmonic 1D monoatomic crystal has shown that the lifetime of phonon modes is not significantly affected by nonlinear interaction forces because it is pathologically difficult to satisfy the conditions for frequency and wave vector conservation. On the contrary, there exist conditions between short wavelength phonons and long wavelength phonons whereby near-resonance peaks emerge in plots of SED-frequency spectra. Satellite peaks materialize when the

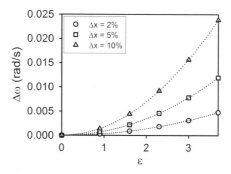

Fig. 9.8 Frequency-shift evaluated at $k = \pi/a$ for 1D anharmonic crystal relative to harmonic case. *Symbols* represent different magnitudes for the maximum initial random displacement imposed upon the masses in the 1D crystal in terms of percentage of the lattice spacing. *Circle*, *square*, and *triangle symbols* represent small, intermediate, and large initial displacements, respectively

anharmonicity of the system is adequately sampled. Lastly, nonlinear interaction forces lead to amplitude-dependent frequency shifts.

9.2.4 Anharmonic One-Dimensional Superlattices

In this section, the insight gained from analysis of the 1D harmonic and anharmonic crystals is extended to a series of superlattice configurations. A characteristic feature offered by periodic media is folded phononic band structures. Band-folding allows the conditions for wave vector and frequency conservation to be easily satisfied thereby greatly impacting three phonon processes because a greater number of phonon mode decay channels are available. Three direct consequences of band folding are (1) modulated eigenfrequencies for vibrational modes, (2) decreased phonon mode group velocities, and (3) altered phonon mode lifetimes. The superlattice configurations considered in this section do not possess the ability to boundary-scatter phonons because the potential describing the inter-action between particles of differing mass is identical to the potential between particles of the same mass. Accordingly, the discussion of phonon mode lifetime is limited to coherent, band-folding effects. The main objective in this section is to illustrate the role superlattice periodicity plays in modulating eigenfrequencies and phonon mode lifetimes at a constant filling fraction. For all superlattices considered, the total number of atoms simulated with MD is $N = 800$. Every plot presented represents an average over a minimum of five unique MD simulations with randomly generated initial conditions. For superlattice unit cells, the mass of the black atom amounts to 50 % of that of the white atom.

To begin with, consider the 1D anharmonic diatomic crystal (superlattice 1:1) as pictured in Fig. 9.9. In comparison to the 1D anharmonic monoatomic crystal, a single fold in the phononic band structure occurs at wave vector mode $k = \pi/2a$.

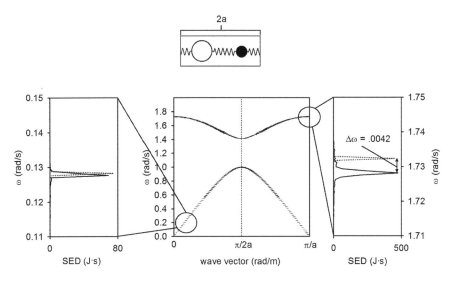

Fig. 9.9 (*Top*) unit cell for diatomic crystal. (*Center*) band structure for 1D anharmonic diatomic crystal. (*Left*) SED-frequency plot at $k = \pi/20a$ with peaks for harmonic (*dashed line*) and anharmonic (*solid line*) cases. (*Right*) SED-frequency plot at $k = \pi/a$ with peaks for harmonic (*dashed line*) and anharmonic (*solid line*) cases

Similar to the monoatomic case, there is a region in the band structure where frequency varies linearly with wave vector. In Fig. 9.9 (right), a SED-frequency plot is highlighted at $k = \pi/a$. This mode, minus a reciprocal space vector, is identical to the mode at the center of the irreducible Brillouin zone. Two peaks are visible in this plot: the dashed line represents a peak for the 1D harmonic diatomic crystal whereas the solid line represents the anharmonic case. There is a noticeable frequency shift as well as a marked difference in peak breadth. Peak broadening is directly associated with satisfaction of conservation of wave vector and frequency conditions; the addition of a second band in the band structure allows these conditions to be met more easily. In the left hand plot of Fig. 9.9, two peaks are apparent. The dashed line corresponds to the diatomic harmonic system and the solid line represents the anharmonic case. There appears to be no significant difference in peak position or width. This result was seen in the anharmonic monoatomic case for long wavelength, low-frequency wave vector modes.

Larger superlattice configurations are now considered to probe the impact superlattice periodicity has on frequency shift and phonon lifetime. In Fig. 9.10, the band structure for a superlattice configuration consisting of a unit cell comprised of two heavy atoms and two light atoms (superlattice 2:2) is displayed. Four distinct bands span the irreducible Brillouin zone. The highest frequency band shows near zero group velocity for all nondegenerate wave vector modes. A SED-frequency plot is highlighted at $k = \pi/2a$. This plot shows information for the harmonic (dashed line) and anharmonic (solid line) cases. Similar to Fig. 9.9, there is a noticeable shift in frequency and the anharmonic peak is significantly broader

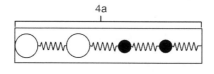

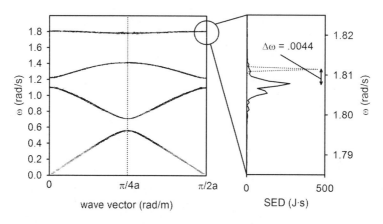

Fig. 9.10 (*Top*) four atom unit cell. (*Left*) anharmonic band structure corresponding to the four atom unit cell. (*Right*) SED-frequency plot at $k = \pi/2a$. *Dashed line* represents the harmonic case whereas the *solid line* represents the anharmonic case

than the harmonic peak. In comparison to the diatomic case, the increased number of bands in the irreducible Brillouin zone allows the conditions for conservation of wave vector and frequency to be met with greater ease. That is, many more three phonon processes satisfy those conditions. Accordingly, the anharmonic peak here shows greater width than the anharmonic peak in the right hand plot of Fig. 9.9.

In Fig. 9.11, the band structure for a superlattice configuration comprised of eight atoms (superlattice 4:4) is displayed. Eight distinct bands span the irreducible Brillouin zone. Of these bands, several show wave vector modes with near zero group velocity. The SED-frequency plot on the right hand side of Fig. 9.11 shows a very wide peak for the anharmonic case. From Figs. 9.9–9.11 it is apparent that anharmonic SED-frequency peaks broaden as the number of bands spanning the irreducible Brillouin zone increases. Accordingly, phonon mode lifetime is significantly reduced by the number of bands available. If the bands spanning the irreducible Brillouin zone are flat bands, then this effect becomes even more pronounced because for a flat band, the conditions for conservation of wave vector can always be satisfied.

With this notion in mind, a final configuration is introduced (Fig. 9.12) with superlattice periodicity 16a (superlattice 8:8). Similar to Figs. 9.9, 9.10, 9.11, Fig. 9.12 shows a frequency shift and peak broadening for the highest frequency anharmonic mode at $k = \pi/8a$.

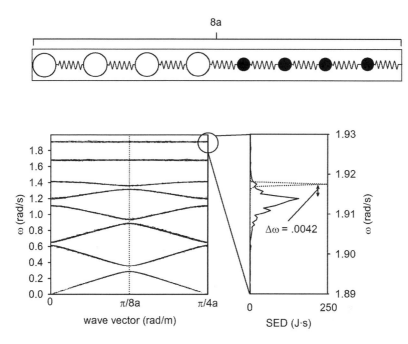

Fig. 9.11 (*Top*) eight atom unit cell. (*Left*) anharmonic band structure corresponding to the eight atom unit cell. (*Right*) SED-frequency plot at $k = \pi/4a$. *Dashed line* represents the harmonic case whereas the *solid line* represents the anharmonic case

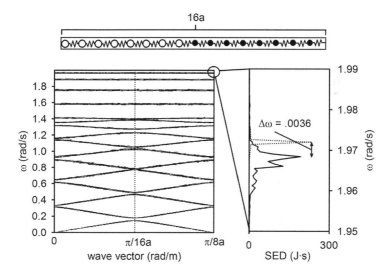

Fig. 9.12 (*Top*) 16 atom unit cell. (*Left*) anharmonic band structure corresponding to the 16 atom unit cell. (*Right*) SED-frequency plot at $k = \pi/8a$. *Dashed line* represents the harmonic case whereas the *solid line* represents the anharmonic case

Fig. 9.13 (a) (*Left* to *right*)
1:1, 2:2, 4:4, and 8:8 SED-
frequency plot with peaks
respectively corresponding to
the superlattice configurations
depicted in Figs. 9.9, 9.10,
9.11, and 9.12. (**b**) Lorentzian
function fits to the SED-
frequency spectra in (**a**).
Lorentzian peaks are labeled
with half-width at half-
maximum values in units of
10^{-6} Hz

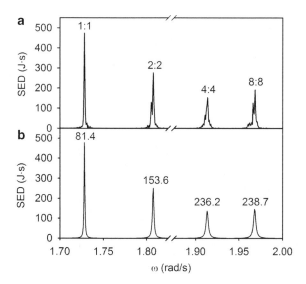

To compare all four superlattice configurations, Fig. 9.13 shows (a) SED-frequency plots and (b) Lorentzian function fits to SED-frequency data corresponding to the mode with highest frequency.

Qualitatively (in Fig. 9.13a) and quantitatively (in Fig. 9.13b), it is observable that as the length of the period decreases a general narrowing occurs for anharmonic SED-frequency peaks. Accordingly, phonon mode lifetime increases when the period of the superlattice is decreased. This observation is consistent with the work presented by Garg et al. [21]. In this study it was shown that the thermal conductivity of a small period Si–Ge superlattice could be higher than that of the constituent materials. In that model, the authors calculated the thermal conductivity using the Boltzmann transport equation within the single mode relaxation time (SMRT) approximation [22]. They modeled the superlattice with harmonic and anharmonic force constants derived from density-functional theory (DFT). In that work the interfaces were treated as perfect(no boundary scattering). It was found that mass mismatch between Si and Ge atoms essentially controls phonon dispersion in the superlattices. The model also considered only three-phonon anharmonic scattering processes. Under these conditions, an increase in lifetime of the transverse acoustic (TA) modes (the majority contributors to thermal conductivity) was responsible for the observed increase in thermal conductivity of the short-period superlattice. This increase in lifetime was explained by the effect of a reduction in periodicity on the band structure of the superlattice that leads to bands that do not allow three phonon scattering events involving TA modes that satisfy the wave vector and frequency conservation rules. Additional lengthening of the phonon lifetime (and increase in thermal conductivity) was further demonstrated by changing the mass mismatch between the constituent materials.

Other authors have addressed the issue of boundary scattering in superlattices; however, these investigations have included interfacial scattering phenomena in addition to coherent band-folding effects. Experimentally and theoretically

[23–25], it has been demonstrated that phonon–boundary collisions play a leading role in decreasing the lifetime of thermal phonons in semiconductor superlattice configurations.

9.3 Phonon Propagation in Two-Dimensional Systems

Having examined the phononic properties of one-dimensional systems, we now turn our attention to two-dimensional systems. In particular, using MD we examine thermal-phonon transport in nanostructured graphene and boron nitride (BN). Both systems are technologically important materials and are characterized by large Debye temperatures; consequently they display distinct harmonic (at low temperatures) and anharmonic regimes, thereby lending themselves well to the study of phonon propagation as a function of temperature and the underlying nanostructure. Specifically, we focus on graphene sheets nanostructured with periodic antidots and boron nitride nanoribbons with aperiodic, spatially asymmetric nanoscale-triangular defects [26]. These contrasting 2D systems, which can also be experimentally synthesized, provide avenues to compare and distinguish the competition between coherent and incoherent phonon scattering and boundary scattering.

The Brenner-Tersoff style potentials [27] are invoked to represent interatomic interactions in graphene and BN as they capture the many-body, covalent nature of the atomic-bonds well, and represent the phonon band structure of the two systems accurately. The Brenner-Tersoff potential includes the anharmonicity of interatomic bonds. The potential parameters for graphene and BN are given in Ref [28] and [29], respectively.

In order to characterize phonon transport in nanostructured graphene and BN, we use relative measures of material parameters such as thermal diffusivity and thermal conductivity as indirect probes to characterize thermal-phonon propagation and lifetimes; it should be noted that it is not our intention to quantify thermal conductivity as well as diffusivity. While in principle the SED method can be invoked for such studies, the mode-by-mode analysis becomes an extremely cumbersome task involving the identification and characterization of the many phonon modes that appear due to the folding of bands within the mini-Brillouin zone corresponding to the periodicity of the phononic crystal.

Strategies to evaluate thermal conductivity and diffusivity include nonequilibrium MD (NEMD) and equilibrium MD (EMD) methods. In the NEMD framework, the thermal conductivity is obtained directly by solving Fourier's law under steady-state conditions, where a temperature gradient is maintained across the modeled material by fixing the temperature of the two ends of the material at different temperatures. Thermal diffusivity is evaluated under transient conditions, by solving for the second-order heat equation. The EMD method is based on the Green-Kubo formulations[29], where NVE (i.e., the microcanonical ensemble) conditions are imposed on the simulated system; based on the equilibrium fluctuations in the heat current (S), the thermal conductivity (κ) is estimated from the time-dependent heat current autocorrelation function (HCAF) as given by (9.26) and (9.27), where V and T are the volume and temperature of the system respectively [30]. The thermal diffusivity (D_T) can then be calculated [see (9.28)], where Cp and ρ are the specific heat and mass density respectively.

$$S(t) = \sum_i E_i v_i + \frac{1}{2} \sum_{i,j} (F_{ij} \cdot v_i) r_{ij} + \frac{1}{6} \sum_{i,j,k} (F_{ijk} \cdot v_i)(r_{ij} + r_{ik}) \qquad (9.26)$$

$$\kappa = \frac{1}{3k_B V T^2} \int_0^\infty \langle S(t) \cdot S(0) \rangle dt \qquad (9.27)$$

$$D_T = \frac{\kappa}{\rho C_P} \qquad (9.28)$$

Here, v_i and E_i represent the velocity and energy of an atom I respectively, while F_{ij} and F_{ijk} represent two body and three-body forces on atom i, due to neighboring atoms j and k.

EMD and NEMD methods have been routinely used to model thermal transport in materials, but care has to be taken in their implementation; in particular, NEMD methods impose extraordinarily large temperature gradients across the material that may not be realized experimentally; further, as discussed by Jiang et al. [31] and as also observed by the authors of this chapter [K. Muralidharan, unpublished work (2011)], the thermostated ends induce spurious vibrational modes characteristic of the size and location of the respective thermostats, which modify the injected heat flux, leading to the possible erroneous estimation of the thermal conductivity. EMD methods, on the other hand, can yield an accurate estimate of the thermal conductivity provided the HCAF is calculated over long time-periods (typically few nanoseconds).

9.3.1 Graphene-Based Phononic Crystals

Here, we report on the thermal-phonon characteristics of antidote graphene comprised of periodic arrangements of holes in a graphene matrix. This system serves as a metaphor for nano-phononic crystal (nano-PC). The lifetime of acoustic and optical phonons is found to be highly sensitive to the filling fraction of the holes in the phononic structure as well as temperature. Results are interpreted in terms of competition between elastic scattering, inelastic phonon–phonon scattering, and boundary scattering.

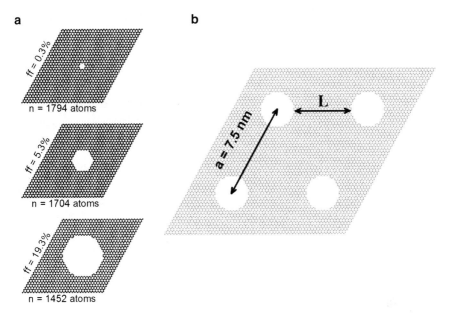

Fig. 9.14 (a) Three examples of unit cells for the nano-PC at different filling fraction. (b) an extended zone representation of the nano-PC with parameter (L), characteristic length, highlighted

9.3.1.1 Simulation Procedure

The nano-PC system of interest is comprised of a graphene matrix with periodically spaced holes. The holes are arranged in a triangular lattice with fixed lattice spacing $a = 7.5$ nm. The radius of the holes varies in size to yield a series of nano-PC unit cells with different filling fractions. Filling fraction (ff) is defined as the atomic fraction of number of atoms removed divided by total number of atoms available. In this study, filling fraction values range from 0.055 % to 20 %. Over this range, EMD calculations using the Green-Kubo method [29] are carried out to extract information on the lifetime of acoustic and optical phonons as a function of temperature (100, 300, 500 K). In Fig. 9.14a, several examples of unit cells for the nano-PC are pictured. A unit cell with ff = 0.0 % represents perfect graphene and contains 1,800 carbon atoms. Two-dimensional PBC are applied with no restrictions in the third dimension. PBC for finite-sized MD simulation cells may constrain some of long-wavelength phonon modes. For every filling fraction, a characteristic length or minimum feature length (L) is identified and is defined as the shortest distance between edges of the holes in the periodic array of the nano-PC. Characteristic length is related to filling fraction through the following relationship: $= a\left(1 - \beta\sqrt{\text{ff}}\right); \beta = \left(\frac{2}{\pi}\right)^{0.5}\sqrt[4]{3}.$

MD simulation cells are initially equilibrated at the temperature of interest by integrating the equations of motion for one million time steps under isothermal conditions using a Berendsen thermostat [32]. Next, the MD system is simulated

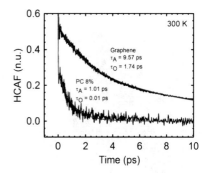

Fig. 9.15 Examples of HCAFs for graphene and nano-PC with ff = 8 %. A sum of two exponential functions is fit to the HCAF to yield estimates for average acoustic and optical phonon lifetime

under constant energy conditions for three million time steps, and the HCAF is calculated over the last two million time steps. In this work, only the last two terms in (9.26) are utilized to calculate the HCAF because these relate strictly to conduction (the convective term (first term) is neglected). Figure 9.15 shows the HCAF at 300 K for (1) perfect graphene and (2) a nano-PC with 8 % filling fraction holes.

The HCAFs exhibit two-stage decay and are, following [30], fit to the sum of two exponential functions of the following form:

$$\frac{S(t) \cdot S(0)}{3} = A_a \, e^{-(t/\tau_a)} + A_o \, e^{-(t/\tau_a)} \tag{9.29}$$

The longer relaxation time is assigned to acoustic modes (τ_a) and the shorter time to optical modes (τ_o). In Fig. 9.15 the average lifetimes for acoustic and optical phonons are also displayed. The decay of the HCAF is extremely rapid in comparison to perfect graphene. The nature of this decay is the subject of the remainder of this section.

The lifetime of a particular phonon mode is well described by Matthiessen's Rule:

$$\frac{1}{\tau} = \frac{1}{\tau_{ph}} + \frac{1}{\tau_e} + \frac{1}{\tau_d} + \frac{1}{\tau_B} \tag{9.30}$$

Here τ represents the total phonon lifetime and τ_{ph}, τ_e, τ_d, and τ_B signify characteristic decay times associated with different types of phonon collision processes, specifically, phonon–phonon, phonon–electron, phonon–defect, and phonon–boundary, respectively. Given the classical nature of the MD simulations, phonon–electron contributions are not included in addition to phonon–defect terms, since the MD simulation-cells are constructed to be defect free. Thus, MD simulation results are interpreted in terms of phonon–phonon and phonon–boundary scattering.

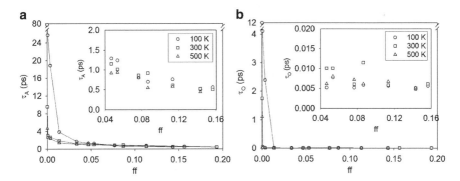

Fig. 9.16 (**a**) average lifetime of acoustic phonons versus filling fraction (100 K, 300 K, 500 K). (**b**) same as (**a**) but for optical phonons. The *insets* are magnifications of the regions of high filling fractions (ff)

9.3.1.2 Phonon–Phonon and Phonon–Boundary Scattering

Figure 9.16 shows plots of average phonon lifetime versus filling fraction of holes for acoustic phonons (a) and optical phonons (b).

For perfect graphene (ff = 0.0 %), the average lifetime of acoustic and optical phonons decreases with increasing temperature. For acoustic phonons, average lifetime decreases from 78.12 ps (100 K) to 9.57 ps (300 K) to 4.69 ps (500 K). For optical phonons, the average lifetime decreases from 11.96 ps (100 K) to 1.74 ps (300 K) to 1.10 ps (500 K). This observation highlights the phonon–phonon collision mechanism embodied in normal and Umklapp phonon processes. The calculated lifetime of optical phonons in perfect graphene at room temperature is consistent with an experimental measurement of 1.2 ps using time-resolved incoherent anti-Stokes Raman scattering [33]. Further, the predicted trend in the estimated acoustic phonon lifetimes matches experimental observations; specifically, using the experimentally measured phonon coherence length in suspended graphene (approximately 800 nm at 300 K [34–37] and 330–400 nm at 400 K [38]) in conjunction with the longitudinal acoustic velocity in graphene (approximately 20,000 m/s [39]), we obtain lifetimes that range between 15 ps (at 400 K) and 40 ps (at 300 K), which compare reasonably well with our predictions.

If a single atom is removed from the MD simulation cell, a nano-PC structure effectively results with filling fraction equal to 0.05 %. Figure 9.16a, b show for all temperatures that the removal of a single atom yields a dramatic decrease in average phonon lifetime. At 100 K the average lifetime of acoustic phonons decreases by 68 %. For 300 K and 500 K, the observed decreases in average lifetime are 63 % and 49 % respectively. This abrupt decrease in phonon lifetime can be attributed to two possible mechanisms: (1) The removal of a single atom offers a superlattice configuration whereby the phononic band structure associated with perfect graphene is folded multiple times thus allowing many more phonon–phonon scattering processes that meet the conditions for conservation of wave vector and frequency, a prerequisite for phonon mode decay; (2) The removal of a single atom creates a boundary/surface that propagating phonons can collide with. Isolating one mechanism from the other is inherently difficult because the two are both present at the same time. However, one may rewrite equation (9.30) in terms of average phonon lifetimes to highlight the dependencies of the different contributions to the total average lifetime:

$$\frac{1}{\tau} = \frac{1}{\tau_{ph}(T,L)} + \frac{1}{\tau_B(L)} \tag{9.31}$$

Here, we have highlighted the dependency of phonon–phonon scattering on temperature (T) as well as the band structure resulting from the periodicity of the structure (L or ff). Boundary scattering depends essentially on the minimum feature length (L) of the structure. As filling fraction increases, for all temperatures, the average lifetime of acoustic phonon modes decreases. For optical phonons, this behavior is less pronounced. For acoustic phonons in the 0.3–3 % filling fraction region, strong temperature dependence suggests that phonon–phonon collisions are the dominant scattering mechanism. The Callaway-Holland model [40, 41] identifies the propensity of a phonon mode (of wave vector k and polarization λ) to undergo normal and Umklapp scattering processes as a function of temperature and frequency:

$$1/\tau_{ph}(k, \lambda, T) = \gamma(k, \lambda)Te^{-\eta/T} \tag{9.32}$$

Here $\gamma(k, \lambda)$ contains the frequency of the specific phonon mode and η is a parameter used to match empirical data. For the purpose of this discussion, (9.32) is adapted by considering average phonon lifetimes by defining $\tau_{ph}(T)$v to represent an average over all polarization branches. We also define a frequency independent average, $\bar{\gamma}$. Equation (9.32) becomes:

$$\frac{1}{\tau_{ph}(T)} = \bar{\gamma}Te^{-\frac{\eta}{T}} \tag{9.33}$$

Figure 9.17a shows a plot of average lifetime (acoustic and optical) versus temperature for perfect graphene. Equation (9.33) is fit to the data points with $\alpha_i = 1/\bar{\gamma}$; $\beta_i = \eta$. This illustrates the temperature dependence of $\tau(T) \sim \tau_{ph}(T)$ in the absence of a periodic array of holes. Figure 9.17b shows a plot of average acoustic lifetime versus temperature for three different filling fractions in the 0.3–3 % range. Similar to Fig. 9.17a, (9.33) is fit to the data points and it is well correlated. This can be interpreted as resulting from the dual dependency of phonon–phonon scattering on temperature and periodicity (through band folding), that is, $\tau(T,L) \sim \tau_{ph}(T,L)$ in the case of the antidot nano-PC structures. This result also implies that $\tau_{ph} < \tau_B$. Therefore, in the 0.3 % - 3 % filling fraction region, phonon–phonon processes appear to be the dominant mechanism behind the lowering of phonon mode lifetimes.

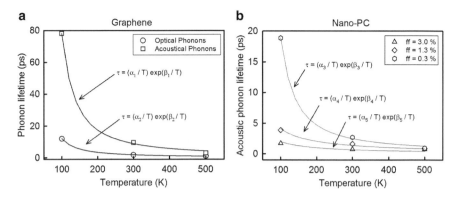

Fig. 9.17 (a) Lifetime versus temperature for acoustic and optical phonons in graphene.(b) Average acoustic phonon lifetime for three different nano-PC filling fractions versus temperature. *Symbols* are calculated values and *solid lines* represent fits using (9.33)

Beyond ff = 3.0 %, the temperature dependency of acoustic phonon lifetime is diminished. The inset in Fig. 9.16a shows, for all temperatures, that average acoustic phonon lifetime follows the same, weak linear trend as filling fraction increases. Temperature dependency of acoustic modes can be diminished if the holes in the graphene matrix have increased in size to a point whereby the characteristic length is such that acoustical phonons have higher probability of getting scattering by the boundary of a hole than with another phonon. This is the case when $\tau_B < \tau_{ph}$ and $\tau \sim \tau_B(L)$. A plot of average acoustical phonon lifetime versus characteristic length (Fig. 9.18) shows for large filling fractions (small L values) the lifetime of acoustical phonons is linearly dependent on L.

The characteristic decay time associated with boundary scattering takes the functional form [40, 41]: $\tau_B = L/v$, where v represents an average speed of sound in graphene. In the small L region of Fig. 9.18 a line is fit where $v = 7000$ m/s (a reasonable value for average speed of sound for acoustic phonons in graphene). In this region, boundary scattering is the dominant scattering mechanism. Beyond $L = 5$ nm, this linear dependence is lost and scattering is attributed to a mix of phonon–boundary and phonon–phonon collisions. As L increases to larger values, the significance of boundary scattering is lost and normal and Umklapp phonon processes dominate.

9.3.2 Phonon Transport in Boron Nitride Nano-Ribbons

Two-dimensional BN structures are isomorphic to their carbon counterparts and capable of demonstrating equally remarkable structure–property relations. Of particular interest are the phonon propagation characteristics in single-layer BN sheets and Boron Nitride nanoribbons (BNNR) containing triangular defects, which have been recently fabricated by Jin et al. [42]. As pointed out by Yang et al. [43], such

Fig. 9.18 Average acoustical phonon lifetime versus characteristic length. Phonon–boundary collisions are the dominant scattering mechanism at low L values. In this region, there is greater probability of phonon-scattering due to the hole-edge than phonon-phonon scattering

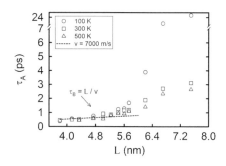

structures open up new avenues for manipulating the thermal properties of defected BNNR, and in particular, the geometric asymmetry of the triangular vacancies/ defects can be exploited to preferentially scatter phonons in BNNR, which can, in turn, lead to spatially dependent thermal properties. A related consequence is the possibility of realizing thermal rectifiers as discussed by Go et al. [44], where it was shown that the thermal conductivity of a material has to be an inseparable function of both space and temperature to exhibit thermal rectification. To examine the interplay between defect-orientation and phonon propagation in BNNR, we employ (1) EMD simulations to correlate the relations between phonon transport and the temporal evolution of spatial HCAF profiles across the simulated system, and (2) a variant of NEMD simulations, where one end of the BNNR is suddenly quenched and held at a fixed temperature; the time taken for the temperature of the rest of the material to equal that of the thermostated end is taken as a measure of the thermal diffusivity and more importantly a measure of transient thermal-phonon characteristics. Note that, in both methods employed, we explicitly avoid estimating heat fluxes, and thereby circumvent problems associated with NEMD as discussed previously.

9.3.2.1 Simulation Procedure

The MD simulations of pristine and defected BNNR employ the Brenner-Tersoff potential as developed by Albe and Moller [29] due to its success in modeling the different hybridization states of BN, an important requirement while modeling defected BNNR. To ensure consistency with experimental observations [42], the arm-chair orientation of BNNR (a-BNNR) was simulated; the length of the simulated a-BNNR was 17.5 nm, while periodic boundary conditions were applied along its 7.1 nm width. Fixed boundary conditions were imposed on the edge atoms (i.e., the thinnest strip consisting of boron and nitrogen atoms at each end). For the defected system, the defect was represented by an equilateral triangle with nitrogen-termination to ensure consistency with experimental observations. The defect orientation is shown in Fig. 9.19, and its dimensions were chosen to be approximately half the BNNR width (corresponding to a filling fraction of 5 %).

Fig. 9.19 Illustration of the pristine and defected BNNR

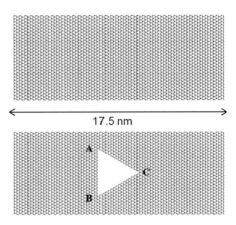

In order to carry out EMD simulations, both pristine as well as defected BNNR were initially equilibrated at 300 K and 900 K respectively, after which NVE simulations were conducted for 7 ns with a time step of 1 fs. The systems were spatially divided into 15 bins along their length to enable the calculations of spatial variations in HCAF, which were obtained over the last 5 ns of each NVE run. Particular attention was paid to the HCAF component along the length of the BNNR, which was primarily used in our data analysis. To ensure better statistics, five different equilibrated starting configurations were used for each case. In the NEMD simulations, the thinnest possible strip of atoms (consisting of equal number of boron and nitrogen), adjacent to the boundary atoms at the opposite ends of the BNNR, were identified to be the thermostated regions which were governed by a Nose-Hoover thermostat [32]. The boundary atoms were not included to avoid edge effects as noted by Jiang et al. [31]. For the 300 K and 900 K systems, the thermostat temperature equaled 150 K and 450 K respectively, and the time for the rest of the unconstrained system to attain the temperature of the thermostated region was calculated when the thermostat was placed at the (1) left and the (2) right edge respectively.

9.3.2.2 Phonon Transport and Rectification

The 300 K spatial variation in HCAF as a function of time for pristine BNNR is given in Fig. 9.20. Interestingly, each spatial-bin is characterized by similar, temporally periodic peaks and valleys, which are systematically displaced with respect to neighboring bins. Since the HCAF is a measure of the material's ability to dissipate thermal fluctuations, and therefore directly related to thermal-phonon energy transport, Fig. 9.20 can be interpreted in terms of phonon propagation. Specifically, the appearance of the first and the second valley in the HCAF for each bin represents phonon-reflection from the nearest and farthest fixed-edge respectively. Clearly, the time-delay between the two valleys is related to the spatial

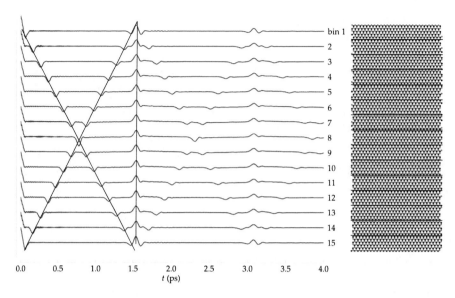

Fig. 9.20 (*Left*)- Time evolution of the spatially resolved HCAF for pristine BNNR at 300 K. Black overlays are an aid to the eye. (*Right*)-Representation of the spatial decomposition of the simulated system into 15 bins

location of the bin. Note that a single reflection from a fixed edge leads to a phase-change as represented by the valley. At approximately 1.5 ps, we see the emergence of a peak for all spatial bins. This corresponds to a 'round-trip' made by the respective phonons, which undergo two reflections (i.e., two phase changes) from either edge; the peak is larger in magnitude than the valley, representing the simultaneous arrival of the two phonons. A similar peak appears at approximately 3 ps, though the magnitude of this peak is reduced as compared to the first peak, implying the role of anharmonicity-induced scattering of phonons that eventually leads to a finite lifetime of phonons as evidenced by the gradual diminishing of the peaks and valleys in the HCAF.

While the spatially decomposed HCAF of pristine BNNR is symmetric (i.e., HCAF of nth bin and (15-n)th bin are similar), this is not observed for the 300 K defected BNNR, as shown in Fig. 9.21.

An inspection of Fig. 9.21 reveals that additional phonon reflection is enabled by the AB-face (see Fig. 9.19) of the triangular defect that is parallel to the BNNR edge, leading to dissimilar HCAF profiles in the two regions that are separated by the triangular vacancy in the defected BNNR.

In particular, consider the first two HCAF valleys/peaks in the bins between the triangle-face AB and the near edge (i.e., bins 1–6). The bins in proximity to the BNNR edge (bins 1–3) are characterized by valleys followed by peaks in HCAF, while the HCAFs in bins closer to the triangle-face AB (bins 4–6) are first described by peaks and subsequently by valleys. This is explained by the fact that the triangle-face AB is not a fixed boundary, and does not lead to a phase change during

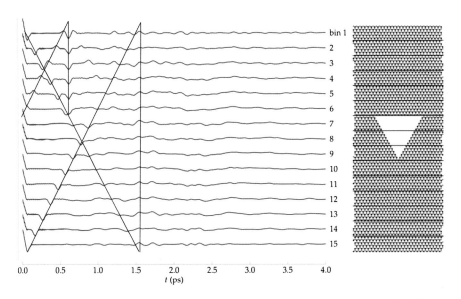

Fig. 9.21 (*Left*)- Time evolution of the spatially resolved HCAF for defected BNNR at 300 K. (*Right*)-Representation of the spatial decomposition of the simulated system into 15 bins

reflection. At approximately 0.7 ps, we see the uniform appearance of valleys for all the bins (9.1–9.6), which is correlated to the simultaneous 'round-trip' arrival of two phonons. Note that the same phenomenon is also observed in the pristine BNNR system at 1.5 ps due to the longer path (almost twice) traversed by the respective phonons. For the region in the defected BNNR between triangle vertex-C and the farther edge, the spatial HCAF profile diverges from that of the other region; all the bins corresponding to this region (9.9–9.15) are characterized by an initial valley (reflection from the farther fixed end), but subsequent features are not well pronounced, a direct consequence of phonon scattering from the sloped edges of the triangle defect, which can be distinguished from the reflection that occurs at the normal AB face. Thus, phonon propagation characteristics in the two regions separated by the geometrically asymmetric triangular defect are indeed different.

Figure 9.22a, b illustrate the HCAF of pristine and defected BNNR at 900 K. A comparison with Figs. 9.20 and 9.21 indicates the role of temperature on the HCAF profile. Clearly, the anharmonic effects become more distinct at the higher temperature, as seen by the absence of higher order HCAF echoes in the respective systems. Thus, by comparing and contrasting the HCAF characteristics of pristine and defected BNNR, one can conclude that geometric asymmetry of the defect leads to distinct spatial- and temperature-dependent thermal-phonon propagation characteristics for the defected BNNR system, indicating the possibility of observing thermal rectification in such systems.

In order to study the transient response of the two systems, the quenching procedure as described earlier was adopted. Figure 9.23a, b illustrates the rate of

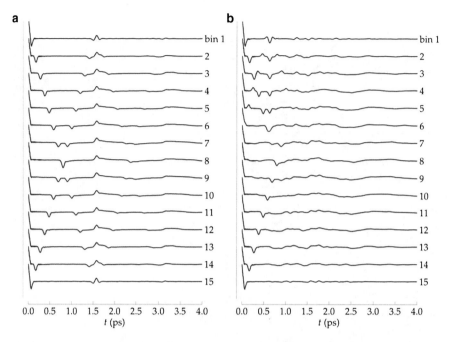

Fig. 9.22 Time evolution of the spatially resolved HCAF for (**a**) pristine and (**b**) defected BNNR at 900 K

temperature change for the pristine BNNR and defected BNNR (both initially at 300 K and quenched to 150 K), when in 'forward' bias (i.e., the thermostated BNNR edge faces the triangle-face AB) and 'reverse' bias (i.e., the thermostated BNNR edge faces the triangle vertex-C). While the pristine BNNR responds identically under both forward and reverse bias, the temperature-time curve do not overlap for the defected BNNR, implying that the thermal diffusivity is position-dependent. Numerical solution of the transient heat equation shows that the reverse-bias apparent thermal diffusivity is higher by a factor of 1.13. A similar result was also observed when the 900 K systems were quenched to 450 K, with the ratio of the reverse-bias to forward-bias thermal diffusivity for defected BNNR equaling 1.07. These results when viewed in conjunction with the HCAF observations clearly indicate that the asymmetric triangular defect plays an important part in the ability of the defected BNNR to respond to external thermal stimuli. Specifically, based on the orientation, specific triangular-faces can impede phonon-energy propagation, thereby allowing defected BNNR systems to exhibit spatially asymmetric thermal transport properties.

Importantly, these results are consistent with past theoretical and experimental investigations, where boundary scattering from arrays of spatially asymmetric triangular holes led to acoustic rectification in the MHz and GHz regimes [45, 46]. An important distinction between these studies and the current work is the explicit inclusion of anharmonic interactions that arise in atomic systems;

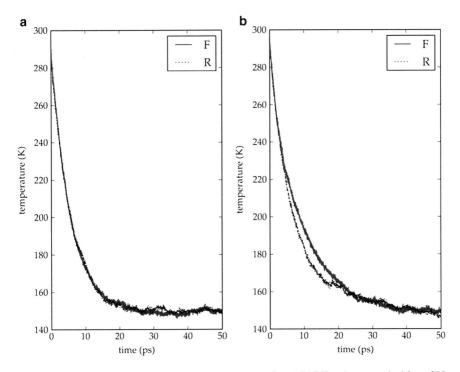

Fig. 9.23 Temperature-time plot of (**a**) pristine and (**b**) defected BNNR, when quenched from 300 to 150 K under forward (F) and reverse (R) bias

nevertheless, in each case it is clear that scattering at the triangular-hole boundary dominates phonon propagation leading to rectification. Other related atomistic investigations include the characterization of interface asperity on the in-plane thermal conductivity of superlattices [47]; here the interface asperity was represented by a series of triangles, and even the in-plane thermal-phonon transport was dictated by the surface roughness (i.e., the size and orientation of interface-triangles) further affirming the effect of boundary scattering on phonon propagation.

Acknowledgments This material is partly based upon work supported by the National Science Foundation under Grant No. 1148936 and Grant No. 0924103.

References

1. P. Hyldgaard, G. D. Mahan,"Phonon Knudsen flow in AlAs/GaAssuperlattices" in *Thermal Conductivity*, vol. 23,(Technomic, Lancaster, PA, 1996)
2. G. Chen, C.L. Tien, X. Wu, J.S. Smith, Thermal diffusivity measurement of GaAs/AlGaAs thin-film structures. J. Heat Transfer **116**, 325 (1994)

3. W.S. Capinski, H. J. Maris, Thermal conductivity of GaAs/AlAssuperlattices. Physica B **219&220**, 699 (1996)
4. E.S. Landry, M.I. Hussein, A.J.H. McGaughey, Complex superlattice unit cell designs for reduced thermal conductivity. Phys. Rev.B **77**, 184302 (2008)
5. A.J.H. McGaughey, M.I. Hussein, E.S. Landry, M. Kaviany, G.M. Hulbert, Phonon band structure and thermal transport correlation in a layered diatomic crystal. Phys. Rev. B **74**, 104304 (2006)
6. T. Gorishnyy, C.K. Ullal, M. Maldovan, G. Fytas, E.L. Thomas, Hypersonicphononiccrystals. Phys. Rev. Lett. **94**, 115501 (2005)
7. J.-N. Gillet, Y. Chalopin, S. Volz, Atomic-scale three-dimensional phononic crystals with a very low thermal conductivity to design crystalline thermoelectric devices. J.Heat Transfer **131**, 043206 (2009)
8. B.L. Davis, M.I. Hussein, Thermal characterization of nanoscalephononic crystals usingsupercell lattice dynamics. AIP Adv. **1**, 041701 (2011)
9. A. Netsch, A. Fleischmann, C. Enss, "Thermal conductivity in glasses with a phononic crystal like structure", PHONONS 2007. J. Phys. Conf. Ser. **92**, 012130 (2007)
10. J.-F. Robillard, K. Muralidharan, J. Bucay, P.A. Deymier, W. Beck, D. Barker, Phononic metamaterials for thermal management: an atomistic computational study. Chin. J. Phys. **49**, 448 (2011)
11. D.C. Wallace, *Thermodynamics of Crystals* (Wiley, New York, 1972)
12. D.C. Wallace, Renormalization and statistical mechanics in many-particle systems. I. Hamiltonian perturbation method. Phys. Rev. **152**, 247 (1966)
13. A.A. Maradudin, A.E. Fein, Scattering of neutrons by an anharmonic crystal. Phys. Rev. **128**, 2589 (1962)
14. R.K. Narisetti, M.J. Leamy, M.J. Ruzzene, A perturbation approach for predicting wave propagation in one-dimensional nonlinear periodic structures. ASME J. Vib. Acoust. **132**, 031001 (2010)
15. K. Manktelow, M.J. Leamy, M. Ruzzene, Multiple scales analysis of wave-wave interactions in a cubically nonlinear monoatomic chain. Nonlinear Dyn. **63**, 193 (2011)
16. N.M. Krylov, N.N. Bogoliubov, *Introduction to Nonlinear Mechanics* trans. by S. Lefshetz (Princeton U.P, Princeton, NJ, 1947)
17. I.C. Khoo, Y.K. Wang, Multiple time scale analysis of an anharmonic crystal. J. Math. Phys. **17**, 222 (1976)
18. J.M. Haile, Molecular dynamics simulation: elementarymethods. (Wiley Inter-Science , 1992)
19. J.A. Thomas, J.E. Turney, R.M. Iutzi, C.H. Amon, A.J.H. McGaughey, Predicting phonon dispersion relations and lifetimes from the spectral energy density. Phys. Rev. B **81**, 081411(R) (2010)
20. G.P. Berman, F.M. Izraileva, The Fermi-Pasta-Ulam problem: fifty years of progress. Chaos **15**, 015104 (2005)
21. J. Garg, N. Bonini, N. Marzani, High thermal conductivity in short-period superlattice. Nano Lett. **11**, 5135 (2011)
22. J.M. Ziman, *Electrons and Phonons* (Oxford University Press, London, 1960)
23. S.M. Lee, D.G. Cahill, R. Vekatasubramanian, Thermal conductivity of Si-Ge superlattices. Appl. Phys. Lett. **70**, 2957 (1997)
24. W.S. Capinski, H.J. Maris, T. Ruf, M. Cardona, K. Ploog, D.S. Latzer, Thermal-conductivity measurements of GaAs/AlAs superlattices using a picoseconds optical pump-and-probe technique. Phys. Rev. B **59**, 8105 (1999)
25. Y. Wang, X. Xu, R. Venkatasubramanian, Reduction in coherent phonon lifetime in Bi2Te3/Sb2Te3 superlattices. Appl. Phys. Lett. **93**, 113114 (2008)
26. K. Muralidharan, R.G. Erdmann, K. runge, P.A. Deymier, Asymmetric energy transport in defected boron nitride nanoribbons: Implications for thermal rectification. AIP Adv. **1**, 041703 (2011)

27. D.W. Brenner, O.A. Shenderova, J.A. Harrison, S.J. Stuart, B. Ni, S.B. Sinnott, A second-generation reactive empirical bond order (REBO) potential energy expression for hydrocarbons. J. Phys. Condens. Matter **14**, 783 (2002)

28. L. Lindsay, D. A. Broido, Optimized Tersoff and Brenner empirical potential parameters for lattice dynamics and phonon thermal transport in carbon nanotubes and graphene. Phys. Rev. B **81**, 205441 (2010)

29. K. Albe, W. Moller, Modelling of boron nitride: atomic scale simulations on thin film growth. Comput. Mater. Sci. **10**, 111 (1998)

30. A.J.H. McGaughey, M. Kaviani, Phonon transport in molecular dynamics simulations: formulation and thermal conductivity prediction. Adv. Heat Transfer **39**, 169 (2006)

31. J.-W. Jiang, J. Chen, J.-S. Wang, B. Li, Edge states induce boundary temperature jump in molecular dynamics simulations of heat conduction. Phys. Rev. B **80**, 052301 (2009)

32. D.C. Rapaport, *The Art of Molecular Dynamics Simulation* (Cambridge University Press , 1995)

33. K. Kang, D. Abdula, D.G. Cahill, M. Shim, Lifetimes of optical phonons in graphene and graphite by time-resolved incoherent anti-Stokes Raman scattering. Phys. Rev. B **81**, 165405 (2010)

34. S. Ghosh, I. Calizo, D. Teweldebrhan, E.P. Pokatilov, D.L. Nika, A.A. Balandin, W. Bao, F. Miao, C.N. Lau, Extremely high thermal conductivity of graphene: prospects for thermal management applications in nanoelectronic circuits. Appl. Phys. Lett. **92**, 151911 (2008)

35. A.A. Balandin, Thermal properties of graphene and nanostructured carbon materials. Nat. Mater. **10**, 569–581 (2011)

36. D.L. Nika, S. Ghosh, E.P. Pokatilov, A.A. Balandin, Lattice thermal conductivity of graphene flakes: comparison with bulk graphene. Appl. Phys. Lett. **94**, 203103 (2009)

37. D.L. Nika, E.P. Pokatilov, A.S. Askerov, A.A. Balandin, Phonon thermal conduction in graphene: role of Umklapp and edge roughness scattering. Phys. Rev. B **79**, 155413 (2009)

38. S. Chen, A.L. Moore, W. Cai, J.W. Suk, J. An, C. Mishra, C. Amos, C.W. Magnuson, J. Kang, L. Shi, R.S. Ruoff, Raman measurements of thermal transport in suspended monolayer graphene of variable sizes in vacuum and gaseous environments. ACS Nano **5**, 321 (2011)

39. H. Suzuura, T. Ando, Phonons and electron–phonon scattering in carbon nanotubes. Phys. Rev. B **65**, 235412 (2002)

40. J. Callaway, Model for lattice thermal conductivity at low temperatures. Phys. Rev. **113**, 1046 (1959)

41. M.G. Holland, Analysis of lattice thermal conductivity. Phys. Rev. **132**, 2461 (1963)

42. C. Jin, F. Lin, K. Suenaga, S. Iijima, Fabrication of a freestanding Boron Nitride single layer and its defect assignments. Phys. Rev. Lett. **102**, 195505 (2009)

43. K. Yang, Y. Chen, Y. Xie, X.L. Wei, T. Ouyang, J. Zhong, Effect of triangle vacancy on thermal transport in boron nitride nanoribbons. Solid State Commun. **151**, 460 (2011)

44. D.B. Go, M. Sen, On the condition for thermal rectification using bulk materials. J. Heat Transfer **132**, 1245021 (2010)

45. R. Krishnan, S. Shirota, Y. Tanaka, N. Nishiguchi, High-efficient acoustic wave rectifier. Solid State Commun. **144**, 194–197 (2007)

46. S. Danworaphong, T.A. Kelf, O. Matsuda, M. Tomoda, Y. Tanaka, N. Nishiguchi, O.B. Wright, Y. Nishijima, K. Ueno, S. Juodkazis, H. Misawa, Real-time imaging of acoustic rectification. Appl. Phys. Lett. **99**, 201910 (2011)

47. A. Rajabpour, S.M. VaezAllaei, Y. Chalopin, F. Kowsary, S. Volz, Tunable superlattice in-plane thermal conductivity based on asperity sharpness at interfaces: Beyond Ziman's model of specularity. J. Appl. Phys. **110**, 113529 (2011)

Chapter 10
Phononic Band Structures and Transmission Coefficients: Methods and Approaches

J.O. Vasseur, Pierre A. Deymier, A. Sukhovich, B. Merheb, A.-C. Hladky-Hennion, and M.I. Hussein

Abstract The purpose of this chapter is first to recall some fundamental notions from the theory of crystalline solids (such as direct lattice, unit cell, reciprocal lattice, vectors of the reciprocal lattice, Brillouin zone, etc.) applied to phononic crystals and second to present the most common theoretical tools that have been developed by several authors to study elastic wave propagation in phononic crystals and acoustic metamaterials. These theoretical tools are the plane wave expansion method, the finite-difference time domain method, the multiple scattering theory, and the finite element method. Furthermore, a model reduction method based on Bloch modal analysis is presented. This method applies on top of any of the numerical methods mentioned above. Its purpose is to significantly reduce the size of the final matrix model and hence enable the computation of the band structure at a very fast rate without any noticeable loss in accuracy. The intention

J.O. Vasseur (✉) • A.-C. Hladky-Hennion
Institut d'Electronique, de Micro-électronique et de Nanotechnologie, UMR CNRS 8520, Cité Scientifique, 59652 Villeneuve d'Ascq Cedex, France
e-mail: Jerome.Vasseur@univ-lille1.fr; anne-christine.hladky@isen.fr

P.A. Deymier • B. Merheb
Department of Materials Science and Engineering, University of Arizona, Tucson, AZ 85721, USA
e-mail: deymier@email.arizona.edu; bassam@merheb.net

A. Sukhovich
Laboratoire Domaines Océaniques, UMR CNRS 6538, UFR Sciences et Techniques, Université de Bretagne Occidentale, Brest, France
e-mail: alexei_suhov@yahoo.co.uk

M.I. Hussein
Department of Aerospace Engineering Sciences, University of Colorado Boulder, Boulder, CO 80309, USA
e-mail: mih@colorado.edu

P.A. Deymier (ed.), *Acoustic Metamaterials and Phononic Crystals*,
Springer Series in Solid-State Sciences 173, DOI 10.1007/978-3-642-31232-8_10,
© Springer-Verlag Berlin Heidelberg 2013

in this chapter is to give to the reader the basic elements necessary for the development of his/her own calculation codes. The chapter does not contain all the details of the numerical methods, and the reader is advised to refer to the large bibliography already devoted to this topic.

10.1 Periodic Structures and Their Properties

Solids possessing crystalline structure are periodic arrays of atoms. The starting point in the description of the symmetry of any periodic arrangement is the concept of a *Bravais* lattice. A Bravais lattice is defined as an infinite array of discrete points with such an arrangement and orientation that it appears exactly the same from whichever of its points the array is viewed [1]. Mathematically, a Bravais lattice in three dimensions is defined as a collection of points with position vectors $\vec{R}$ of the form

$$\vec{R} = n\vec{a}_1 + m\vec{a}_2 + k\vec{a}_3 \tag{10.1}$$

where $\vec{a}_1, \vec{a}_2, \vec{a}_3$ are any three vectors not all in the same plane and n, m, k are any three integer numbers. Vectors $\vec{a}_1, \vec{a}_2, \vec{a}_3$ are called *primitive* vectors of a given Bravais lattice. When any of the primitive vectors are zero, (10.1) also defines a two-dimensional (2D) Bravais lattice, one example of which is shown in Fig. 10.1.

It is also worth mentioning that for any given Bravais lattice, the set of primitive vectors is not unique, and there are very many different choices, as shown in Fig. 10.1.

In three dimensions, there exist a total of 14 different Bravais lattices. The symmetry of any physical crystal is described by one of the Bravais lattices plus a *basis*. The basis consists of identical units (usually made by group of atoms), which are attached to every point of the underlying Bravais lattice. A crystal, whose basis consists of a single atom or ion, is said to have a monatomic Bravais lattice.

Another important concept widely used in the study of crystals is that of a *primitive cell*. The primitive cell is a volume of space that contains precisely one lattice point and can be translated through all the vectors of a Bravais lattice to fill all the space without overlapping itself or leaving voids. Just as in the case of primitive vectors, there is no unique way of choosing a primitive cell. The most common choice, however, is the *Wigner–Seitz* cell, which has the full symmetry of the underlying Bravais lattice. The Wigner–Seitz cell about a lattice point also has a property of being closer to that point than to any other lattice point. It can be constructed by drawing lines connecting a given point to nearby lying points, bisecting each line with a plane and taking the smallest polyhedron bounded by these planes.

The Bravais lattice, which is defined in *real* space, is sometimes referred to as a *direct* lattice. At the same time, there exist the concepts of a *reciprocal* space

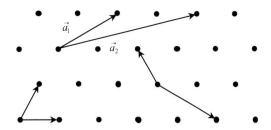

Fig. 10.1 A 2D triangular Bravais lattice. Several possible choices of the primitive vectors $\vec{a_1}$ and $\vec{a_2}$ are indicated

and a *reciprocal* lattice, which play an extremely important role in virtually any study of wave propagation, diffraction, and other wave phenomena in crystals. For any Bravais lattice, given by a set of vectors $\vec{R}$ [see (10.1)], and a plane wave $\exp(i\vec{k} \cdot \vec{r})$, the reciprocal lattice is defined as a set of all wave vectors $\vec{G}$ that yield plane waves with the periodicity of a given Bravais lattice [1]. Mathematically, a wave vector $\vec{G}$ belongs to the reciprocal lattice of a Bravais lattice with vectors $\vec{R}$, if the equation

$$\exp(i\vec{G} \cdot (\vec{r} + \vec{R})) = \exp(i\vec{G} \cdot \vec{r}) \tag{10.2}$$

is true for any $\vec{r}$ and $\vec{R}$ of the given Bravais lattice. It follows from (10.2) that a reciprocal lattice can also be viewed as a set of points, whose positions are given by a set of wave vectors $\vec{G}$ satisfying the condition:

$$\exp(\vec{G} \cdot \vec{R}) = 1 \tag{10.3}$$

for all $\vec{R}$ in the Bravais lattice. The reciprocal lattice itself is a Bravais lattice. The primitive vectors $\vec{b_1}, \vec{b_2}, \vec{b_3}$ of the reciprocal lattice are constructed from the primitive vectors $\vec{a_1}, \vec{a_2}, \vec{a_3}$ of the direct lattice and given in three dimensions by the following expressions:

$$\begin{aligned}
\vec{b_1} &= 2\pi \frac{\vec{a_2} \times \vec{a_3}}{\vec{a_1} \cdot (\vec{a_2} \times \vec{a_3})} \\
\vec{b_2} &= 2\pi \frac{\vec{a_3} \times \vec{a_1}}{\vec{a_2} \cdot (\vec{a_3} \times \vec{a_1})} \\
\vec{b_3} &= 2\pi \frac{\vec{a_1} \times \vec{a_2}}{\vec{a_3} \cdot (\vec{a_1} \times \vec{a_2})}
\end{aligned} \tag{10.4}$$

As an example, Fig. 10.2 shows a simple-cubic Bravais lattice with a lattice constant a as well as its reciprocal lattice, which is also a simple-cubic one with a lattice constant $2\pi/a$ (as follows from relations (10.4)).

Since the reciprocal lattice is a Bravais lattice, one can also find its Wigner–Seitz cell. The Wigner–Seitz cell of a reciprocal lattice is conventionally called a *first Brillouin* zone (BZ). Planes in k-space, which bisect the lines joining a particular point of a reciprocal lattice with all other points, are known as *Bragg* planes.

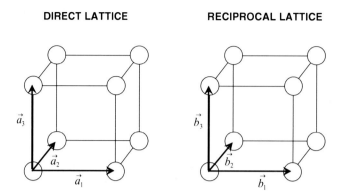

Fig. 10.2 Simple-cubic direct lattice and its reciprocal lattice. The primitive vectors of both lattices are also indicated

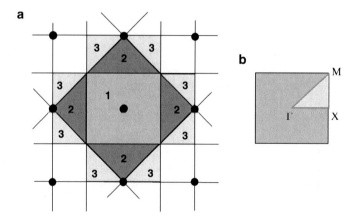

Fig. 10.3 (a) The first three Brillouin zones of the reciprocal lattice of the 2D square Bravais lattice. The *dots* indicate reciprocal lattice points, the *solid lines* indicate Bragg planes, and the digits indicate the order of the corresponding Brillouin zone. (b) The first Brillouin zone with the two high-symmetry directions commonly referred to as ΓX and ΓM. The triangle ΓXM is named the irreducible Brillouin zone

Therefore, the first BZ can also be defined as the set of all points in k-space that can be reached from the origin without crossing *any* Bragg plane. The BZs of higher orders also exist, with the *n*th BZ defined as the set of points that can be reached from the origin by crossing $(n-1)$ Bragg planes [1]. The first BZ is of great importance in the theory of solids with periodic structures, since the periodicity of the structure allows the description of the properties of the solids within the first BZ. Figure 10.3 shows the first three BZs of the 2D square Bravais lattice. The first BZ has a shape of a square with two high-symmetry directions, which are commonly referred to as ΓX and ΓM.

It is well known from quantum mechanics that the energy of an electron in an atom assumes discrete values. However, when the atomic orbitals overlap as the atoms come close together in a solid, the energy levels of the electrons broaden and form continuous regions, also known as energy *bands*. At the same time, because of the periodicity of the crystal structure, the electronic wave functions undergo strong Bragg reflections at the boundaries of the BZs. The destructive interference of the Bragg-scattered wave functions gives rise to the existence of the energy regions, in which no electronic energy levels exist. Since these regions are not accessible by the electrons, they are also known as *forbidden* bands. If the forbidden band occurs along the particular direction inside the crystal, it is conventionally called a *stop* band. If it happens to span *all* the directions inside the crystal, the term "complete *band gap*," or simply *band gap*, is used instead. The electronic properties of crystalline solids are conveniently described with the help of the *band structure* plots, which represent energy levels of the electrons of the solid as a function of the direction inside the solid.

The concepts of the direct and reciprocal lattices, BZs and energy bands discussed in this section, are of general nature and can be applied to *any* periodic system without being limited to atomic crystals. These concepts appear throughout the different chapters of this book.

10.2   Plane Wave Expansion methods

10.2.1   Plane Wave Expansion Method for Bulk Phononic Crystals

We first present with many details the plane wave expansion (PWE) method used for the calculation of the band structures of bulk phononic crystals, i.e., assumed of infinite extent along the three spatial directions. For the sake of simplicity, we limit ourselves to 2D phononic crystals, but the method can be easily extended to 3D structures. Two-dimensional phononic crystals are modeled as periodic arrays of infinite cylinders of different shape (circular, square, etc.) made up of a material A embedded in an infinite matrix B. Elastic materials A and B may be isotropic or of specific crystallographic symmetry. The elastic cylinders are assumed parallel to the z axis of the Cartesian coordinates system (O, x, y, z). The intersections of the cylinders axis with the (xOy) transverse plane form a 2D periodic array and the nearest neighbor distance between cylinders is a. The 2D primitive unit cell may contain one cylinder, or more. The filling factor, f_i, of each inclusion is defined as the ratio between the cross-sectional area of a cylinder and the surface of the primitive unit cell (see Fig. 10.4).

In absence of an external force, the equation of propagation of the elastic waves in any composite material is given as

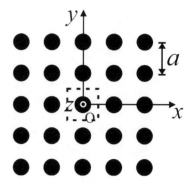

Fig. 10.4 Transverse cross section of the (*square*) array of inclusions . The cylinders are parallel to the *z* direction. The *dotted lines* represent the primitive unit cell of the 2D array

$$\rho(\vec{r})\frac{\partial^2 u_i(\vec{r},t)}{\partial t^2} = \sum_j \frac{\partial}{\partial x_j}\left(\sum_{m,n} C_{ijmn}(\vec{r})\frac{\partial u_n(\vec{r},t)}{\partial x_m}\right) \tag{10.5}$$

where $u_i(\vec{r},t)(i=1,2,3)$) is a component of the elastic displacement field. The elements $C_{ijmn}(i,j,m,n=1,2,3)$ of the elastic stiffness tensor and the mass density ρ are periodic functions of the position vector, $\vec{r}=(\vec{r}_{//},z)=(x,y,z)$.

In (10.5), x_1, x_2, x_3, u_1, u_2 and u_3 are equivalent to x, y, z, u_x, u_y, and u_z respectively.

For the sake of clarity, we consider constituent materials of cubic symmetry (but the method could be applied for lower crystallographic symmetry) characterized by the following stiffness tensor:

$$\overline{\overline{C}} = \begin{pmatrix} C_{11} & C_{12} & C_{12} & 0 & 0 & 0 \\ C_{12} & C_{11} & C_{12} & 0 & 0 & 0 \\ C_{12} & C_{12} & C_{11} & 0 & 0 & 0 \\ 0 & 0 & 0 & C_{44} & 0 & 0 \\ 0 & 0 & 0 & 0 & C_{44} & 0 \\ 0 & 0 & 0 & 0 & 0 & C_{44} \end{pmatrix}, \tag{10.6}$$

where the Voigt notation has been used. In this case, (10.5) becomes

$$\begin{cases} \rho\frac{\partial^2 u_x}{\partial t^2} = \frac{\partial}{\partial x}\left(C_{11}\frac{\partial u_x}{\partial x}+C_{12}\left(\frac{\partial u_y}{\partial y}+\frac{\partial u_z}{\partial z}\right)\right) + \frac{\partial}{\partial y}\left(C_{44}\left(\frac{\partial u_x}{\partial y}+\frac{\partial u_y}{\partial x}\right)\right) + \frac{\partial}{\partial z}\left(C_{44}\left(\frac{\partial u_x}{\partial z}+\frac{\partial u_z}{\partial x}\right)\right) \\ \rho\frac{\partial^2 u_y}{\partial t^2} = \frac{\partial}{\partial x}\left(C_{44}\left(\frac{\partial u_x}{\partial y}+\frac{\partial u_y}{\partial x}\right)\right) + \frac{\partial}{\partial y}\left(C_{11}\frac{\partial u_y}{\partial y}+C_{12}\left(\frac{\partial u_x}{\partial x}+\frac{\partial u_z}{\partial z}\right)\right) + \frac{\partial}{\partial z}\left(C_{44}\left(\frac{\partial u_y}{\partial z}+\frac{\partial u_z}{\partial y}\right)\right). \\ \rho\frac{\partial^2 u_z}{\partial t^2} = \frac{\partial}{\partial x}\left(C_{44}\left(\frac{\partial u_z}{\partial x}+\frac{\partial u_x}{\partial z}\right)\right) + \frac{\partial}{\partial y}\left(C_{44}\left(\frac{\partial u_z}{\partial z}+\frac{\partial u_y}{\partial y}\right)\right) + \frac{\partial}{\partial z}\left(C_{11}\frac{\partial u_z}{\partial z}+C_{12}\left(\frac{\partial u_y}{\partial y}+\frac{\partial u_x}{\partial x}\right)\right) \end{cases} \tag{10.7}$$

For bulk phononic crystals, *the elastic constants and the mass density do not depend on z*. Then taking advantage of the 2D periodicity in the (xOy) plane, they can be expanded in Fourier series in the form:

$$C_{ij}(\vec{r}_{//}) = \sum_{\vec{G}''_{//}} C_{ij}(\vec{G}''_{//}) e^{i\vec{G}''_{//}\cdot\vec{r}_{//}} \tag{10.8}$$

$$\rho(\vec{r}_{//}) = \sum_{\vec{G}''_{//}} \rho(\vec{G}''_{//}) e^{i\vec{G}''_{//}\cdot\vec{r}_{//}} \tag{10.9}$$

where $\vec{G}''_{//}$ is a 2D reciprocal lattice vector. One writes, with the help of the Bloch theorem, the elastic displacement field as

$$\vec{u}(\vec{r}) = e^{i(\omega t - \vec{K}_{//}\cdot\vec{r}_{//} - K_z z)} \sum_{\vec{G}'_{//}} \vec{u}_{\vec{K}}(\vec{G}'_{//}) e^{i\vec{G}'_{//}\cdot\vec{r}_{//}} \tag{10.10}$$

where $\vec{K} = (\vec{K}_{//}, K_z) = (K_x, K_y, K_z)$ is a wave vector, $\vec{G}'_{//}$, a 2D reciprocal lattice vector, and ω, an angular frequency. Substituting (10.8), (10.9), and (10.10) into (10.5) and posing $\vec{G}_{//} = \vec{G}'_{//} + \vec{G}''_{//}$ leads to a set of three coupled equations

$$
\begin{cases}
\omega^2 \sum_{\vec{G}'_{//}} B^{(11)}{}_{\vec{G}_{//},\vec{G}'_{//}} u_{x_{\vec{K}}}\left(\vec{G}'_{//}\right) \\
\quad = \sum_{\vec{G}'_{//}} \left\{ u_{x_{\vec{K}}}\left(\vec{G}'_{//}\right) A^{(11)}{}_{\vec{G}_{//},\vec{G}'_{//}} + u_{y_{\vec{K}}}\left(\vec{G}'_{//}\right) A^{(12)}{}_{\vec{G}_{//},\vec{G}'_{//}} + u_{z_{\vec{K}}}\left(\vec{G}'_{//}\right) A^{(13)}{}_{\vec{G}_{//},\vec{G}'_{//}} \right\} \\
\omega^2 \sum_{\vec{G}'_{//}} B^{(22)}{}_{\vec{G}_{//},\vec{G}'_{//}} u_{y_{\vec{K}}}\left(\vec{G}'_{//}\right) \\
\quad = \sum_{\vec{G}'_{//}} \left\{ u_{x_{\vec{K}}}\left(\vec{G}'_{//}\right) A^{(21)}{}_{\vec{G}_{//},\vec{G}'_{//}} + u_{y_{\vec{K}}}\left(\vec{G}'_{//}\right) A^{(22)}{}_{\vec{G}_{//},\vec{G}'_{//}} + u_{z_{\vec{K}}}\left(\vec{G}'_{//}\right) A^{(23)}{}_{\vec{G}_{//},\vec{G}'_{//}} \right\} \\
\omega^2 \sum_{\vec{G}'_{//}} B^{(33)}{}_{\vec{G}_{//},\vec{G}'_{//}} u_{z_{\vec{K}}}\left(\vec{G}'_{//}\right) \\
\quad = \sum_{\vec{G}'_{//}} \left\{ u_{x_{\vec{K}}}\left(\vec{G}'_{//}\right) A^{(31)}{}_{\vec{G}_{//},\vec{G}'_{//}} + u_{y_{\vec{K}}}\left(\vec{G}'_{//}\right) A^{(32)}{}_{\vec{G}_{//},\vec{G}'_{//}} + u_{z_{\vec{K}}}\left(\vec{G}'_{//}\right) A^{(33)}{}_{\vec{G}_{//},\vec{G}'_{//}} \right\}
\end{cases}
$$

$$\tag{10.11}$$

where

$$
\begin{cases}
B^{(11)}{}_{\vec{G}_{//},\vec{G}'_{//}} = B^{(22)}{}_{\vec{G}_{//},\vec{G}'_{//}} = B^{(33)}{}_{\vec{G}_{//},\vec{G}'_{//}} = \rho\left(\vec{G}_{//} - \vec{G}'_{//}\right) \\[4pt]
A^{(11)}{}_{\vec{G}_{//},\vec{G}'_{//}} = C_{11}\left(\vec{G}_{//} - \vec{G}'_{//}\right)(G_x + K_x)\left(G'_x + K_x\right) \\[2pt]
\quad + C_{44}\left(\vec{G}_{//} - \vec{G}'_{//}\right)\left[(G_y + K_y)\left(G'_y + K_y\right) + (K_z)^2\right] \\[2pt]
A^{(12)}{}_{\vec{G}_{//},\vec{G}'_{//}} = C_{12}\left(\vec{G}_{//} - \vec{G}'_{//}\right)(G_x + K_x)\left(G'_y + K_y\right) \\[2pt]
\quad + C_{44}\left(\vec{G}_{//} - \vec{G}'_{//}\right)\left(G'_x + K_x\right)(G_y + K_y) \\[2pt]
A^{(13)}{}_{\vec{G}_{//},\vec{G}'_{//}} = C_{12}\left(\vec{G}_{//} - \vec{G}'_{//}\right)(G_x + K_x)(K_z) + C_{44}\left(\vec{G}_{//} - \vec{G}'_{//}\right)\left(G'_x + K_x\right)(K_z) \\[2pt]
A^{(21)}{}_{\vec{G}_{//},\vec{G}'_{//}} = C_{12}\left(\vec{G}_{//} - \vec{G}'_{//}\right)(G'_x + K_x)(G_y + K_y) \\[2pt]
\quad + C_{44}\left(\vec{G}_{//} - \vec{G}'_{//}\right)\left(G'_y + K_y\right)(G_x + K_x) \\[2pt]
A^{(22)}{}_{\vec{G}_{//},\vec{G}'_{//}} = C_{11}\left(\vec{G}_{//} - \vec{G}'_{//}\right)(G_y + K_y)(G'_y + K_y) \\[2pt]
\quad + C_{44}\left(\vec{G}_{//} - \vec{G}'_{//}\right)\left[(G_x + K_x)(G'_x + K_x) + (K_z)^2\right] \\[2pt]
A^{(23)}{}_{\vec{G}_{//},\vec{G}'_{//}} = C_{12}\left(\vec{G}_{//} - \vec{G}'_{//}\right)(K_z)(G_y + K_y) + C_{44}\left(\vec{G}_{//} - \vec{G}'_{//}\right)(G'_y + K_y)(K_z) \\[2pt]
A^{(31)}{}_{\vec{G}_{//},\vec{G}'_{//}} = C_{12}\left(\vec{G}_{//} - \vec{G}'_{//}\right)(G'_x + K_x)(K_z) + C_{44}\left(\vec{G}_{//} - \vec{G}'_{//}\right)(G_x + K_x)(K_z) \\[2pt]
A^{(32)}{}_{\vec{G}_{//},\vec{G}'_{//}} = C_{12}\left(\vec{G}_{//} - \vec{G}'_{//}\right)(G'_y + K_y)(K_z) + C_{44}\left(\vec{G}_{//} - \vec{G}'_{//}\right)(G_y + K_y)(K_z) \\[2pt]
A^{(33)}{}_{\vec{G}_{//},\vec{G}'_{//}} = C_{11}\left(\vec{G}_{//} - \vec{G}'_{//}\right)(K_z)^2 + C_{44}\left(\vec{G}_{//} - \vec{G}'_{//}\right)[(G_x + K_x)(G'_x + K_x) \\[2pt]
\quad + (G_y + K_y)(G'_y + K_y)]
\end{cases}
$$

$$(10.12)$$

and G_x, G_y (resp. G'_x, G'_y) are the components of the $\vec{G}_{//}$ (resp. $\vec{G}'_{//}$) vectors.

Equation (10.12) can be rewritten as a standard generalized eigenvalue equation in the form

$$
\omega^2 \begin{pmatrix}
B^{(11)}{}_{\vec{G}_{//},\vec{G}'_{//}} & 0 & 0 \\
0 & B^{(22)}{}_{\vec{G}_{//},\vec{G}'_{//}} & 0 \\
0 & 0 & B^{(33)}{}_{\vec{G}_{//},\vec{G}'_{//}}
\end{pmatrix}
\begin{pmatrix}
u_{x_{\vec{K}}}\left(\vec{G}'_{//}\right) \\
u_{y_{\vec{K}}}\left(\vec{G}'_{//}\right) \\
u_{z_{\vec{K}}}\left(\vec{G}'_{//}\right)
\end{pmatrix}
$$

$$
= \begin{pmatrix}
A^{(11)}{}_{\vec{G}_{//},\vec{G}'_{//}} & A^{(12)}{}_{\vec{G}_{//},\vec{G}'_{//}} & A^{(13)}{}_{\vec{G}_{//},\vec{G}'_{//}} \\
A^{(21)}{}_{\vec{G}_{//},\vec{G}'_{//}} & A^{(22)}{}_{\vec{G}_{//},\vec{G}'_{//}} & A^{(23)}{}_{\vec{G}_{//},\vec{G}'_{//}} \\
A^{(31)}{}_{\vec{G}_{//},\vec{G}'_{//}} & A^{(32)}{}_{\vec{G}_{//},\vec{G}'_{//}} & A^{(33)}{}_{\vec{G}_{//},\vec{G}'_{//}}
\end{pmatrix}
\begin{pmatrix}
u_{x_{\vec{K}}}\left(\vec{G}'_{//}\right) \\
u_{y_{\vec{K}}}\left(\vec{G}'_{//}\right) \\
u_{z_{\vec{K}}}\left(\vec{G}'_{//}\right)
\end{pmatrix}. \qquad (10.13)
$$

Equation (10.13) is equivalent to $\omega^2 \overset{\leftrightarrow}{B}\vec{u}_{\vec{K}} = \overset{\leftrightarrow}{A}\vec{u}_{\vec{K}}$, where $\overset{\leftrightarrow}{A}$ and $\overset{\leftrightarrow}{B}$ are square matrices whose size depends on the number of 2D $\vec{G}_{//}$ vectors taken into account in the Fourier series. The numerical resolution of this eigenvalue equation is performed along the principal directions of propagation of the 2D irreducible BZ of the array of inclusions.

If one assumes that the elastic waves propagate only in the transverse plane (xOy), i.e., $K_z=0$, then the elements of the sub-matrices $A^{(13)}_{\vec{G}_{//},\vec{G}'_{//}}$, $A^{(23)}_{\vec{G}_{//},\vec{G}'_{//}}$, $A^{(31)}_{\vec{G}_{//},\vec{G}'_{//}}$, and $A^{(32)}_{\vec{G}_{//},\vec{G}'_{//}}$ vanish and (10.13) can be rewritten as

$$\omega^2 \begin{pmatrix} B^{(11)}_{\vec{G}_{//},\vec{G}'_{//}} & 0 & 0 \\ 0 & B^{(22)}_{\vec{G}_{//},\vec{G}'_{//}} & 0 \\ 0 & 0 & B^{(33)}_{\vec{G}_{//},\vec{G}'_{//}} \end{pmatrix} \begin{pmatrix} u_{x_{\vec{K}}}\left(\vec{G}'_{//}\right) \\ u_{y_{\vec{K}}}\left(\vec{G}'_{//}\right) \\ u_{z_{\vec{K}}}\left(\vec{G}'_{//}\right) \end{pmatrix}$$

$$= \begin{pmatrix} A^{(11)}_{\vec{G}_{//},\vec{G}'_{//}} & A^{(12)}_{\vec{G}_{//},\vec{G}'_{//}} & 0 \\ A^{(21)}_{\vec{G}_{//},\vec{G}'_{//}} & A^{(22)}_{\vec{G}_{//},\vec{G}'_{//}} & 0 \\ 0 & 0 & A^{(33)}_{\vec{G}_{//},\vec{G}'_{//}} \end{pmatrix} \begin{pmatrix} u_{x_{\vec{K}}}\left(\vec{G}'_{//}\right) \\ u_{y_{\vec{K}}}\left(\vec{G}'_{//}\right) \\ u_{z_{\vec{K}}}\left(\vec{G}'_{//}\right) \end{pmatrix} \quad (10.14)$$

The matrices involved in (10.14) are super-diagonal, and one can separate this equation into two independent uncoupled eigenvalues equations as follows:

$$\omega^2 \begin{pmatrix} B^{(11)}_{\vec{G}_{//},\vec{G}'_{//}} & 0 \\ 0 & B^{(22)}_{\vec{G}_{//},\vec{G}'_{//}} \end{pmatrix} \begin{pmatrix} u_{x_{\vec{K}}}\left(\vec{G}'_{//}\right) \\ u_{y_{\vec{K}}}\left(\vec{G}'_{//}\right) \end{pmatrix}$$

$$= \begin{pmatrix} A^{(11)}_{\vec{G}_{//},\vec{G}'_{//}} & A^{(12)}_{\vec{G}_{//},\vec{G}'_{//}} \\ A^{(21)}_{\vec{G}_{//},\vec{G}'_{//}} & A^{(22)}_{\vec{G}_{//},\vec{G}'_{//}} \end{pmatrix} \begin{pmatrix} u_{x_{\vec{K}}}\left(\vec{G}'_{//}\right) \\ u_{y_{\vec{K}}}\left(\vec{G}'_{//}\right) \end{pmatrix} \quad (10.15)$$

$$\omega^2 \sum_{\vec{G}'_{//}} B^{(33)}_{\vec{G}_{//},\vec{G}'_{//}} u_{z_{\vec{K}}}\left(\vec{G}'_{//}\right) = \sum_{\vec{G}'_{//}} A^{(33)}_{\vec{G}_{//},\vec{G}'_{//}} u_{z_{\vec{K}}}\left(\vec{G}'_{//}\right) \quad (10.16)$$

Equation (10.15) leads to XY vibration modes polarized in the transverse plane (xOy) and (10.16) corresponds to Z modes with a displacement field along the z direction. Decoupling of the propagation modes in bulk phononic crystals leads to the diagonalization of matrices of reduced size and then to save computation time.

In order to evaluate the Fourier transform of the elastic constants and the density defined by (10.8) and (10.9), we need to specify the symmetry of the array of inclusions, the shape, and the cross-sectional area of the cylinder inclusion. For example, one considers a square array of cylinders of circular cross section of radius R with a lattice parameter a. Then one inclusion of filling factor $f = \pi \left(\frac{R}{a}\right)^2$ is located at the center of the 2D primitive unit cell (Wigner–Seitz cell) and the Fourier coefficients in (10.8) and (10.9) are given as

$$\zeta(\vec{G}_{//}) = \frac{1}{A_u} \iint\limits_{\binom{\text{primitive}}{\text{unit cell}}} \zeta(\vec{r}_{//}) e^{-i\vec{G}_{//}\cdot\vec{r}_{//}} d^2\vec{r}_{//} \tag{10.17}$$

where $\zeta \equiv \rho, C_{ij}$ and A_u is the area of the 2D primitive unit cell. These Fourier coefficients can be calculated as follows:

$$
\begin{aligned}
\zeta(\vec{G}_{//}) &= \frac{1}{A_u} \iint\limits_{\binom{\text{primitive}}{\text{unit cell}}} \zeta(\vec{r}_{//}) e^{-i\vec{G}_{//}\cdot\vec{r}_{//}} d^2\vec{r}_{//} \\
&= \frac{1}{A_u} \left\{ \iint\limits_{(A_{u.c.})} \zeta_A e^{-i\vec{G}_{//}\cdot\vec{r}_{//}} d^2\vec{r}_{//} + \iint\limits_{(B_{u.c.})} \zeta_B e^{-i\vec{G}_{//}\cdot\vec{r}_{//}} d^2\vec{r}_{//} \right\} \\
&= \frac{1}{A_u} \iint\limits_{(A_{u.c.})} \zeta_A e^{-i\vec{G}_{//}\cdot\vec{r}_{//}} d^2\vec{r}_{//} - \frac{1}{A_u} \iint\limits_{(A_{u.c.})} \zeta_B e^{-i\vec{G}_{//}\cdot\vec{r}_{//}} d^2\vec{r}_{//} \\
&\quad + \frac{1}{A_u} \iint\limits_{(A_{u.c.})} \zeta_B e^{-i\vec{G}_{//}\cdot\vec{r}_{//}} d^2\vec{r}_{//} + \frac{1}{A_u} \iint\limits_{(B_{u.c.})} \zeta_B e^{-i\vec{G}_{//}\cdot\vec{r}_{//}} d^2\vec{r}_{//} \\
&= \frac{1}{A_u} \iint\limits_{(A_{u.c.})} \zeta_A e^{-i\vec{G}_{//}\cdot\vec{r}_{//}} d^2\vec{r}_{//} - \frac{1}{A_u} \iint\limits_{(A_{u.c.})} \zeta_B e^{-i\vec{G}_{//}\cdot\vec{r}_{//}} d^2\vec{r}_{//} \\
&\quad + \zeta_B \left\{ \frac{1}{A_u} \iint\limits_{\binom{\text{primitive}}{\text{unit cell}}} e^{-i\vec{G}_{//}\cdot\vec{r}_{//}} d^2\vec{r}_{//} \right\} = \frac{1}{A_u}(\zeta_A - \zeta_B) \iint\limits_{(A_{u.c.})} e^{-i\vec{G}_{//}\cdot\vec{r}_{//}} d^2\vec{r}_{//} \\
&\quad + \zeta_B \left\{ \frac{1}{A_u} \iint\limits_{\binom{\text{primitive}}{\text{unit cell}}} e^{-i\vec{G}_{//}\cdot\vec{r}_{//}} d^2\vec{r}_{//} \right\}.
\end{aligned}
\tag{10.18}
$$

But $\frac{1}{A_u} \iint\limits_{\binom{\text{primitive}}{\text{unit cell}}} e^{-i\vec{G}_{//}\cdot\vec{r}_{//}} d^2\vec{r}_{//} = \delta_{\vec{G}_{//},\vec{0}}$ where δ is the Dirac distribution and (10.18) can be rewritten as

$$\zeta(\vec{G}_{//}) = (\zeta_A - \zeta_B).F\left(\vec{G}_{//}\right) + \zeta_B.\delta_{\vec{G}_{//},\vec{0}} \tag{10.19}$$

where $F(\vec{G}_{//})$ is the structure factor defined as

$$F(\vec{G}_{//}) = \frac{1}{A_u} \iint\limits_{(A_{u.c})} e^{-i\vec{G}_{//}\cdot\vec{r}_{//}} d^2\vec{r}_{//} \tag{10.20}$$

In (10.20), integration is performed over the cross section of the cylindrical inclusion denoted by $(A_{u.c.})$. Using the polar coordinates $r_{//}, \theta$, one shows that

$$
\begin{aligned}
F(\vec{G}_{//}) &= \frac{1}{a^2} \int_0^R \int_0^{2\pi} e^{-iG_{//} \cdot r_{//} \cos\theta} r_{//} dr_{//} d\theta = \frac{1}{a^2} \int_0^R 2\pi r_{//} dr_{//} J_0(G_{//}r_{//}) \\
&= \frac{2\pi}{a^2 G_{//}^2} \int_0^{G_{//}R} (G_{//}r_{//}) J_0(G_{//}r_{//}) d(G_{//}r_{//}) \\
&= \frac{2\pi}{a^2 G_{//}^2} G_{//} R . J_1(G_{//}R) = 2f \frac{J_1(G_{//}R)}{G_{//}R}
\end{aligned}
\tag{10.21}
$$

where J_0 and J_1 are Bessel functions of the first kind of orders 0 and $1, f = \pi R^2/a^2$ and $0 \le f \le \pi/4$. The maximum value of f corresponds to the close-packed structure where one cylinder touches another one. Similar calculations lead, for rods of square cross section of width d, to $F(\vec{G}_{//}) = f \left(\frac{\sin(G_x d/2)}{(G_x d/2)} \right) \left(\frac{\sin(G_y d/2)}{(G_y d/2)} \right)$ where $f = d^2/a^2$ and $0 \le f \le 1$.

Note that for $\vec{G}_{//} = \vec{0}, F(\vec{G}_{//} = \vec{0}) = f$ and

$$
\zeta(\vec{G}_{//} = \vec{0}) = (\zeta_A - \zeta_B)f + \zeta_B = f\zeta_A + (1-f)\zeta_B
\tag{10.22}
$$

and $\zeta(\vec{G}_{//} = \vec{0})$ corresponds to the average value of ζ.

The components of the 2D reciprocal lattice vectors $\vec{G}_{//}$ are $G_x = \frac{2\pi}{a} n_x$ and $G_y = \frac{2\pi}{a} n_y$ where n_x and n_y are integers. In the course of the numerical resolution of (10.13), we consider $-M_x \le n_x \le +M_x$ and $-M_y \le n_y \le +M_y$ (with M_x and M_y positive integers), i.e., $(2M_x + 1)(2M_y + 1)$ 2D $\vec{G}_{//}$ vectors (G_x and G_y have $(2M_x + 1)$ and $(2M_y + 1)$ different values, respectively) are taken into account. This gives $3(2M_x + 1)(2M_y + 1)$ real eigenfrequencies $\omega(\vec{K})$ for a given wave vector $\vec{K}$ describing the principal directions of propagation in the irreducible BZ. Following the same process, the PWE method can be applied to other symmetries of the array (triangular, honeycomb, etc.) and other shapes of the inclusion (square, rotated square, etc.). The choice of the values of the integers M_x, M_y is of crucial importance for insuring the convergency of the Fourier series. The convergency is fast when considering constituent materials with closed physical properties but is slower when materials A and B present very different densities and elastic moduli [2]. The PWE method is also useful for computing band structures of phononic crystals made of fluid constituents [3]. In this case, the Fourier transform of the equation of propagation of longitudinal acoustic waves in a heterogeneous periodic fluid leads to a generalized eigenvalue equation similar to (10.16). But the PWE method fails to predict accurately the band structures of mixed phononic crystals made of solid (resp. fluid) inclusions surrounded by a fluid (resp. solid) [4]. Nevertheless, in some particular cases, the PWE method is very well adapted for the calculations of band structures of mixed systems, provided the inclusions can be assumed to be infinitely rigid as it happens in arrays of solid inclusions surrounded with air [5]. On the other hand, the PWE method assumes the phononic crystal to be

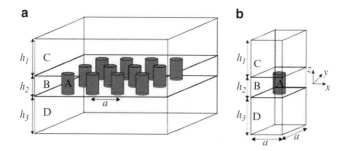

Fig. 10.5 (a) 2D phononic crystal plate sandwiched between two slabs of homogeneous materials and (b) 3D super-cell considered in the course of the super-cell PWE computation

of infinite extent in the three spatial directions and does not allow the calculation of the reflection and transmission coefficients of elastic waves through phononic crystals of finite thickness.

10.2.2 PWE Method for Phononic Crystal Plates: The Super-Cell Method

To calculate the elastic band structures of 2D phononic crystal plates, one modifies the PWE method presented in Sect. 10.2.1. The phononic crystal plate of thickness, h_2, is assumed to be infinite in the (xOy) plane of the Cartesian coordinates system (O, x, y, z). The plate is sandwiched between two slabs of thicknesses h_1 and h_3, made of elastic homogeneous materials C and D (see Fig. 10.5a). In the course of the numerical calculations, one considers the parallelepipedic super-cell depicted in Fig. 10.5b.

The basis of the super-cell in the (xOy) plane includes that of the 2D primitive unit cell (which may contain one cylinder or more) of the array of inclusions, and its height along the z direction is $\ell = h_1 + h_2 + h_3$. This super-cell is repeated periodically along the x, y, and z directions. This triple periodicity allows one to develop the elastic constants and the mass density of the constituent materials as Fourier series as

$$\zeta(\vec{r}) = \sum_{\vec{G}} \zeta(\vec{G}) e^{i\vec{G}\cdot\vec{r}} \tag{10.23}$$

where $\vec{r} = (\vec{r}_{//}, z) = (x, y, z)$ and $\vec{G} = (\vec{G}_{//}, G_z) = (G_x, G_y, G_z)$ are *3D position vectors and reciprocal lattice vectors*, respectively. Moreover, the elastic displacement field can be written as

$$\vec{u}(\vec{r}) = e^{i(\omega t - \vec{K}_{//}\cdot\vec{r}_{//} - K_z z)} \sum_{\vec{G}} \vec{u}_{\vec{K}}(\vec{G}) e^{i\vec{G}\cdot\vec{r}}. \tag{10.24}$$

The components in the (xOy) plane of the $\vec{G}$ vectors depend on the geometry of the array of inclusions while along the z direction, $G_z = \frac{2\pi}{\ell} n_z$, where n_z is an integer. The Fourier coefficients in (10.23) are now given as

$$\zeta(\vec{G}) = \frac{1}{V_u} \iiint_{\text{(super cell)}} \zeta(\vec{r})e^{-i\vec{G}\cdot\vec{r}}d^3\vec{r} \tag{10.25}$$

with $V_u = A_u.\ell$ is the volume of the super-cell.

For a square array of inclusions, the Fourier coefficients become

$$\zeta(\vec{G}) = \begin{cases} f\zeta_A\left(\frac{h_2}{\ell}\right) + (1-f)\zeta_B\left(\frac{h_2}{\ell}\right) + \zeta_C\left(\frac{h_1}{\ell}\right) + \zeta_D\left(\frac{h_3}{\ell}\right), & \text{if } \vec{G} = \vec{0} \\ (\zeta_A - \zeta_B)F_I^s(\vec{G}) + (\zeta_C - \zeta_B)F_{II}^s(\vec{G}) + (\zeta_D - \zeta_B)F_{III}^s(\vec{G}), & \text{if } \vec{G} \neq \vec{0} \end{cases} \tag{10.26}$$

with

$$F_I^s(\vec{G}) = \frac{1}{V_u} \iiint_{(A)} e^{-i\vec{G}\cdot\vec{r}}d^3\vec{r} = F(\vec{G}_{//}) \left(\frac{\sin\left(G_z \frac{h_2}{2}\right)}{G_z \frac{h_2}{2}}\right) \cdot \left(\frac{h_2}{\ell}\right) \tag{10.27}$$

$$F_{II}^s(\vec{G}) = \frac{1}{V_u} \iiint_{(C)} e^{-i\vec{G}\cdot\vec{r}}d^3\vec{r}$$

$$= \left(\frac{\sin\left(G_x \frac{a}{2}\right)}{\left(G_x \frac{a}{2}\right)}\right) \cdot \left(\frac{\sin\left(G_y \frac{a}{2}\right)}{\left(G_y \frac{a}{2}\right)}\right) \cdot \left(\frac{\sin\left(G_z \frac{h_1}{2}\right)}{\left(G_z \frac{h_1}{2}\right)}\right) \cdot \left(\frac{h_1}{\ell}\right) \cdot e^{-iG_z\left(\frac{h_1+h_2}{2}\right)} \tag{10.28}$$

$$F_{III}^s(\vec{G}) = \frac{1}{V_u} \iiint_{(D)} e^{-i\vec{G}\cdot\vec{r}}d^3\vec{r}$$

$$= \left(\frac{\sin(G_x \frac{a}{2})}{(G_x \frac{a}{2})}\right) \cdot \left(\frac{\sin(G_y \frac{a}{2})}{(G_y \frac{a}{2})}\right) \cdot \left(\frac{\sin(G_z \frac{h_3}{2})}{(G_z \frac{h_3}{2})}\right) \cdot \left(\frac{h_3}{\ell}\right) \cdot e^{-iG_z\left(\frac{h_2+h_3}{2}\right)} \tag{10.29}$$

In (10.27), (10.28), and (10.29), the integration is performed over the volume occupied by each material A, C, or D inside the unit cell. In (10.27), $F(\vec{G}_{//})$ is the structure factor defined by (10.21) for cylindrical inclusions.

As for the bulk phononic crystals, the equation of motion is Fourier transformed by substituting (10.23) and (10.24) in (10.5), and this leads to the following generalized eigenvalue equation:

$$
\omega^2 \begin{pmatrix} B^{(11)}{}_{\vec{G},\vec{G}'} & 0 & 0 \\ 0 & B^{(22)}{}_{\vec{G},\vec{G}'} & 0 \\ 0 & 0 & B^{(33)}{}_{\vec{G},\vec{G}'} \end{pmatrix} \begin{pmatrix} u_{x_{\vec{K}}}(\vec{G}') \\ u_{y_{\vec{K}}}(\vec{G}') \\ u_{z_{\vec{K}}}(\vec{G}') \end{pmatrix}
$$

$$
= \begin{pmatrix} A^{(11)}{}_{\vec{G},\vec{G}'} & A^{(12)}{}_{\vec{G},\vec{G}'} & A^{(13)}{}_{\vec{G},\vec{G}'} \\ A^{(21)}{}_{\vec{G},\vec{G}'} & A^{(22)}{}_{\vec{G},\vec{G}'} & A^{(23)}{}_{\vec{G},\vec{G}'} \\ A^{(31)}{}_{\vec{G},\vec{G}'} & A^{(32)}{}_{\vec{G},\vec{G}'} & A^{(33)}{}_{\vec{G},\vec{G}'} \end{pmatrix} \begin{pmatrix} u_{x_{\vec{K}}}(\vec{G}') \\ u_{y_{\vec{K}}}(\vec{G}') \\ u_{z_{\vec{K}}}(\vec{G}') \end{pmatrix}
$$

where

$$
\begin{cases}
B^{(11)}{}_{\vec{G},\vec{G}'} = B^{(22)}{}_{\vec{G},\vec{G}'} = B^{(33)}{}_{\vec{G},\vec{G}'} = \rho\left(\vec{G}-\vec{G}'\right) \\[2mm]
A^{(11)}{}_{\vec{G},\vec{G}'} = C_{11}\left(\vec{G}-\vec{G}'\right)(G_x+K_x)(G'_x+K_x) \\[1mm]
\qquad + C_{44}\left(\vec{G}-\vec{G}'\right)\left[(G_y+K_y)(G'_y+K_y)+(G_z+K_z)(G'_z+K_z)\right] \\[2mm]
A^{(12)}{}_{\vec{G},\vec{G}'} = C_{12}\left(\vec{G}-\vec{G}'\right)(G_x+K_x)(G'_y+K_y)+C_{44}\left(\vec{G}-\vec{G}'\right)(G'_x+K_x)(G_y+K_y) \\[2mm]
A^{(13)}{}_{\vec{G},\vec{G}'} = C_{12}\left(\vec{G}-\vec{G}'\right)(G_x+K_x)(G'_z+K_z)+C_{44}\left(\vec{G}_{//}-\vec{G}'_{//}\right)(G'_x+K_x)(G_z+K_z) \\[2mm]
A^{(21)}{}_{\vec{G},\vec{G}'} = C_{12}\left(\vec{G}-\vec{G}'\right)(G'_x+K_x)(G_y+K_y)+C_{44}\left(\vec{G}-\vec{G}'\right)(G'_y+K_y)(G_x+K_x) \\[2mm]
A^{(22)}{}_{\vec{G},\vec{G}'} = C_{11}\left(\vec{G}-\vec{G}'\right)(G_y+K_y)(G'_y+K_y) \\[1mm]
\qquad + C_{44}\left(\vec{G}-\vec{G}'\right)\left[(G_x+K_x)(G'_x+K_x)+(G_z+K_z)(G'_z+K_z)\right] \\[2mm]
A^{(23)}{}_{\vec{G},\vec{G}'} = C_{12}\left(\vec{G}-\vec{G}'\right)(G'_z+K_z)(G_y+K_y)+C_{44}\left(\vec{G}-\vec{G}'\right)(G'_y+K_y)(G_z+K_z) \\[2mm]
A^{(31)}{}_{\vec{G},\vec{G}'} = C_{12}\left(\vec{G}-\vec{G}'\right)(G'_x+K_x)(G_z+K_z)+C_{44}\left(\vec{G}-\vec{G}'\right)(G_x+K_x)(G'_z+K_z) \\[2mm]
A^{(32)}{}_{\vec{G},\vec{G}'} = C_{12}\left(\vec{G}-\vec{G}'\right)(G'_y+K_y)(G_z+K_z)+C_{44}\left(\vec{G}-\vec{G}'\right)(G_y+K_y)(G'_z+K_z) \\[2mm]
A^{(33)}{}_{\vec{G},\vec{G}'} = C_{11}\left(\vec{G}-\vec{G}'\right)(G_z+K_z)(G'_z+K_z) \\[1mm]
\qquad + C_{44}\left(\vec{G}-\vec{G}'\right)\left[(G_x+K_x)(G'_x+K_x)+(G'_y+K_y)(G_y+K_y)\right]
\end{cases}
$$

$$(10.30)$$

The numerical resolution of this eigenvalue equation is performed along the principal directions of propagation of the 2D irreducible BZ of the array of inclusions while K_z is fixed to any value lower than $\frac{\pi}{l}$. In the course of the numerical calculations, G_x, G_y, and G_z take respectively $(2M_x + 1)$, $(2M_y + 1)$, and $(2M_z + 1)$ discrete values, and this leads to $3(2M_x + 1)(2M_y + 1)(2M_z + 1)$ eigenfrequencies ω for a given wave vector $\vec{K}$.

The super-cell method requires an interaction as low as possible between the vibrational modes of neighboring periodically repeated phononic crystal plates. Then, in order to allow the top surface of the plate to be free of stress, medium C should behave, for instance, like vacuum [6]. But as already observed by various authors [6–8], the choice of the physical parameters characterizing vacuum in the course of the PWE computations is of critical importance. Indeed, in the framework of the PWE method, taking abruptly $C_{ij} = 0$ and $\rho = 0$ for vacuum leads to numerical instabilities and unphysical results [6–8]. Then vacuum must be modeled as a pseudo-solid material with very low C_{ij} and ρ. For the sake of simplicity, this low impedance medium (LIM) is supposed to be elastically isotropic and is characterized by a longitudinal speed of sound C_1, and a transversal speed of sound C_t or equivalently by two elastic moduli expressed with the Voigt notation as $C_{11} = \rho C_\ell^2$ and $C_{44} = \rho C_t^2$. The choice of the values of these parameters is governed by the boundary condition between any solid material and vacuum. Indeed, one knows that this interface must be free of stress, and this requires that $C_{11} = 0$ and $C_{44} = 0$ rigorously in vacuum [6]. Then, using the LIM to model vacuum in the PWE computations, the nonvanishing values of these parameters must be as small as possible, and we consider that the ratio between the elastic moduli of the LIM and those of any other solid material constituting the phononic crystal must approach zero. We choose C_1 and C_t to be much larger than the speeds of sound in usual solid materials in order to limit propagation of acoustic waves to the solid. Large speeds of sound and small elastic moduli impose a choice of a very low mass density for the LIM. More specifically, we choose $\rho = 10^{-4}\,\mathrm{kg\,m^{-3}}$ and $C_1 = C_t = 10^5\,\mathrm{m\,s^{-1}}$, i.e., the acoustic impedances of the LIM are equal to $10\,\mathrm{kg\,m^{-2}\,s^{-1}}$. With these values, $C_{11} = C_{44} = 10^6\,\mathrm{N\,m^{-2}}$ and the elastic constants of the LIM are approximately 10^4 times lower than those of any usual solid material that are typically on the order of $10^{10}\,\mathrm{N\,m^{-2}}$. The values we choose for C_{11} and C_{44} are a compromise to achieve satisfactory convergence of the SC-PWE method and still satisfy boundary conditions. Values of the elastic constants of the LIM lower than $10^4\,\mathrm{N\,m^{-2}}$ can have, in some cases, effects on the numerical convergence. We choose $C_{11} = C_{44}$ for convenience. In the course of the PWE calculations, these values of the LIM physical characteristics allow one to model vacuum without numerical difficulties.

In the super-cell, medium D can be either vacuum or a homogeneous material depending on whether one wants to model a phononic crystal plate or a structure made of a phononic crystal plate deposited on a substrate of finite thickness. Computations of dispersion curves of phononic crystal plates with $K_z = 0$ and with any other nonvanishing value of K_z, lower than $\frac{\pi}{\ell}$, lead to nearly the same result. Indeed, the eigenvalues computed with $K_z = 0$ and $K_z \neq 0$ differ only in their third decimal. This indicates that the homogeneous slabs C and D made of the LIM modeling vacuum rigorously provide appropriate decoupling of the plate modes of vibration in the z direction. Then, the value of K_z may be fixed to zero. Due to this 3D nature, the numerical convergency of the super-cell PWE (SC-PWE) method is relatively slow, and it has been shown that this method is suitable for voids/solid matrix plates but is not reliable for constituent materials with very different physical properties [9]. The SC-PWE method does not to require to write and to satisfy explicitly the boundary conditions at the free surfaces. Nevertheless, other authors have proposed PWE schemes for phononic crystals plates where these boundary conditions are satisfied, but these methods also suffer from convergence difficulties [10].

10.2.3   PWE Method for Complex Band Structures

In classical PWE methods (see Sect. 10.2.1), one calculates a set of real eigenfrequencies $\omega(\vec{K})$ for a specific wave vector $\vec{K}$. That means that only propagating modes with a real wave vector can be deduced from $\omega(\vec{K})$ PWE methods. Then an extended PWE method has been proposed that allows the calculation of not only the propagating modes but also the evanescent modes. The wave vector for evanescent waves possesses a nonvanishing imaginary part. We have seen previously that the Fourier transform of the equation of propagation of elastic waves in a phononic crystal leads to the resolution of a generalized eigenvalue equation in the form $\omega^2 \overset{\leftrightarrow}{B}$ $\vec{U} = \overset{\leftrightarrow}{A}\vec{U}$. The matrix elements of $\overset{\leftrightarrow}{A}$ and $\overset{\leftrightarrow}{B}$ involve terms depending on the components of the wave vector $\vec{K}$. It is always possible to rewrite matrix $\overset{\leftrightarrow}{A}$ as $\overset{\leftrightarrow}{A}$ $= K_\alpha^2 \overset{\leftrightarrow}{A}_1 + K_\alpha \overset{\leftrightarrow}{A}_2 + \overset{\leftrightarrow}{A}_3$, where K_α is one of the components of the wave vector, and $\overset{\leftrightarrow}{A}_1$, $\overset{\leftrightarrow}{A}_2$, and $\overset{\leftrightarrow}{A}_3$ are matrices of the same size as $\overset{\leftrightarrow}{A}$. The generalized eigenvalue equation $\omega^2 \overset{\leftrightarrow}{B}\vec{U} = \overset{\leftrightarrow}{A}\vec{U}$ may be recast as $K_\alpha^2 \overset{\leftrightarrow}{A}_1 \vec{U} = \omega^2 \overset{\leftrightarrow}{B}\vec{U} - \overset{\leftrightarrow}{A}_3 \vec{U} - K_\alpha \overset{\leftrightarrow}{A}_2 \vec{U}$ and one can write

$$K_\alpha \begin{pmatrix} \overset{\leftrightarrow}{I} & \overset{\leftrightarrow}{0} \\ \overset{\leftrightarrow}{0} & \overset{\leftrightarrow}{A}_1 \end{pmatrix} \begin{pmatrix} \vec{U} \\ K_\alpha \vec{U} \end{pmatrix} = \begin{pmatrix} \overset{\leftrightarrow}{0} & \overset{\leftrightarrow}{I} \\ \omega^2 \overset{\leftrightarrow}{B} - \overset{\leftrightarrow}{A}_3 & -\overset{\leftrightarrow}{A}_2 \end{pmatrix} \begin{pmatrix} \vec{U} \\ K_\alpha \vec{U} \end{pmatrix} \tag{10.31}$$

where $\overset{\leftrightarrow}{I}$ is the identity matrix. Equation (10.31) is nothing else than a generalized eigenvalue equation where the eigenvalues are the component K_α of the wave vector. For a specific value of the circular frequency ω, one calculates a set of complex eigenvalues K_α. This method is named $\vec{K}(\omega)$ PWE method. The size of the matrices occurring on the left and right sides of (10.31) is twice that of matrices $\overset{\leftrightarrow}{A}$ and $\overset{\leftrightarrow}{B}$. One may illustrate these general ideas by considering the peculiar case of the Z elastic modes propagating in a bulk 2D phononic crystal made of a square array of lattice parameter a, of cylindrical inclusions embedded in a solid matrix. If one assumes $K_z = 0$, then these modes are given by (10.16), where ω depends on the two variables K_x and K_y. Consider the propagation of elastic waves along the ΓX

direction of the irreducible BZ for which $K_y = 0$ and $0 \leq \mathrm{Re}(K_x) \leq \frac{\pi}{a}$. Equation (10.16) leads to

$$K_x^2 \sum_{\vec{G}'_{//}} C_{44}\left(\vec{G}_{//} - \vec{G}'_{//}\right) u_{z_K}\left(\vec{G}'_{//}\right)$$

$$= \sum_{\vec{G}'_{//}} \left\{ \omega^2 \rho\left(\vec{G}_{//} - \vec{G}'_{//}\right) - (G_x.G'_x + G_y.G'_y)C_{44}\left(\vec{G}_{//} - \vec{G}'_{//}\right)\right\} u_{z_K}\left(\vec{G}'_{//}\right)$$

$$- K_x \sum_{\vec{G}'_{//}} (G_x + G'_x)C_{44}\left(\vec{G}_{//} - \vec{G}'_{//}\right) u_{z_K}\left(\vec{G}'_{//}\right) \tag{10.32}$$

and can be rewritten as

$$K_x \begin{pmatrix} \overset{\leftrightarrow}{I} & \overset{\leftrightarrow}{0} \\ \overset{\leftrightarrow}{0} & \overset{\leftrightarrow}{A_1} \end{pmatrix} \begin{pmatrix} \vec{U} \\ K_x\vec{U} \end{pmatrix} = \begin{pmatrix} \overset{\leftrightarrow}{0} & \overset{\leftrightarrow}{I} \\ \omega^2\overset{\leftrightarrow}{B} - \overset{\leftrightarrow}{A_3} & -\overset{\leftrightarrow}{A_2} \end{pmatrix} \begin{pmatrix} \vec{U} \\ K_x\vec{U} \end{pmatrix} \tag{10.33}$$

$$\text{where} \begin{cases} B_{\vec{G}_{//},\vec{G}'_{//}} = \rho\left(\vec{G}_{//} - \vec{G}'_{//}\right) \\ A_{1\vec{G}_{//},\vec{G}'_{//}} = C_{44}\left(\vec{G}_{//} - \vec{G}'_{//}\right) \\ A_{2\vec{G}_{//},\vec{G}'_{//}} = C_{44}\left(\vec{G}_{//} - \vec{G}'_{//}\right)(G_x + G'_x) \\ A_{3\vec{G}_{//},\vec{G}'_{//}} = C_{44}\left(\vec{G}_{//} - \vec{G}'_{//}\right)\left[(G_xG'_x + G_yG'_y)\right] \end{cases} \tag{10.34}$$

Numerical resolution of (10.34) leads to $2N$ (if $N \times N$ is the size of matrices $\overset{\leftrightarrow}{A}$ and $\overset{\leftrightarrow}{B}$) complex values of $K_x = \mathrm{Re}(K_x) - i\mathrm{Im}(K_x)$ for any value of ω. Eigenvalues belonging to the irreducible BZ and corresponding to waves with a vanishing amplitude when $x \to +\infty$ may be taken into account, i.e., $0 \leq \mathrm{Re}(K_x) \leq \frac{\pi}{a}$ and $\mathrm{Im}(K_x) \geq 0$. Figure 10.6 presents the band structures calculated by both $\omega(\vec{K})$ and $\vec{K}(\omega)$ methods. This figure shows the ability of the $\vec{K}(\omega)$ method to calculate the evanescent modes. Of particular interest is the existence of *additional bands* (see right panel of Fig. 10.6 for reduced frequency around 1.1) not predicted by the classical $\omega(\vec{K})$ PWE method (red dots). These vibrational modes are characterized by a nonvanishing $\mathrm{Im}(K_x)$.

To apply this, $\vec{K}(\omega)$ PWE method requires to consider only one component of the wave vector $\vec{K}$ as eigenvalue. That needs to keep fixed the other component or to write a linear relation between them. For example, along the ΓM direction in the irreducible BZ of the square array, one can write $K_x = K_y$ and consider K_x as the eigenvalue. In the same way, one can deal with any direction of propagation and not only with the high-symmetry directions. Plotting all the values of K_x and K_y corresponding to a specific frequency leads to the equi-frequency contour (EFC) of the phononic crystal. Knowing precisely the shape of these EFCs is of fundamental interest when studying focusing or self-collimating of elastic waves by

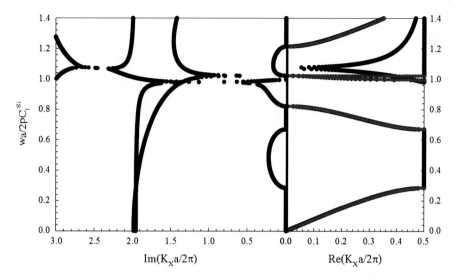

Fig. 10.6 Band structures along the ΓX direction of the irreducible Brillouin zone for a square array of holes drilled in a Silicon matrix: *Red dots*: $\omega(\vec{K})$ method; *Black dots*: $\vec{K}(\omega)$ method

phononic crystals [11]. Moreover, the $\vec{K}(\omega)$ PWE method allows to take into account elastic moduli depending on the frequency and should be applied for calculating the band structures of phononic crystals made of viscoelastic materials.

10.3 Finite-Difference Time Domain Method

10.3.1 Calculation of Transmission Coefficients

We present here the basic principles of the finite-difference time domain (FDTD) method applied to the calculation of transmission coefficients of elastic waves through phononic crystals made of nonviscous or nonviscoelastic constituents. The method is based on discretizations of the differential equations of motion on both spatial and time domains. As previously and for the sake of simplicity, we limit ourselves to 2D phononic crystals.

We consider a 2D phononic crystal containing cylindrical inclusions surrounded by a host matrix. Constituent materials are supposed to be isotropic solids or fluids. The inclusions are parallel to the z direction and are arranged periodically in the transverse (x,y) plane. A phononic crystal of finite thickness along the y direction is realized by considering a small number of periods in this direction. The "sample" is bounded by semi-infinite homogeneous media on both sides. The system is infinite in the vertical direction z, and all its physical properties do not depend on z. That means that we propose a strictly 2D FDTD scheme. The probing signal

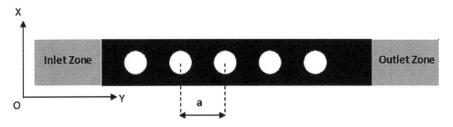

Fig. 10.7 Two-dimensional cross section of the FDTD model structure. The cylinders are parallel to the z axis of the Cartesian coordinate system $(Oxyz)$. The lattice parameter is a

corresponding to a longitudinal wave that propagates along the y direction is launched from the left homogeneous medium (inlet zone) and detected in the right one (outlet zone) (see Fig. 10.7). We just describe here a 2D FDTD scheme just as it has been reported in [12].

The elastic wave equation is given by

$$\frac{\partial \vec{v}}{\partial t} = \frac{1}{\rho(x,y)} \overrightarrow{\nabla}.\bar{\bar{\sigma}} \tag{10.35}$$

with

$$\vec{v} = \frac{\partial \vec{u}}{\partial t} \tag{10.36}$$

where t is time, $\rho(x, y)$ is the mass density, $\vec{u}(x, y, t)$ is the displacement field, $\vec{v}(x, y, t)$ is the velocity vector, and $\bar{\bar{\sigma}}(x, y)$ is the total stress tensor. The nonzero Cartesian components of the 2D stress tensor $\bar{\bar{\sigma}}$ are

$$\sigma_{xx} = C_{11} \frac{\partial u_x}{\partial x} + C_{12} \frac{\partial u_y}{\partial y} \tag{10.37}$$

$$\sigma_{yy} = C_{44} \left(\frac{\partial u_y}{\partial x} + \frac{\partial u_x}{\partial y} \right) \tag{10.38}$$

$$\sigma_{xy} = C_{11} \frac{\partial u_y}{\partial y} + C_{12} \frac{\partial u_x}{\partial x} \tag{10.39}$$

with $C_{11}(x,y)$, $C_{44}(x,y)$, $C_{12}(x,y) = C_{11}(x,y) - 2C_{44}(x,y)$, the position-dependent elastic moduli. For a given isotropic medium, C_{11} and C_{44} are related to the longitudinal C_l and transverse C_t speeds of sound as $C_{11} = \rho C_l^2$ and $C_{44} = \rho C_t^2$. A fluid is treated as a solid with zero transverse speed of sound in this 2D FDTD scheme. From (10.37), (10.38), and (10.39), one notes that we consider only modes of vibration analog to XY modes as defined by (10.15) in the preceding section. The

FDTD method involves transforming the governing differential equations given by (10.35) and (10.36) in the time domain into finite differences and solving them as one progress in time in small increments. For the implementation of the FDTD method, we divide the computational domain into $N_x \times N_y$ sub-domains (grids) with dimensions Δx, Δy. For the time derivative, we use forward difference, with a time interval Δt, and the displacement field is calculated at multiple integers of Δt, whereas the velocity is calculated on a time grid shifted by half the step. The probing signal is launched from the left homogeneous medium and corresponds to a longitudinal wave that propagates along the y direction for increasing y. This can be written as $F(y,t) = F(y - C_l t)$, where C_l is the longitudinal speed of sound in the inlet medium. The initial conditions on the displacement field and the speed vector are such as $\vec{u}(t = 0) = \begin{pmatrix} u_x = 0 \\ u_y = F(y) \end{pmatrix}$ and $\vec{v}\left(t = \frac{\Delta t}{2}\right) = \begin{pmatrix} v_x = 0 \\ v_y = -C_l \frac{dF(y,t)}{dt}\Big|_{t=+\Delta t/2} \end{pmatrix}$.

The stress component σ_{xx} is calculated at time $(n+1)$ from the components of the displacement field calculated at time t by discretizing (10.37), then

$$\sigma_{xx}^{n+1}(i,j) = C_{11}\left(i + \frac{1}{2}, j\right)\left(\frac{u_x^n(i+1,j) - u_x^n(i,j)}{\Delta x}\right)$$
$$+ C_{12}\left(i + \frac{1}{2}, j\right)\left(\frac{u_y^n(i,j) - u_y^n(i,j-1)}{\Delta y}\right) \tag{10.40}$$

where we define $C_{11}\left(i + \frac{1}{2}, j\right) = \sqrt{C_{11}(i+1,j)C_{11}(i,j)}$ and $C_{12}\left(i + \frac{1}{2}, j\right) = \sqrt{C_{12}(i+1,j)C_{12}(i,j)}$.

Similarly, the components σ_{xy} and σ_{yy} are obtained in discretized form as

$$\sigma_{xy}^{n+1}(i,j) = C_{11}\left(i + \frac{1}{2}, j\right)\left(\frac{u_y^n(i,j) - u_y^n(i,j-1)}{\Delta y}\right) + C_{12}\left(i + \frac{1}{2}, j\right)$$
$$\times \left(\frac{u_x^n(i+1,j) - u_x^n(i,j)}{\Delta y}\right) \tag{10.41}$$

$$\sigma_{yy}^{n+1}(i,j) = C_{44}\left(i, j + \frac{1}{2}\right)\left(\frac{u_x^n(i,j+1) - u_x^n(i,j)}{\Delta y} + \frac{u_y^n(i,j) - u_y^n(i-1,j)}{\Delta x}\right) \tag{10.42}$$

where we define $C_{44}\left(i, j + \frac{1}{2}\right) = \sqrt{C_{44}(i,j+1)C_{44}(i,j)}$.

Using expansions at point (i,j) and time n, (10.35) in component form becomes

$$v_x^{n+1}(i,j) = v_x^n(i,j) + \frac{\Delta t}{\rho(i,j)} \left(\frac{\sigma_{xx}^{n+1}(i,j) - \sigma_{xx}^{n+1}(i-1,j)}{\Delta x} + \frac{\sigma_{xy}^{n+1}(i,j) - \sigma_{xy}^{n+1}(i,j-1)}{\Delta y} \right)$$

(10.43)

$$v_y^{n+1}(i,j) = v_y^n(i,j) + \frac{\Delta t}{\rho\left(i + \frac{1}{2}, j + \frac{1}{2}\right)}$$
$$\times \left(\frac{\sigma_{yy}^{n+1}(i,j+1) - \sigma_{yy}^{n+1}(i,j)}{\Delta x} + \frac{\sigma_{xy}^{n+1}(i+1,j) - \sigma_{xy}^{n+1}(i,j)}{\Delta y} \right) \quad (10.44)$$

where we define $\rho\left(i + \frac{1}{2}, j + \frac{1}{2}\right) = \sqrt[4]{\rho(i,j)\rho(i+1,j)\rho(i,j+1)\rho(i+1,j+1)}$.

Finally, the components of the displacement field at time $(n+1)$ are deduced from the same component but evaluated at time n as $u_x^{n+1}(i,j) = u_x^n(i,j) + \Delta t.v_x^n(i,j)$ and $u_y^{n+1}(i,j) = u_y^n(i,j) + \Delta t.v_y^n(i,j)$.

Using this iterative procedure, the elastic wave equation is solved numerically, and the components of the time-dependent displacement field are calculated at the exit of the outlet. The component $u_y(t)$ is then averaged on a period of the slab along the x direction and Fourier transformed with respect to time. The same procedure is applied when the phononic crystal slab is replaced by a homogeneous medium identical to the inlet and the outlet media. The ratio between the two Fourier-transformed signals (with and without the PC slab) leads to the transmission coefficient. A reliable calculation of the transmission coefficient strongly depends on the choice of the function $F(y,t)$ corresponding to the probing signal. In particular, when considering the propagation through a homogeneous structure, i.e., without the PC slab, the Fourier-transformed signal must vary smoothly with the frequency on a specific frequency range $[0,\omega_{max}]$. This condition can be satisfied by taken into account a sinusoidal function weighted by a Gaussian profile such as

$$F(y,t) = F(y - C_l t) = F(Y) = A\cos[k_0 Y].\exp\left[-\frac{(k_0 Y)^2}{2}\right], \quad \text{where} \quad k_0 \sim \frac{\omega_{max}}{C_l}. \quad \text{The}$$

choice of this kind of function also allows to mimic the frequency response of a transducer generating pressure waves with a pass band $[0,\omega_{max}]$ usually used in ultrasonic measurements.

Periodic boundary conditions are applied along the x direction. That means that the elastic displacement is imposed to be the same on $x = 0$ and $x = L$, where L is the width of the FDTD mesh along the x direction for any value of y. For example, one must satisfy for any time step that $u_y(i_{max}+1, j) = u_y(1, j)$, where the integer i denoting the number of the spatial discretization step along the x direction varies between 1 and i_{max}. For closing the FDTD mesh along the y direction, it is necessary to impose absorbing boundary conditions on y_{min} and y_{max}, where y_{min} and y_{max} denote the entry of the inlet zone and the exit of the outlet zone. Absorbing boundary conditions are implemented in order to prevent reflection from the end elements of the FDTD mesh. First-order Mur's absorbing conditions [13] are usually used and can be implemented in the FDTD code by satisfying the following formula:

$$u_y^{n+1}(i, j_{max}) = u_y^n(i, j_{max} - 1) + \left(\frac{C_l \Delta t - \Delta y}{C_l \Delta t + \Delta y}\right) \left[u_y^{n+1}(i, j_{max} - 1) - u_y^n(i, j_{max})\right]$$

$$(10.45)$$

$$u_y^{n+1}(i, 1) = u_y^n(i, 2) + \left(\frac{C_l \Delta t - \Delta y}{C_l \Delta t + \Delta y}\right) \left[u_y^{n+1}(i, 2) - u_y^n(i, 1)\right] \qquad (10.46)$$

where the integer j denoting the number of the spatial discretization step along the y direction varies between 1 and j_{max}. Same formula should be satisfied for the x component of the displacement field.

Finally, for insuring the numerical stability of the FDTD code, it must be checked that the time step Δt and the discretization meshes Δx and Δy satisfy the following stability criterion [14]:

$$\Delta t \leq \frac{0.5}{C_l^{max} \sqrt{\left(\frac{1}{\Delta x}\right)^2 + \left(\frac{1}{\Delta y}\right)^2}} \qquad (10.47)$$

where C_l^{max} stands for the largest longitudinal speed of sound of the constituent materials involved in the structure.

10.3.2 Band Structure Calculation

In some cases, the PWE method fails to predict accurately the band structure of phononic crystals especially for mixed composites where one of the constituent is a fluid. Tanaka et al. [8] have reported an extension of the FDTD method for the calculation of dispersion relations of acoustic waves in 2D phononic crystals. In contrast with the standard FDTD approach presented in Sect. 10.3.1, the band structure FDTD technique implies a periodic system in the transverse plane xy.

The displacement field, the velocity vector, and the stress tensor must satisfy the Bloch theorem, i.e.,

$$\vec{u}(\vec{r}, t) = e^{i \overrightarrow{K} \cdot \vec{r}} \vec{U}(\vec{r}, t) \qquad (10.48)$$

$$\vec{v}(\vec{r}, t) = e^{i \overrightarrow{K} \cdot \vec{r}} \vec{V}(\vec{r}, t) \qquad (10.49)$$

$$\bar{\bar{\sigma}}(\vec{r}, t) = e^{i \overrightarrow{K} \cdot \vec{r}} \bar{\bar{\Sigma}}(\vec{r}, t) \qquad (10.50)$$

where $\vec{r}(x, y)$ is the position vector in the xy plane and $\vec{K}(K_x, K_y)$ is the Bloch wave vector.

$\vec{U}(\vec{r}, t)$, $\vec{V}(\vec{r}, t)$, and $\overline{\overline{\Sigma}}(\vec{r}, t)$ are spatial periodic functions satisfying $\vec{U}(\vec{r} + \vec{a}) = \vec{U}(\vec{r})$, $\vec{V}(\vec{r} + \vec{a}) = \vec{V}(\vec{r})$, and $\overline{\overline{\Sigma}}(\vec{r} + \vec{a}) = \overline{\overline{\Sigma}}(\vec{r})$, where $\vec{a}$ is the lattice translation vector. One inserts (10.48), (10.49), and (10.50) into the equations of propagation of the elastic waves, i.e., (10.35) and (10.36), and these later become

$$\frac{d\vec{V}}{dt} = \frac{1}{\rho(\vec{r})} i \vec{K} . \overline{\overline{\Sigma}}(\vec{r}, t) \quad \text{with} \quad \vec{V} = \frac{d\vec{U}}{dt}. \tag{10.51}$$

To solve (10.51), one first specifies a 2D wave vector, $\vec{K}(K_x, K_y)$, along the principal direction of the irreducible BZ. An assumption on the initial displacement $\vec{U}(\vec{r}, t = 0)$ in the form of a delta stimulus at some random location within the unit cell is then made. The equations of motion are then solved by discretizing both space and time. The time evolution of $\vec{U}(\vec{r_i}, t)$ at several predetermined locations $\vec{r_i}$ within the unit cell is recorded. Peaks in the frequency space of the Fourier-transformed signals are identified as the eigenfrequencies of the normal modes of the system for a given wave vector, $\vec{K}$.

10.3.3   Viscoelastic Media

The FDTD method reported in Sect. 10.3.1 is suitable for the calculation of transmission coefficient through phononic crystals made of non-lossy purely elastic material. Nevertheless several experimental studies were devoted to phononic crystals made of viscoelastic materials such as rubber, epoxy. Taking into account the effects of viscoelasticity on the propagation of elastic waves in phononic crystals is of fundamental as well as of practical interest in many areas. In this section, an alternate FDTD scheme where the viscoelastic properties, i.e., time-dependent elastic moduli, are rigorously taken into account is presented. As visco-elastic materials, we consider the general linear viscoelastic fluid (GLVF).

10.3.3.1   Viscoelastic Model

When the GLVF material also is compressible, the components of the total stress tensor are given by

$$\sigma(t) = 2 \int_{-\infty}^{t} G(t - t') D(t') dt' + \int_{-\infty}^{t} \left[K(t - t') - \frac{2}{3} G(t - t') \right] \left[\vec{\nabla} \vec{v}(t') \right] I dt' \tag{10.52}$$

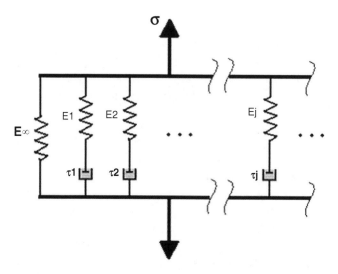

Fig. 10.8 Spring and dashpot illustration of the generalized Maxwell model

where t is time, $\vec{v}(\vec{r}, t)$ is the velocity vector, $\vec{D}(\vec{r}, t)$ is the rate of deformation tensor given by

$$\vec{D} = \frac{1}{2}\left[\left(\vec{\nabla}\vec{v}\right) + \left(\vec{\nabla}\vec{v}\right)^{\mathrm{T}}\right] \tag{10.53}$$

and $G(t)$ and $K(t)$ are the steady shear and bulk moduli, respectively.

These moduli can be experimentally determined through rheometry, and the data can be fit in a variety of ways, including the use of mechanical-analog models. A viscoelastic model, or in effect, the behavior pattern it describes, may be illustrated schematically by combinations of springs and dashpots, representing elastic and viscous factors, respectively. Hence, a spring is assumed to reflect the properties of an elastic deformation and similarly a dashpot to depict the characteristics of viscous flow. The generalized Maxwell model, also known as the Maxwell– Weichert model, takes into account the fact that the relaxation does not occur with a single time constant, but with a distribution of relaxation times. The Weichert model shows this by having as many spring–dashpot Maxwell elements as are necessary to accurately represent the distribution (Fig. 10.8). A multiple element Maxwell model is therefore more apt to represent the numerous timescales associated with relaxation in real viscoelastic materials.

For an n-element generalized Maxwell solid model, the extensional modulus $E(t)$ is calculated to be

$$E(t) = E_\infty + \sum_i^n E_i e^{-\frac{t}{\tau_i}} \tag{10.54}$$

where $\{E_i, \tau_i = 1, 2, \ldots, n\}$ are the moduli and relaxation times of the elements, and $E_\infty = E(\infty)$ is the equilibrium extensional modulus.

Introducing $\alpha(t) = \alpha_0 + \sum\limits_{i=1}^{n} \alpha_i e^{-t/\tau_i}$ where $\alpha_0 = \frac{E_\infty}{E_{\mathrm{sum}}}$, $\alpha_i = \frac{E_i}{E_{\mathrm{sum}}}$ ($i = 1, 2, .., n$), and $\sum\limits_{i=0}^{n} \alpha_i = 1$, $E_{\mathrm{sum}} = \sum\limits_{i=1}^{n} E_i$, we obtain $E(t) = E_{\mathrm{sum}}\alpha(t)$.

Consequently, we assume that

$$E(t) = 2G(t)(1 + v) = 3K(t)(1 - 2v) \tag{10.55}$$

$$\text{with}\begin{cases} G(t) = G_{\mathrm{sum}}\alpha(t) \\ K(t) = K_{\mathrm{sum}}\alpha(t) \end{cases} \quad \text{and} \quad \begin{cases} G_{\mathrm{sum}} = \mu \\ K_{\mathrm{sum}} - \dfrac{2}{3}G_{\mathrm{sum}} = \lambda \end{cases}. \tag{10.56}$$

In (10.55) and (10.56), v is the Poisson's ratio and λ and μ are the Lamé constant and shear modulus, respectively.

Now we consider a 2D elastic/viscoelastic material, where the system is infinite in the vertical direction z, and none of its properties depends on z (translational invariance). In this case, the Cartesian components of the 2D stress tensor deduced from (10.52) become

$$\sigma_{xx}(t) = 2\int_{-\infty}^{t} G(t - t')\frac{\partial v_x(t')}{\partial x}\mathrm{d}t' + \int_{-\infty}^{t}\left(K(t - t') - \frac{2}{3}G(t - t')\right)$$
$$\times \left(\frac{\partial v_x(t')}{\partial x} + \frac{\partial v_y(t')}{\partial y}\right)\mathrm{d}t' \tag{10.57}$$

$$\sigma_{yy}(t) = 2\int_{-\infty}^{t} G(t - t')\frac{\partial v_y(t')}{\partial y}\mathrm{d}t' + \int_{-\infty}^{t}\left(K(t - t') - \frac{2}{3}G(t - t')\right)$$
$$\times \left(\frac{\partial v_x(t')}{\partial x} + \frac{\partial v_y(t')}{\partial y}\right)\mathrm{d}t' \tag{10.58}$$

$$\sigma_{xy}(t) = \sigma_{yx}(t) = \int_{-\infty}^{t} G(t - t')\left(\frac{\partial v_x(t')}{\partial y} + \frac{\partial v_y(t')}{\partial x}\right)\mathrm{d}t' \tag{10.59}$$

For the sake of illustration, let us insert (10.56) into (10.57). Using $C_{11} = 2\mu + \lambda$, $C_{12} = \lambda$, and $C_{44} = \mu$, $\sigma_{xx}(t)$ becomes

$$\sigma_{xx}(t) = \alpha_0\left[C_{11}\frac{\partial u_x(t)}{\partial x} + C_{12}\frac{\partial u_y(t)}{\partial y}\right]$$
$$+ C_{11}\sum_{1}^{n}\alpha_i\int_{-\infty}^{t}\frac{\partial v_x(t')}{\partial x}e^{-\frac{(t-t')}{\tau_i}}\mathrm{d}t' + C_{12}\sum_{1}^{n}\alpha_i\int_{-\infty}^{t}\frac{\partial v_y(t')}{\partial y}e^{-\frac{(t-t')}{\tau_i}}\mathrm{d}t'$$
$$\tag{10.60}$$

Equation (10.60) involves integrals of the type

$$Ixx_i(t) = \int_{-\infty}^{t} \frac{\partial v_x(t')}{\partial x} e^{-\frac{(t-t')}{\tau_i}} dt' \tag{10.61}$$

in which calculations can be achieved by the following recursive method.

First we assume that for an incident wave that arrives from an elastic medium, we have $\int_{-\infty}^{t} \approx \int_{0}^{t}$. Then the following variable $w = t - t'$, ($\Rightarrow dw = -dt'$) leads to

$$Ixx_i(t) = \int_{0}^{t} \frac{\partial v_x(t - w)}{\partial x} e^{-\frac{w}{\tau_i}} dw \tag{10.62}$$

Now we calculate $Ixx_i(t + \Delta t)$.

$$Ixx_i(t + \Delta t) = \int_{0}^{t+\Delta t} \frac{\partial v_x(t + \Delta t - w)}{\partial x} e^{-\frac{w}{\tau_i}} dw \tag{10.63}$$

$$Ixx_i(t + \Delta t) = \int_{0}^{\Delta t} \frac{\partial v_x(t + \Delta t - w)}{\partial x} e^{-\frac{w}{\tau_i}} dw + \int_{\Delta t}^{t+\Delta t} \frac{\partial v_x(t + \Delta t - w)}{\partial x} e^{-\frac{w}{\tau_i}} dw \tag{10.64}$$

By changing $s = w - \Delta t => ds = dw$

$$Ixx_i(t + \Delta t) = \int_{-\Delta t}^{0} \frac{\partial v_x(t - s)}{\partial x} e^{-\frac{(s+\Delta t)}{\tau_i}} ds + \int_{0}^{t} \frac{\partial v_x(t - s)}{\partial x} e^{-\frac{(s+\Delta t)}{\tau_i}} ds \tag{10.65}$$

$$Ixx_i(t + \Delta t) = \left[\frac{\frac{\partial v_x(t)}{\partial x} e^{-\frac{\Delta t}{\tau_i}} + \frac{\partial v_x(t + \Delta t)}{\partial x}}{2} \Delta t \right] + e^{-\frac{\Delta t}{\tau_i}} \int_{0}^{t} \frac{\partial v_x(t - s)}{\partial x} e^{-\frac{s}{\tau_i}} ds \tag{10.66}$$

And finally a recursive form for the integral calculation is obtained as

$$Ixx_i(t + \Delta t) = \left[\frac{\frac{\partial v_x(t)}{\partial x} e^{-\frac{\Delta t}{\tau_i}} + \frac{\partial v_x(t + \Delta t)}{\partial x}}{2} dt \right] + e^{-\frac{\Delta t}{\tau_i}} Ixx_i(t) \tag{10.67}$$

where $Ixx_i(0) = 0$

Similar equations are obtained for the yy and xy components.

$$Iyy_i(t + \Delta t) = \left[\frac{\frac{\partial v_y(t)}{\partial y} e^{-\frac{\Delta t}{\tau_i}} + \frac{\partial v_y(t + \Delta t)}{\partial y}}{2} \Delta t \right] + e^{-\frac{\Delta t}{\tau_i}} Iyy_i(t) \tag{10.68}$$

$$Ixy_i(t + \Delta t) = \left[\frac{\frac{\partial v_x(t)}{\partial y} e^{-\frac{\Delta t}{\tau_i}} + \frac{\partial v_x(t + \Delta t)}{\partial y}}{2} \cdot \Delta t \right] + e^{-\frac{\Delta t}{\tau_i}} Ixy_i(t) \qquad (10.69)$$

$$Iyx_i(t + \Delta t) = \left[\frac{\frac{\partial v_y(t)}{\partial x} e^{-\frac{\Delta t}{\tau_i}} + \frac{\partial v_y(t + \Delta t)}{\partial x}}{2} \cdot \Delta t \right] + e^{-\frac{\Delta t}{\tau_i}} Iyx_i(t) \qquad (10.70)$$

We can now develop the FDTD method for the generalized Maxwell model.

10.3.3.2 FDTD Method for the Generalized Maxwell Model

As in Sect. 10.3.1, (10.35) stands for the basis equation for implementing the FDTD scheme taking into account the viscoelastic properties of the constituent materials of the 2D phononic crystal. The components of the velocity vector are given in discretized form by (10.43) and (10.44).

The stress component σ_{xx} is calculated by discretizing (10.60), using expansion at point (i, j) and time (n):

$$\sigma_{xx}^{n+1}(i,j) = \alpha_0\left(i+\frac{1}{2},j\right)C_{11}\left(i+\frac{1}{2},j\right)\left(\frac{u_x^n(i+1,j) - u_x^n(i,j)}{\Delta x}\right)$$
$$+ \alpha_0\left(i+\frac{1}{2},j\right)C_{12}\left(i+\frac{1}{2},j\right)\left(\frac{u_y^n(i,j) - u_y^n(i,j-1)}{\Delta y}\right)$$
$$+ C_{11}\left(i+\frac{1}{2},j\right)\sum_{p=1}^{n}\alpha_p\left(i+\frac{1}{2},j\right)$$
$$\cdot \left[\frac{v_x^n(i+1,j) - v_x^n(i,j)}{2\Delta x} + \frac{v_x^{n-1}(i+1,j) - v_x^{n-1}(i,j)}{2\Delta x} e^{-\frac{\Delta t}{\tau_p\left(i+\frac{1}{2},j\right)}} + e^{-\frac{\Delta t}{\tau_p(i,j)}}I_{xx_p}^n\right]$$
$$+ C_{12}\left(i+\frac{1}{2},j\right)\sum_{p=1}^{n}\alpha_p\left(i+\frac{1}{2},j\right)$$
$$\cdot \left[\frac{v_y^n(i,j) - v_y^n(i,j-1)}{2\Delta x} + \frac{v_y^{n-1}(i,j) - v_y^{n-1}(i,j-1)}{2\Delta x} e^{-\frac{\Delta t}{\tau_p\left(i+\frac{1}{2},j\right)}} + e^{-\frac{\Delta t}{\tau_p(i,j)}}I_{yy_p}^n\right]$$

$$(10.71)$$

where we define $C_{11}(i+1/2,j) = \sqrt{C_{11}(i+1,j)C_{11}(i,j)}$, $C_{12}(i+1/2,j) = \sqrt{C_{12}(i+1,j)C_{12}(i,j)}$, and $\alpha_p(i+1/2,j) = \sqrt{\alpha_p(i+1,j)\alpha_p(i,j)}$, $p = 0,1,2,\ldots,n$.

Similarly, the components σ_{yy} and σ_{xy} are obtained in discretized form:

$$\sigma_{yy}^{n+1}(i,j) = \alpha_0\left(i+\frac{1}{2},j\right)C_{11}\left(i+\frac{1}{2},j\right)\left(\frac{u_y^n(i,j)-u_y^n(i,j-1)}{\Delta y}\right)$$

$$+ \alpha_0\left(i+\frac{1}{2},j\right)C_{12}\left(i+\frac{1}{2},j\right)\left(\frac{u_x^n(i+1,j)-u_x^n(i,j)}{\Delta y}\right)$$

$$+ C_{11}\left(i+\frac{1}{2},j\right)\sum_{p=1}^{n}\alpha_p\left(i+\frac{1}{2},j\right)$$

$$\cdot\left[\frac{v_y^n(i,j)-v_y^n(i,j-1)}{2\Delta y}+\frac{v_y^{n-1}(i,j)-v_y^{n-1}(i,j-1)}{2\Delta y}e^{-\frac{\Delta t}{\tau_p\left(i+\frac{1}{2},j\right)}}+e^{-\frac{\Delta t}{\tau_p(i,j)}}I_{yy_p}^n\right]$$

$$+ C_{12}\left(i+\frac{1}{2},j\right)\sum_{p=1}^{n}\alpha_p\left(i+\frac{1}{2},j\right)$$

$$\cdot\left[\frac{v_x^n(i+1,j)-v_x^n(i,j-1)}{2\Delta x}+\frac{v_x^{n-1}(i+1,j)-v_x^{n-1}(i,j-1)}{2\Delta x}e^{-\frac{\Delta t}{\tau_p\left(i+\frac{1}{2},j\right)}}+e^{-\frac{\Delta t}{\tau_p(i,j)}}I_{xx_p}^n\right].$$

$$(10.72)$$

$$\sigma_{xy}^{n+1}(i,j) = \alpha_0\left(i,j+\frac{1}{2}\right)C_{44}\left(i,j+\frac{1}{2}\right)$$

$$\times\left(\frac{u_x^n(i,j+1)-u_x^n(i,j)}{\Delta y}+\frac{u_y^n(i,j)-u_y^n(i-1,j-1)}{\Delta x}\right)$$

$$+ C_{44}\left(i,j+\frac{1}{2}\right)\sum_{p=1}^{n}\alpha_p\left(i,j+\frac{1}{2}\right)$$

$$\cdot\left[\frac{v_x^n(i,j+1)-v_x^n(i,j)}{2\Delta y}+\frac{v_x^{n-1}(i,j+1)-v_x^{n-1}(i,j)}{2\Delta y}e^{-\frac{\Delta t}{\tau_p\left(i,j+\frac{1}{2}\right)}}+e^{-\frac{\Delta t}{\tau_p(i,j)}}I_{xy_p}^n\right]$$

$$+ C_{44}\left(i,j+\frac{1}{2}\right)\sum_{p=1}^{n}\alpha_p\left(i,j+\frac{1}{2}\right)$$

$$\cdot\left[\frac{v_y^n(i,j)-v_y^n(i-1,j)}{2\Delta x}+\frac{v_y^{n-1}(i,j)-v_y^{n-1}(i-1,j)}{2\Delta x}e^{-\frac{\Delta t}{\tau_p\left(i,j+\frac{1}{2}\right)}}+e^{-\frac{\Delta t}{\tau_p(i,j)}}I_{yx_p}^n\right]\quad(10.73)$$

where $\quad C_{44}(i,j+1/2) = \sqrt{C_{44}(i,j+1)C_{44}(i,j)}\quad$ and $\quad \alpha_p(i,j+1/2) = \sqrt{\alpha_p(i,j+1)\alpha_p(i,j)}, p = 0,1,2,\ldots,n$.

It has to be mentioned that the above way of discretizing the equations ensures second-order accurate central difference for the space derivatives. The field components u_x and u_y have to be centered in different space points. Calculations of transmission coefficients through 2D phononic crystals made of viscoelastic constituents follow the same procedures as in Sect. 10.3.1. These calculations must be done considering the structure depicted in Fig. 10.7 and applying periodic boundary conditions in the x direction and Mur's absorbing boundary conditions on the two extremities of the discretization mesh along the y direction.

Such calculations for 2D phononic crystals made of steel cylinders embedded in rubber modeled as GLVF were reported in [15, 16]. Results have shown the very good agreement between the numerical predictions and the experimental measurements.

10.4   Multiple Scattering Theory

The multiple scattering theory (MST) was introduced for 3D phononic crystals by three different groups at about the same time [17–19], and its 2D version was developed 3 years later by Prof. Liu's group in the theoretical work by Mei et al. [20]. The MST is essentially an extension of the Korringa–Kohn–Rostoker (KKR) theory (which is a well-known method used by the solid-state community for electronic band structure calculations) to the case of elastic/acoustic waves. The MST is ideally suited for phononic crystals (both 2D and 3D) in which scattering units have simple symmetries, such as spheres or cylinders. It is also a quickly converging method that takes into account the full vector character of the elastic field and is able to deal with the phononic crystals of any type (e.g., liquid/solid crystals, for which the PWE method fails). We present in this sub-section the main points of the MST in case of the 3D phononic crystals by following the steps along which it was developed by Liu et al. in [18].

In a homogeneous isotropic medium, the elastic wave equation may be written as

$$(\lambda + 2\mu)\vec{\nabla}(\vec{\nabla} \cdot \vec{u}) - \mu\vec{\nabla} \times \vec{\nabla} \times \vec{u} + \rho\omega^2\vec{u} = 0 \tag{10.74}$$

where ρ is the density of the medium, λ, μ are its Lamé constants, and $\vec{u}$ is the displacement field. Because of the spherical symmetry of the scatterers, it is natural to work with the general solution of (10.74) expressed in the spherical coordinates:

$$\vec{u}(\vec{r}) = \sum_{lm\sigma}[a_{lm\sigma}\vec{J}_{lm\sigma}(\vec{r}) + b_{lm\sigma}\vec{H}_{lm\sigma}(\vec{r})] \tag{10.75}$$

where $\vec{J}_{lm\sigma}(\vec{r}), \vec{H}_{lm\sigma}(\vec{r})$ are defined as follows:

$$\vec{J}_{lm1}(\vec{r}) = \frac{1}{\alpha}\vec{\nabla}[j_l(\alpha r)Y_{lm}(\hat{r})]$$

$$\vec{J}_{lm2}(\vec{r}) = \frac{1}{\sqrt{l(l+1)}}\vec{\nabla} \times [\vec{r}j_l(\beta r)Y_{lm}(\hat{r})]$$

$$\vec{J}_{lm3}(\vec{r}) = \frac{1}{\beta\sqrt{l(l+1)}}\vec{\nabla} \times \vec{\nabla} \times [\vec{r}j_l(\beta r)Y_{lm}(\hat{r})] \tag{10.76}$$

and

$$\vec{H}_{lm1}(\vec{r}) = \frac{1}{\alpha}\vec{\nabla}\left[h_l(\alpha r)Y_{lm}(\hat{r})\right]$$

$$\vec{H}_{lm2}(\vec{r}) = \frac{1}{\sqrt{l(l+1)}}\vec{\nabla}\times\left[\vec{r}h_l(\beta r)Y_{lm}(\hat{r})\right]$$

$$\vec{H}_{lm3}(\vec{r}) = \frac{1}{\beta\sqrt{l(l+1)}}\vec{\nabla}\times\vec{\nabla}\times\left[\vec{r}h_l(\beta r)Y_{lm}(\hat{r})\right] \qquad (10.77)$$

where $\alpha = \omega\sqrt{\rho/(\lambda+2\mu)}, \beta = \omega\sqrt{\rho/\mu}, j_l(x)$ is the spherical Bessel function, $h_l(x)$ is the spherical Hankel function of the first kind, and $Y_{lm}(\hat{r})$ is the usual spherical harmonic with $\hat{r}$ denoting angular coordinates (θ, φ) of $\vec{r}$ in spherical coordinate system. In (10.75), index σ assumes values from 1 to 3, where $\sigma = 1$ indicates the longitudinal wave and $\sigma = 2, 3$ indicates two transverse waves of different polarizations. In the case when the coefficients $b_{lm\sigma}$ in (10.75) are equal to zero, $\vec{u}(\vec{r})$ represents an *incident* wave, and in the case of $a_{lm\sigma} = 0$, $\vec{u}(\vec{r})$ represents a *scattered* wave. Therefore, the wave incident on an ith scatterer is expressed as

$$\vec{u}_i^{in}(\vec{r}_i) = \sum_{lm\sigma} a_{lm\sigma}^i \vec{J}_{lm\sigma}^i(\vec{r}_i) \qquad (10.78)$$

where $\vec{r}_i$ indicates some point in space as measured from the center of the ith scatterer. The wave scattered by scatterer i can be expressed as

$$\vec{u}_i^{sc}(\vec{r}_i) = \sum_{lm\sigma} b_{lm\sigma}^i \vec{H}_{lm\sigma}^i(\vec{r}_i). \qquad (10.79)$$

The first key point of MST is the idea that the wave (10.78) incident on a given scatterer i can be viewed as a sum of the externally incident wave $\vec{u}_i^{(0)}(\vec{r}_i)$ expressed as

$$\vec{u}_i^{(0)}(\vec{r}_i) = \sum_{lm\sigma} a_{lm\sigma}^{i(0)} \vec{J}_{lm\sigma}^i(\vec{r}_i) \qquad (10.80)$$

and *all* other scattered waves *except* the one scattered by the ith scatterer, which can be expressed as

$$\sum_{j\neq i} \vec{u}_j^{sc}(\vec{r}_j) = \sum_{j\neq i}\sum_{lm\sigma} b_{lm\sigma}^j \vec{H}_{lm\sigma}^j(\vec{r}_j) \qquad (10.81)$$

so that (10.78) can also be written as

$$\vec{u}_i^{in}(\vec{r}_i) = \vec{u}_i^{(0)}(\vec{r}_i) + \sum_{j\neq i} \vec{u}_j^{sc}(\vec{r}_j) \qquad (10.82)$$

Here $\vec{r}_i$ and $\vec{r}_j$ refer to the position of the same point in space and are measured from the centers of scatterers i and j, respectively.

Another crucial point of MST is that for a given scatterer, the scattered field is *completely* determined from the incident field with the help of the scattering matrix T. In other words, the expansion coefficients $A = \{a^j_{lm\sigma}\}$ and $B = \{b^j_{lm\sigma}\}$ are related through $T = \{t_{lm\sigma l'm'\sigma'}\}$ as follows:

$$B = TA$$

or more explicitly

$$b^j_{lm\sigma} = \sum_{l'm'\sigma'} t_{lm\sigma l'm'\sigma'} a^j_{l'm'\sigma'}. \tag{10.83}$$

For objects of simple geometry, such as spheres or cylinders, the calculation of the scattering matrix T is an *exactly* solvable boundary-value problem, and this is the origin of MST's reliability and precision when handling arrangements of scatterers of spherical symmetry. In short, the coefficients $t_{lm\sigma l'm'\sigma'}$ are found by applying the boundary conditions that require the continuity of the normal components of both the displacement and the stress vectors at the scatterer–matrix interface. The explicit expressions of the T matrix coefficients for an elastic sphere can be found in [17] (liquid matrix) and in [19] (elastic matrix), and in [20] for an elastic cylinder in an elastic matrix.

The final MST equation is obtained by substituting (10.78), (10.80), (10.81), and (10.83) into (10.82) and reads

$$\sum_{jl'm'\sigma'} \left(\delta_{ij}\delta_{ll'}\delta_{mm'}\delta_{\sigma\sigma'} - \sum_{l''m'\sigma''} t^j_{l'm''\sigma''l'm'\sigma'} G^{ij}_{l''m''\sigma''lm\sigma} \right) a^j_{l'm'\sigma'} = a^{i(0)}_{lm\sigma} \tag{10.84}$$

where $G_{lm\sigma l'm'\sigma'}$ is the so-called vector structure constant, which relates $\vec{H}^j_{lm\sigma}(\vec{r}_j)$ in (10.81) and $\vec{J}^i_{lm\sigma}(\vec{r}_i)$ through the relation

$$\vec{H}^j_{lm\sigma}(\vec{r}_j) = \sum_{l'm'\sigma'} G^{ij}_{lm\sigma l'm'\sigma'} \vec{J}^i_{l'm'\sigma'}(\vec{r}_i)$$

(more details can be found in [18]). The normal modes of the system may be obtained by solving the secular equation that follows from (10.84) in the absence of an external incident wave (i.e., when all $a^{i(0)}_{lm\sigma}$ are zero):

$$\det \left| \delta_{ij}\delta_{ll'}\delta_{mm'}\delta_{\sigma\sigma'} - \sum_{l''m''\sigma''} t^j_{l''m''\sigma''l'm'\sigma'} G^{ij}_{l''m''\sigma''lm\sigma} \right| = 0. \tag{10.85}$$

In case of the periodic system, $G_{lm\sigma l'm'\sigma'}$ is modified to take into account the symmetry of the structure. The solutions of (10.85) give the band structure of an elastic periodic system.

Fig. 10.9 Geometry of the
layer MST. Vectors $\vec{a_1}$ and $\vec{a_2}$
are the primitive vectors of
the corresponding 2D Bravais
lattice

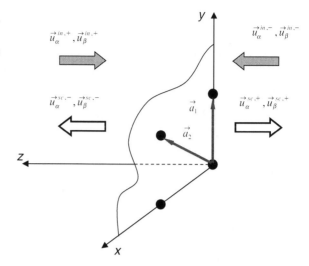

To facilitate the *direct* comparison with the real samples, a successful theory
must also be able to calculate the quantities that one measures in a typical experiment, e.g., transmission and reflection coefficients. This is accomplished in the
framework of the *layer* MST, which allows one to calculate the transmission of an
elastic wave through a *finite* slab (with an arbitrary number of layers) of periodically arranged scatterers. The approach starts by calculating the field of the elastic
wave scattered (or transmitted) by a *single* layer of scatterers. Let us assume that the
layer of scatterers (elastic spheres) lies completely in the x–y plane and that
positions of the scatterers are given by vectors $\{\vec{R}_n\}$ of a 2D Bravais lattice,
which is generated by two primitive vectors $\vec{a}_1, \vec{a}_2$, i.e.,

$$\vec{R}_n = n_1 \vec{a}_1 + n_2 \vec{a}_2 \tag{10.86}$$

where n_1, n_2 are integers. The positive direction of the z-axis is chosen to be to the
left of the layer as explained by Fig. 10.9.

A plane elastic wave $\vec{u}^{\text{in}}(\vec{r})$ incident on the layer can be expressed in general as

$$\vec{u}^{\text{in}}(\vec{r}) = \sum_s \vec{u}_\alpha^{\text{in},s}(\vec{r}) + \sum_s \vec{u}_\beta^{\text{in},s}(\vec{r}) \tag{10.87}$$

where $s = +/-$ indicates waves incident from the left (positive z) and from the
right (negative z) respectively, while $\alpha = 1$ and $\beta = 2, 3$ are identical to index σ in
(10.75) and distinguish between the longitudinal and the transverse (with two
polarizations) waves . Each term in (10.87) can be expressed in terms of the
primitive vectors $\vec{b}_1, \vec{b}_2$ of the 2D reciprocal lattice as follows:

$$\vec{u}_{\alpha}^{\text{in},\pm}(\vec{r}) = \sum_g \vec{u}_{\alpha g}^{\text{in},\pm}(\vec{r}) = \sum_g \vec{U}_{\alpha g}^{\text{in},\pm}\exp(i\vec{k}_{\alpha g}^{\pm}\cdot\vec{r}) \tag{10.88a}$$

$$\vec{u}_{\beta}^{\text{in},\pm}(\vec{r}) = \sum_{\vec{g}} \vec{u}_{\beta g}^{\text{in},\pm}(\vec{r}) = \sum_{\vec{g}} \vec{U}_{\beta g}^{\text{in},\pm}\exp(i\vec{k}_{\beta g}^{\pm}\cdot\vec{r}) \tag{10.88b}$$

where wave vectors $\vec{k}_{\alpha g}^{\pm}$ and $\vec{k}_{\beta g}^{\pm}$ are given by the expressions

$$\vec{k}_{\alpha g}^{\pm} = \left(\vec{k}_{\|} + \vec{g}, \pm\sqrt{\alpha^2 - \left|\vec{k}_{\|} + \vec{g}\right|^2}\right) \tag{10.89a}$$

$$\vec{k}_{\beta g}^{\pm} = \left(\vec{k}_{\|} + \vec{g}, \pm\sqrt{\beta^2 - \left|\vec{k}_{\|} + \vec{g}\right|^2}\right) \tag{10.89b}$$

Here $\vec{g}$ is the 2D reciprocal lattice vector ($\vec{g} = m_1\vec{b}_1 + m_2\vec{b}_2$, where m_1, m_2 are integers), and $\vec{k}_{\|}$ is a reduced wave vector in the first BZ of the reciprocal lattice. In (10.89a) and (10.89b), ($\vec{k}_{\|} + \vec{g}$) simply represents components of wave vectors $\vec{k}_{\alpha g}^{\pm}$ and $\vec{k}_{\beta g}^{\pm}$ that are parallel to the layer of scatterers. These expressions are chosen to simplify subsequent calculations.

Much in the same way, the wave $\vec{u}^{\text{sc}}(\vec{r})$ scattered by the layer can be expressed as follows:

$$\vec{u}^{\text{sc}}(\vec{r}) = \sum_s \vec{u}_{\alpha}^{\text{sc},s}(\vec{r}) + \sum_s \vec{u}_{\beta}^{\text{sc},s}(\vec{r})$$

$$= \sum_{s,\vec{g}} \vec{U}_{\alpha g}^{\text{sc},s}\exp(i\vec{k}_{\alpha g}^{s}\cdot\vec{r}) + \sum_{s,\vec{g}} \vec{U}_{\beta g}^{\text{sc},s}\exp(i\vec{k}_{\beta g}^{s}\cdot\vec{r}) \tag{10.90}$$

Indices α and β have the same meaning as in case of incident wave (10.87). The index $s = +/-$, however, reverses its meaning and now indicates the scattered waves propagating away from the layer on its right (negative z) and on its left (positive z) correspondingly (see Fig. 10.9).

After lengthy and complicated calculations, one can show (see Ref. [18]) that amplitudes $\vec{U}_{\alpha g}^{\text{sc},\pm}$ and $\vec{U}_{\beta g}^{\text{sc},\pm}$ of the scattered wave are related to the amplitudes $\vec{U}_{\alpha g}^{\text{in},\pm}$ and $\vec{U}_{\beta g}^{\text{in},\pm}$ of the incident wave with the help of matrices $\mathbf{M}_{\kappa\kappa'}^{ss'}$ ($s, s' = +/-$ and $\kappa, \kappa' = \alpha, \beta$) as follows:

$$\begin{bmatrix} \mathbf{U}_{\alpha}^{\text{sc},+} \\ \mathbf{U}_{\beta}^{\text{sc},+} \end{bmatrix} = \begin{bmatrix} \mathbf{M}_{\alpha\alpha}^{++} & \mathbf{M}_{\alpha\beta}^{++} \\ \mathbf{M}_{\beta\alpha}^{++} & \mathbf{M}_{\beta\beta}^{++} \end{bmatrix}\begin{bmatrix} \mathbf{U}_{\alpha}^{\text{in},+} \\ \mathbf{U}_{\beta}^{\text{in},+} \end{bmatrix} + \begin{bmatrix} \mathbf{M}_{\alpha\alpha}^{+-} & \mathbf{M}_{\alpha\beta}^{+-} \\ \mathbf{M}_{\beta\alpha}^{+-} & \mathbf{M}_{\beta\beta}^{+-} \end{bmatrix}\begin{bmatrix} \mathbf{U}_{\alpha}^{\text{in},-} \\ \mathbf{U}_{\beta}^{\text{in},-} \end{bmatrix}$$

$$\begin{bmatrix} \mathbf{U}_{\alpha}^{\text{sc},-} \\ \mathbf{U}_{\beta}^{\text{sc},-} \end{bmatrix} = \begin{bmatrix} \mathbf{M}_{\alpha\alpha}^{-+} & \mathbf{M}_{\alpha\beta}^{-+} \\ \mathbf{M}_{\beta\alpha}^{-+} & \mathbf{M}_{\beta\beta}^{-+} \end{bmatrix}\begin{bmatrix} \mathbf{U}_{\alpha}^{\text{in},+} \\ \mathbf{U}_{\beta}^{\text{in},+} \end{bmatrix} + \begin{bmatrix} \mathbf{M}_{\alpha\alpha}^{--} & \mathbf{M}_{\alpha\beta}^{--} \\ \mathbf{M}_{\beta\alpha}^{--} & \mathbf{M}_{\beta\beta}^{--} \end{bmatrix}\begin{bmatrix} \mathbf{U}_{\alpha}^{\text{in},-} \\ \mathbf{U}_{\beta}^{\text{in},-} \end{bmatrix} \tag{10.91}$$

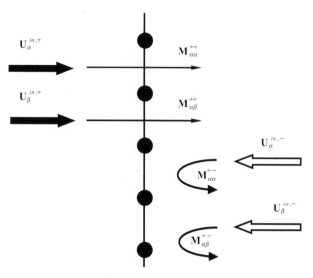

Fig. 10.10 Schematic illustration of the physical significance of the matrices $\mathbf{M}_{\kappa\kappa'}^{ss'}$

In the above equations, $\mathbf{U}_\kappa^{\mathrm{sc},\pm}$ and $\mathbf{U}_\kappa^{\mathrm{in},\pm}$ are column vectors defined as

$$\mathbf{U}_\kappa^{\mathrm{sc},\pm} = [\, \mathbf{U}_{\kappa g_1}^{\mathrm{sc},\pm} \quad \mathbf{U}_{g_2}^{\mathrm{sc},\pm} \quad \ldots \quad \mathbf{U}_{\kappa g_{N-1}}^{\mathrm{sc},\pm} \quad \mathbf{U}_{\kappa g_N}^{\mathrm{sc},\pm} \,]^{\mathrm{Tr}} \tag{10.92a}$$

$$\mathbf{U}_\kappa^{\mathrm{in},\pm} = [\, \mathbf{U}_{\kappa g_1}^{\mathrm{in},\pm} \quad \mathbf{U}_{g_2}^{\mathrm{in},\pm} \quad \ldots \quad \mathbf{U}_{\kappa g_{N-1}}^{\mathrm{in},\pm} \quad \mathbf{U}_{\kappa g_N}^{\mathrm{in},\pm} \,]^{\mathrm{Tr}} \tag{10.92b}$$

where the Tr superscript denotes the operation of transposing. The explicit expressions for the elements of the matrices $\mathbf{M}_{\kappa\kappa'}^{ss'}$ are given by Liu et al. [18]. Being very complicated mathematical objects, matrices $\mathbf{M}_{\kappa\kappa'}^{ss'}$ nevertheless have simple physical meaning (Fig. 10.10). They are transmission and reflection matrices for incident waves $\mathbf{U}_\alpha^{\mathrm{in},\pm}$ and $\mathbf{U}_\beta^{\mathrm{in},\pm}$. For example, by expanding first line in the first matrix equation in (10.91), one obtains

$$\mathbf{U}_\alpha^{\mathrm{sc},+} = \mathbf{M}_{\alpha\alpha}^{++}\mathbf{U}_\alpha^{\mathrm{in},+} + \mathbf{M}_{\alpha\beta}^{++}\mathbf{U}_\beta^{\mathrm{in},+} + \mathbf{M}_{\alpha\alpha}^{+-}\mathbf{U}_\alpha^{\mathrm{in},-} + \mathbf{M}_{\alpha\beta}^{+-}\mathbf{U}_\beta^{\mathrm{in},-}$$

Figure 10.10 shows a schematic diagram explaining the physical meaning of matrices contained in the above equation.

Having found transmission and reflection matrices through the *single* layer, one needs to find a way to calculate similar matrices for a phononic crystal with an *arbitrary* number of layers. Figure 10.10 shows a schematic diagram explaining the physical meaning of matrices $\mathbf{M}_{\kappa\kappa'}^{ss'}$ contained in the above equation. This is accomplished by calculating matrices $\mathbf{Q}_{\kappa\kappa'}^{ss'}$ for each of two single layers that are displaced with respect to the x–y plane by vectors $\vec{a}_3/2$ and $-\vec{a}_3/2$, where $\vec{a}_3$ is a third primitive vector of the Bravais lattice of the phononic crystal. In other words,

$\vec{a}_3$ is a vector by which a single 2D layer of scatterers should be repeated to form the 3D phononic crystal. Matrices $\mathbf{Q}^{ss'}_{\kappa\kappa'}$ have the same physical meaning as $\mathbf{M}^{ss'}_{\kappa\kappa'}$ and are connected with matrices $\mathbf{M}^{ss'}_{\kappa\kappa'}$ by another translation matrix φ^s_κ, whose elements are explicitly expressed in [18]. The transmission and reflection matrices for the pair of two successive layers (denoted by N and $N+1$) are obtained by combining corresponding matrices $\mathbf{Q}^{ss'}_{\kappa\kappa'}(N)$ and $\mathbf{Q}^{ss'}_{\kappa\kappa'}(N + 1)$. The essential physics here is that two sets of matrices are combined by taking into account all *multiple* reflections that the incident wave undergoes between two layers as it propagates through the two-layer system. By repeating this procedure, the transmission and reflection matrices through the slab consisting of 2^n layers can be found. The corresponding matrices for the crystal with an arbitrary number of layers can be obtained by combining matrices for the slab with even number of layers and one extra layer.

It also should be noted that in addition to the band structure, which displays normal modes of the system along high-symmetry directions, the MST also allows calculation of the modes along *any* direction inside the crystal. The geometrical set of all points belonging to a particular mode (which is characterized by a certain frequency) is referred to as an equi-frequency surface or equi-frequency contour for 3D or 2D structures correspondingly.

10.5   Finite Element Method

The finite element (FE) method is suitable for the calculation of band structures of phononic crystals, containing several phases or materials. To present the model, a doubly periodic structure is considered. Square-, rectangular-, triangular-, or honeycomb-type structures can be considered, but for the sake of simplicity, only the square array is presented in this section with a 2D mesh. The phononic crystal contains two or more different phases and consists, for instance, in a periodic array of holes in a solid matrix or a periodic array of cylindrical rods or tubes in a solid matrix. The formalism is the same when the periodic structure is all fluid. The structure is supposed to be infinite and periodic in the x-y plane and is infinite and uniform in the third direction. Consequently, the problem is strictly bidimensional, depending only on the x and y coordinates, using plane strain conditions. The whole domain is split into successive cells (Fig. 10.11). Due to the periodicity of the structure, the A1 and A2 lines, parallel to the y axis, and the B1 and B2 lines, parallel to the x axis, limit the unit cell, which is $2d_1$ wide in the x direction and $2d_2$ wide in the y direction. In Fig. 10.11, corners are marked by letter C.

Then the structure is excited by a plane monochromatic wave, the direction of incidence of which is marked by an angle θ with respect to the positive y axis. The incident wave is characterized by a real wave vector $\vec{k}$, of modulus k, the wave number.

Because the structure is assumed to extend from $-\infty$ to $+\infty$ in the x and y directions and to be periodic, any space function F (pressure, displacement, electrical potential, etc.) has to satisfy the classical Bloch relation:

Fig. 10.11 Schematic description of one unit cell of the doubly periodic structure, used to define the $A1$, $A2$, $B1$, and $B2$ lines, the $C1$, $C2$, $C3$, and $C4$ corners, and the phase relation between the lines

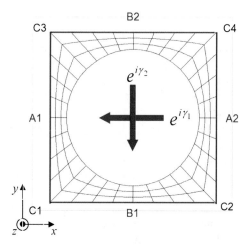

$$F(x + 2d_1, y + 2d_2) = F(x, y)\mathrm{e}^{\mathrm{j}2d_1 k\sin\theta}\mathrm{e}^{\mathrm{j}2d_2 k\cos\theta} = F(x, y)\mathrm{e}^{\mathrm{j}\gamma_1}\mathrm{e}^{\mathrm{j}\gamma_2}. \qquad (10.93)$$

Using relation (10.93) allows reducing the model to only one unit cell, which can be meshed using FEs (Fig. 10.11). Writing relation (10.93) between the displacement values for nodes separated by one period provides the boundary conditions between adjacent cells. Using the FE method, a modal analysis is considered, and the whole system of equations is classically

$$\left([K_{uu}] - \omega^2[M]\right)\vec{U} = \vec{F} \qquad (10.94)$$

where the unknown is the vector of nodal values of the displacement $\vec{U} \cdot [K_{uu}]$ and $[M]$ are, respectively, the structure stiffness and coherent mass matrices. ω is the angular frequency. $\vec{F}$ contains the nodal values of the applied forces.

The application of the periodic boundary conditions implies that the phase relation (10.93) between nodal values belonging to the $A1$ and $A2$ lines, on the one hand, to the $B1$ and $B2$ lines on the other hand, has to be incorporated in the matrix equation (10.94). The unit cell is divided into nine parts: the four lines $A1$, $A2$, $B1$, and $B2$; the four corners $C1$, $C2$, $C3$, and $C4$; and the inner domain I. Displacement vector $\vec{U}$ and force vector $\vec{F}$ are then split into the corresponding nine parts. Due to relation (10.93), their components have to verify

$$\vec{U}_{A2} = \mathrm{e}^{\mathrm{j}\gamma_1}\vec{U}_{A1}; \vec{U}_{B2} = \mathrm{e}^{\mathrm{j}\gamma_2}\vec{U}_{B1}; \vec{U}_{C2} = \mathrm{e}^{\mathrm{j}\gamma_1}\vec{U}_{C1}; \vec{U}_{C3} = \mathrm{e}^{\mathrm{j}\gamma_2}\vec{U}_{C1}; \vec{U}_{C4} = \mathrm{e}^{\mathrm{j}\gamma_1 + \mathrm{j}\gamma_2}\vec{U}_{C1}.$$
$$(10.95)$$

Then owing to the equilibrium of interconnecting forces between two adjacent cells, relation (10.93) leads to analogous relations for the force vector. $\vec{F}_I$, which corresponds to forces applied to inner nodes, is equal to zero. Defining the reduced vector $\vec{U}_R$ as a vector containing values of the displacement on the $A1$ and $B1$ lines, on the $C1$ corner, and in the inner domain I, relations given in (10.95) imply a simple matrix relation between $\vec{U}$ and $\vec{U}_R$, which can be written as

$$\vec{U} = [P_U]\vec{U}_R = [P_U]\begin{pmatrix} \vec{U}_{A1} \\ \vec{U}_{B1} \\ \vec{U}_{C1} \\ \vec{U}_I \end{pmatrix}. \tag{10.96}$$

In the same way, a matrix relation can be defined between the vector $\vec{F}$ and the reduced vector $\vec{F}_R$:

$$\vec{F} = [P_F]\vec{F}_R = [P_F]\begin{pmatrix} \vec{F}_{A1} \\ \vec{F}_{B1} \\ \vec{F}_{C1} \\ \vec{0} \end{pmatrix}. \tag{10.97}$$

Thus, the equation to be solved can be reduced to

$$[P_U]^{*T}\left([K_{uu}] - \omega^2[M]\right)[P_U]\vec{U}_R = \left([K_R] - \omega^2[M_R]\right)\vec{U}_R = [P_U]^{*T}[P_F]\vec{F}_R. \tag{10.98}$$

Finally, the matrices $[K_R]$ and $[M_R]$ are divided into following four parts, *A1*, *B1*, *C1*, and *I* and the resulting equation is

$$\left([K_R] - \omega^2[M_R]\right)\vec{U}_R = \vec{0}. \tag{10.99}$$

A detailed expression of $[K_R]$ and $[M_R]$ are presented in Appendix 2 of [21].

For a given value of the wave number k, the phase shifts of (10.93) and (10.95) are deduced and incorporated in relations (10.96) and (10.97). The resolution of the system (10.99) gives the corresponding eigenvalues ω that are real because the reduced matrices $[K_R]$ and $[M_R]$ are hermitians.

The angular frequency ω is a periodical function of wave vector $\vec{k}$. Thus, the problem can be reduced to the first BZ. The dispersion curves are built varying $\vec{k}$ on the first BZ, for a given propagation direction. The whole diagram is deduced using symmetries.

A particular interest is the study of phononic crystal plates, made for instance of arrays of air inclusions drilled in a plate. In that case, a 3D mesh is considered and the structure is supposed to be of finite size along the thickness of the plate, periodic and infinite in the two other directions. Only one unit mesh is considered, and a phase relation is applied on only the four faces of the mesh, defining boundary conditions between adjacent cells. The FE method is accurate for the study of phononic crystal plates because it does not introduce hypothesis on the displacement field or on the characteristics of the medium surrounding the plate [9, 22].

Another way to characterize periodic structures is the scattering or the radiation of plane acoustic waves from immersed passive or active periodic structures at any incidence. Therefore, the calculation of the transmission and reflection coefficients is performed, when N unit cells are taken into account (Fig. 10.12). For this study, the mesh of the N unit cells of the periodic structure is enough, with a small part of

Fig. 10.12 Schematic
description of N unit cells
($N=6$) of the periodic
structure, for the calculation
of the transmission and
reflection coefficients.
A small part of the fluid
domain is meshed before and
after the periodic structure

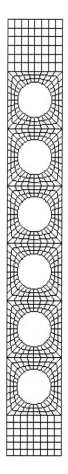

the surrounding fluid domain, which can be air, and a harmonic analysis is
performed at a given frequency. The general system of equation is

$$\begin{pmatrix} [K_{uu}] - \omega^2 [M] & -[L] \\ -\rho^2 c^2 \omega^2 [L]^T & [H] - \omega^2 [M_1] \end{pmatrix} \begin{pmatrix} \vec{U} \\ \vec{P} \end{pmatrix} = \begin{pmatrix} \vec{F} \\ \vec{\psi} \end{pmatrix} \tag{10.100}$$

where the unknown is the vector of nodal values of the displacement $\vec{U}$ and of the
pressure field $\vec{P}$. $[H]$ and $[M_1]$ are, respectively, the compressibility and mass
matrices for the fluid. $[L]$ is the connectivity matrix at the interface and ρ and c
are the density and the sound velocity in the fluid, respectively. $\vec{\psi}$ contains the
nodal values of the pressure normal gradient on the fluid boundaries, on the top
and bottom surfaces. In this system, the periodic boundary conditions are
introduced as previously by the phase relations between nodes separated by the
periodic spacing. Then, the effects of the remaining fluid domain are accounted for

by matching the pressure field in the FE mesh with simple PWEs of the incoming and outgoing waves. Writing the continuity equations introduces matrix relations between the nodal values of the pressure on the bottom and top surfaces, which are then incorporated into system (10.100). The resolution of the system gives the pressure in the fluid domain. Then, the transmission and reflection coefficients are calculated.

10.6   Model Reduction for Band Structure Calculations

10.6.1   Background

As thoroughly discussed in previous chapters and sections, the study of wave propagation in phononic crystals, or periodic media in general, utilizes Bloch's theorem, which allows for the calculation of dispersion curves (frequency band structure) and density of states. Due to crystallographic symmetry, the Bloch wave solution needs to be applied only to a single unit cell in the reciprocal lattice space covering the first BZ [23]. Further utilization of symmetry reduces the solution domain, even more, to the irreducible Brillouin zone (IBZ). As mentioned in previous sections, there are several techniques for band structure calculations for phononic crystals and acoustic metamaterials (which are also applicable to photonic crystals and electromagnetic metamaterials). Some of the methods involve expanding the periodic domain and the wave field using a truncated basis. This provides a means of classification in terms of the type of basis, e.g., the plane wave method (Sect. 10.2) involves a Fourier basis expansion and the FE method (Sect. 10.5) involves a real space basis expansion. The pros and cons of the various methods are discussed in depth in the literature [24].

Regardless of the type of system and type of method used for band structure calculations, the computational effort is usually high because it involves solving a complex eigenvalue problem and doing so numerous times as the value of the wave vector, $\mathbf{k}$, is varied. The size of the problem, and hence the computational load, is particularly high for the following cases: (a) when the unit cell configuration requires a large number of degrees of freedom to be adequately described; (b) when the presence of defects is incorporated in the calculations, thus requiring the modeling of large super-cells; and (c) when a large number of calculations; are needed such as in band structure optimization [25, 26]. All these cases suggest that a fast technique for band structure calculation would be very beneficial.

Some techniques have been developed to expedite band structure calculations; examples include utilization of the multigrid concept [27], development of fast iterative solvers for the Bloch eigenvalue problem [28, 29], and extension of homogenization methods to capture dispersion [30, 31]. In this section, we provide a model reduction method that is based on modal transformation [32, 33]. This method, which is referred to as the *reduced Bloch mode expansion (RBME)* method, involves carrying out an expansion employing a natural basis composed of a selected reduced set of Bloch eigenfunctions[1]. This reduced basis is selected within the IBZ at high-symmetry points determined by the crystal structure and group theory (and possibly at additional related points). At each of these high-symmetry points, a number of Bloch eigenfunctions are selected up to the frequency range of interest for the band structure calculations. As mentioned above, it is common to initially discretize the problem at hand using some choice of basis. In this manner, RBME constitutes a secondary expansion using a set of Bloch eigenvectors and hence keeps and builds on any favorable attributes the primary expansion approach might exhibit. The proposed method is in line with the well-known concept of modal analysis, which is widely used in various fields in the physical sciences and engineering[2].

In the next section, a description of the RBME process and its application in a discrete setting (e.g., using FEs) is given for a phononic crystal problem. Some results from a case study are also presented to demonstrate the application of the method.

10.6.2   Reduced Bloch Mode Expansion method

The starting point for the RBME method is a discrete generalized eigenvalue problem emerging from the application of Bloch's theorem applied to a standard periodic unit cell model. This yields an equation of the form

$$(\mathbf{K}(\mathbf{k}) - \omega^2 \mathbf{M})\tilde{\mathbf{U}} = \mathbf{0}, \tag{10.101}$$

where $\mathbf{M}$ and $\mathbf{K}(\mathbf{k})$ are the global mass and stiffness matrices, respectively; $\tilde{\mathbf{U}}$ is the discrete Bloch vector, which is periodic in the unit cell domain; $\mathbf{k}$ is the wave vector; and ω is the frequency. Equation (10.101) is then solved at a reduced set of selected wave vector points (i.e., reduced set of $\mathbf{k}$-points), providing the eigenvectors from which a reduced Bloch modal matrix, denoted Ψ, is formed. Several schemes are available for this selection, the simplest of which is the set of eigenvectors corresponding to the first few branches at the high-symmetry points Γ, X, M for a 2D model and Γ, X, M, R for a 3D model, as illustrated in Fig. 10.13 for square and simple-cubic cells (more details on selection schemes are given in [32]). The matrix Ψ is then used to expand the eigenvectors $\tilde{\mathbf{U}}$, i.e.,

---

[1] The same mode selection concept, but in the context of a multiscale two-field variational method, was presented in [31, 34].

[2] The concept of modal analysis is rooted in the idea of extracting a reduced set of representative information on the dynamical nature of a complex system. This practice is believed to have originated by the Egyptians in around 4700 B.C. in their quest to find effective ways to track the flooding of the Nile and predict celestial events [35].

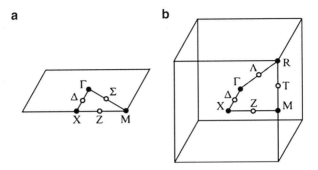

Fig. 10.13 Unit cell in reciprocal lattice space with the irreducible Brillouin zone, high-symmetry **k**-points (*solid circles*) and intermediate **k**-points (*hollow* ciircles) shown. (**a**) 2D square unit cell, (**b**) 3D simple-cubic unit cell

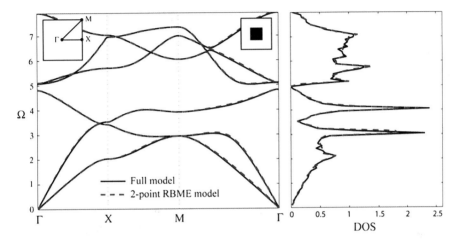

Fig. 10.14 Phononic band structure and density of states (DOS) calculated using full model (matrix size: 4,050 × 4,050) and reduced Bloch mode expansion model (matrix size: 24 × 24). The IBZ and eigenvector selection points are shown in the *left inset*. The 2D unit cell is shown in the *right inset*; the stiff/dense material phase is in *black*, and the compliant/light material phase is in *white*. The finite element method was used for the primary expansion

$$\tilde{\mathbf{U}}_{(n\times1)} = \mathbf{\Psi}_{(n\times m)}\tilde{\mathbf{V}}_{(m\times1)}, \tag{10.102}$$

where $\tilde{\mathbf{V}}$ is a vector of modal coordinates for the unit cell Bloch mode shapes. In (10.102), n and m refer to the number of rows and number of columns for the matrix equation. To enable significant model reduction, the chosen **k**-point selection scheme has to ensure that $m \ll n$. Substituting (10.102) into (10.101), and premultiplying by the complex transpose of $\mathbf{\Psi}$,

$$\mathbf{\Psi}^*\mathbf{K}(\mathbf{k})\mathbf{\Psi}\tilde{\mathbf{V}} - \omega^2\mathbf{\Psi}^*\mathbf{M}\mathbf{\Psi}\tilde{\mathbf{V}} = \mathbf{0}, \tag{10.103}$$

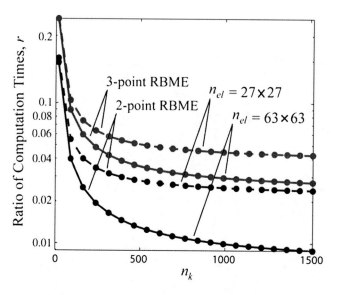

Fig. 10.15 Computational efficiency: ratio of reduced Bloch mode expansion model to full model calculation times, r, versus number of sampled k-points along the border of the IBZ, n_k (for two 2D finite element meshes). The number of elements is denoted by n_{el}

yields a reduced eigenvalue problem of size $m \times m$,

$$\bar{\mathbf{K}}(\mathbf{k})\tilde{\mathbf{V}} - \omega^2\bar{\mathbf{M}}\tilde{\mathbf{V}} = \mathbf{0}, \tag{10.104}$$

where $\bar{\mathbf{M}}$ and $\bar{\mathbf{K}}(\mathbf{k})$ are reduced generalized mass and stiffness matrices. The eigenvalue problem given in (10.104) can then be solved for the entire region of interest within the IBZ at a significantly lower cost compared to using the full model given in (10.101).

To demonstrate the RBME approach, we consider a linear elastic, isotropic, continuum model of a 2D phononic crystal under plain strain conditions. As an example, a square lattice is considered with a bi-material unit cell. One material phase is chosen to be stiff and dense and the other compliant and light. In particular, a ratio of Young's moduli of $E_2/E_1 = 16$ and a ratio of densities of $\rho_2/\rho_1 = 8$ are chosen. The topology of the material phase distribution in the unit cell is shown in the inset of Fig. 10.14. The unit cell is discretized into 45×45 uniformly sized four-node bilinear quadrilateral FEs, i.e., 2,025 elements. With the application of periodic boundary conditions, the number of degrees of freedom is $n = 4050$. Figure. 10.14 shows the calculated band structure and density of states using two-point expansion, that is, the selection is carried out at the Γ, X, M points in **k**-space. In the calculations, eight modes were utilized at each of these selection points. As such, a total of 24 eigenvectors ($m = 24$) were used to form the Bloch modal matrix. The results for the full model are overlaid for comparison indicating excellent agreement, despite a reduction of model size from 4050 to 24 degrees of freedom. For models with a larger number of degrees of freedom, and a calculation with high **k**-point sampling, two orders of magnitude or greater reduction in computational expense will be achieved (as shown in Fig. 10.15).

While the focus in this section has been on phononic crystals, the RBME method is also applicable to acoustic metamaterials, to discrete lattice dynamics calculations, and to photonic and electronic band structure calculations. Furthermore, the method is applicable to any type of lattice symmetry.

References

1. N.W. Ashcroft, N.D. Mermin, *Solid State Physics* (Saunders College, Philadelphia, 1976)
2. M. Sigalas, E.N. Economou, Band structure of elastic waves in two dimensional systems. Solid State Commun. **86**, 141–143 (1993)
3. J.O. Vasseur, B. Djafari-Rouhani, L. Dobrzynskiand, P.A. Deymier, Acoustic band gaps in fibre composite materials of boronnitride structure. J. Phys. Condens Matter **9**, 7327–7341 (1997)
4. ZhilinHou' Xiujun Fu, and Youyan Liu, Singularity of the Bloch theorem in the fluid/solid phononic crystal. Phys. Rev. B **73**, 024304–024308 (2006)
5. J.O. Vasseur, P.A. Deymier, A. Khelif, P. Lambin, B. Djafari-Rouhani, A. Akjouj, L. Dobrzynski, N. Fettouhi, J. Zemmouri, Phononic crystal with low filling fraction and absolute acoustic band gap in the audible frequency range: a theoretical and experimental study. Phys. Rev. E **65**, 056608 (2002)
6. B. Manzanares-Martinez, F. Ramos-Mendieta, Surface elastic waves in solid composites of two-dimensional periodicity. Phys. Rev. B **68**, 134303 (2003)
7. C. Goffaux, J.P. Vigneron, Theoretical study of a tunable phononic band gap system. Phys. Rev. B **64**, 075118 (2001)
8. Y. Tanaka, Y. Tomoyasu, S.I. Tamura, Band structure of acoustic waves in phononic lattices: Two-dimensional composites with large acoustic mismatch. Phys. Rev. B **62**, 7387 (2000)
9. J.O. Vasseur, P.A. Deymier, B. Djafari-Rouhani, Y. Pennec, A.-C. Hladky-Hennion, Absolute forbidden bands and waveguiding in two-dimensional phononic crystal plates. Phys. Rev. B **77**, 085415 (2008)
10. C. Charles, B. Bonello, F. Ganot, Propagation of guided elastic waves in 2D phononic crystals. Ultrasonics **44**, 1209(E) (2006)
11. C. Croënne, E.D. Manga, B. Morvan, A. Tinel, B. Dubus, J. Vasseur, A.-C. Hladky-Hennion, Negative refraction of longitudinal waves in a two-dimensional solid-solid phononic crystal. Phys. Rev. B **83**, 054301 (2011)
12. P. Lambin, A. Khelif, J.O. Vasseur, L. Dobrzynski, B. Djafari-Rouhani, Stopping of acoustic waves by sonic polymer-fluid composites. Phys. Rev. E **63**, 066605 (2001)
13. G. Mur, Absorbing boundary conditions for the finite difference approximation of the time-domain electromagnetic field equations. IEEE Trans. Electromagn. Compatibility **23**, 377 (1981)
14. A. Taflove, *Computational electrodynamics: the finite difference time domain method* (Artech House, Boston, 1995)
15. B. Merheb, P.A. Deymier, M. Jain, M. Aloshyna-Lesuffleur, S. Mohanty, A. Berker, R.W. Greger, Elastic and viscoelastic effects in rubber/air acoustic band gap structures: a theoretical and experimental study. J. Appl. Phys. **104**, 064913 (2008)
16. B. Merheb, P.A. Deymier, K. Muralidharan, J. Bucay, M. Jain, M. Aloshyna-Lesuffleur, R.W. Greger, S. Mohanty, A. Berker, Viscoelastic effect on acoustic band gaps in polymer-fluid composites. Model. Simul. Mater. Sci. Eng. **17**, 075013 (2009)

17. M. Kafesaki, E.N. Economou, Multiple-scattering theory for three-dimensional periodic acoustic composites. Phys. Rev. B. **60**, 11993 (1999)
18. Z. Liu, C.T. Chan, P. Sheng, A.L. Goertzen, J.H. Page, Elastic wave scattering by periodic structures of spherical objects: theory and experiment. Phys. Rev. B. **62**, 2446 (2000)
19. I.E. Psarobas, N. Stefanou, A. Modinos, Scattering of elastic waves by periodic arrays of spherical bodies. Phys. Rev. B. **62**, 278 (2000)
20. J. Mei, Z. Liu, J. Shi, D. Tian, Theory for elastic wave scattering by a two-dimensional periodical array of cylinders: an ideal approach for band-structure calculations. Phys. Rev. B **67**, 245107 (2003)
21. P. Langlet, A.-C. Hladky-Hennion, J.N. Decarpigny, Analysis of the propagation of plane acoustic waves in passive periodic materials using the finite element method. J. Acoust. Soc. Am. **95**, 1792 (1995)
22. J.O. Vasseur, A.-C. Hladky-Hennion, B. Djafari-Rouhani, F. Duval, B. Dubus, Y. Pennec, Waveguiding in two-dimensional piezoelectric phononic crystal plates. J. Appl. Phys. **101**, 114904 (2007)
23. L. Brillouin, *Wave Propagation in Periodic Structures* (Dover, New York, 1953)
24. K. Busch, G. von Freymann, S. Linden, S.F. Mingaleev, L. Tkeshelashvili, M. Wegener, Periodic nanostructures for photonics. Phys. Rep. **444**, 101–202 (2007)
25. O. Sigmund, J.S. Jensen, Systematic design of phononic band-gap materials and structures by topology optimization. Philos. Trans. R. Soc. Lond. A361, 1001–1019 (2003)
26. O.R. Bilal, M.I. Hussein, Ultrawidephononic band gap for combined in-plane and out-of-plane waves. Phys. Rev. E **84**, 065701(R) (2011)
27. R.L. Chern, C.C. Chang, R.R. Hwang, Large full band gaps for photonic crystals in two dimensions computed by an inverse method with multigrid acceleration. Phys. Rev. E **68**, 026704 (2003)
28. D.C. Dobson, An efficient method for band structure calculations in 2D photonic crystals. J. Comput. Phys. **149**, 363–376 (1999)
29. S.G. Johnson, J.D. Joannopoulos, Photonic crystals: putting a new twist on light. Opt. Express **8**, 173 (2001)
30. T.W. McDevitt, G.M. Hulbert, N. Kikuchi, An assumed strain method for the dispersive global-local modeling of periodic structures. Comput. Methods Appl. Mech. Eng. **190**, 6425–6440 (2001)
31. M.I. Hussein, G.M. Hulbert, Mode-enriched dispersion models of periodic materials within a multiscale mixed finite element framework. Finite Elem. Anal. Des. **42**, 602–612 (2006)
32. M.I. Hussein, Reduced Bloch mode expansion for periodic media band structure calculations. Proc. R. Soc. Lond. **A465**, 2825–2848 (2009)
33. Q. Guo, O.R. Bilal, M.I. Hussein, Convergence of the reduced Bloch mode expansion method for electronic band structure calculations," in *Proceedings of Phononics 2011,* Paper PHONONICS-2011-0176, Santa Fe, New Mexico, USA, May 29–June 2, 2011, pp. 238–239
34. M.I. Hussein, Dynamics of banded materials and structures: analysis, design and computation in multiple scales, Ph.D. Thesis, University of Michigan, Ann Arbor, USA, 2004.
35. O. Døssing, IMAC-XIII keynote address: going beyond modal analysis, or IMAC in a new key. Modal Anal. Int. J. Anal. Exp. Modal Anal. **10**, 69 (1995)

Index

P.A. Deymier (ed.), *Acoustic Metamaterials and Phononic Crystals*,
Springer Series in Solid-State Sciences 173, DOI 10.1007/978-3-642-31232-8,
© Springer-Verlag Berlin Heidelberg 2013

Printed by Publishers' Graphics LLC
SO20130116.19.19.42